Turbulence in the Free Atmosphere

Turbulence in the Free Atmosphere

N. K. Vinnichenko, N. Z. Pinus,
S. M. Shmeter, and G. N. Shur

Central Aerological Observatory
Dolgoprudnyi, USSR

Second Edition

Translated from Russian by
Frank L. Sinclair

Consultants Bureau · New York and London

Library of Congress Cataloging in Publication Data

Main entry under title:

Turbulence in the free atmosphere.

Translation of Turbulentnost' v svobodnoĭ atmosfere.
Includes indexes.
1. Atmospheric turbulence. I. Vinnichenko, Nikolaĭ Konstantinovich.
QC880.4.T8T8713 551.5'17 80-23743
ISBN 0-306-10959-X

A Division of Plenum Publishing Corporation
227 West 17th Street, New York, N.Y. 10011

Printed in the United States of America

Preface to the Second American Edition

This book, like the first edition published in 1968 in the USSR and in 1973 in the USA, is devoted mainly to the results of experimental, rather than theoretical, investigations of atmospheric turbulence affecting the flight of aircraft. It should be noted that the basic theoretical principles have not undergone any radical changes in recent years, whereas definite progress has been made in the experimental investigation of turbulence, particularly in clouds and in the vicinity of the tropopause, since the publication of the first edition of our book. This induced us to revise and supplement a considerable part of the text of the first edition of the monograph. The changes made to the text also reflect, of course, the more thorough understanding, acquired by the authors in recent years, of the principles of development of turbulence in the free atmosphere.

We sincerely thank Plenum Press for the translation and publication in English of our revised monograph. We value particularly highly the interest shown in our book by American readers. We hope that the second edition of our book will be a useful textbook for all specialists concerned with problems of atmospheric turbulence itself and its diverse technical applications.

Moscow

N. Z. Pinus
S. M. Shmeter
G. N. Shur
N. K. Vinnichenko

Preface

Turbulence—the randomly disordered movement of volumes of air of widely varying size—is one of the characteristic features of atmospheric air flows; its investigation is essential for the solution of several theoretical and practical problems. Until recently, owing to experimental difficulties, research on turbulence was confined mainly to the lower half of the troposphere. Theoretical investigations have consequently been based on these data.

The rapid development of high-altitude aviation and cases of aircraft encountering hazardous turbulence led to a sharp intensification of research on turbulence in the atmosphere up to 10–12 km, and subsequently at greater altitudes. Such research was confined initially to the characterization of the frequency of occurrence of gusts of different speeds, their relation to altitude, geographical conditions, time of day and year, and so on. At the end of the fifties, when the required measuring equipment and experimental techniques had been developed, it became possible to investigate the complete statistical characteristics of turbulence: the spectral densities of the velocity fluctuations of air flows, structure functions, etc. These data stimulated the further development of theory related to the specific conditions of the free atmosphere.

The first edition of this monograph was published in 1968. It consisted mainly of an account of the results of experimental investigations. Since the theory of atmospheric turbulence is the subject of the comprehensive monograph *Statistical Hydromechanics* by A. S. Monin and A. M. Yaglom (Nauka Press, 1965 and 1967), the authors considered it advisable to deal only with those aspects of theory that facilitate the physical analysis of the presented data. The book also gives a short description of instruments for measuring various parameters of turbulence in the free atmosphere. Since the majority of instruments used for scientific research are constantly and rapidly modified, it was decided not to give details of the design of instruments, but simply to indicate the physical principles on which their operation is based. In a special chapter we give the techniques of statistical analysis of the results of measurements and the errors involved.

This book discusses in detail the origin of turbulence in the free atmosphere, the variation of its structure in relation to stratification of the atmosphere, and turbulence in relation to waves and convection. Considerable attention is given to these mesoscale movements since they are important not only as sources of turbulent energy, but also per se, particularly in connection with their effect on aircraft. In addition to the general characteristics of turbulence, this monograph describes special features of its structure in clouds, jet streams, the tropopause, and so on. A separate chapter is devoted to the effect of turbulence on aircraft.

Particular attention has been paid, of course, to aspects of the problem relating to the authors' own scientific interest and research, and the fullest account is given of the authors' own results, obtained mainly in the course of numerous research flights conducted by the Central Aerological Observatory (TsAO) in collaboration with the State Scientific-Research Institute for Civil Aviation.

This book does not deal with the microscale turbulence that affects the propagation of radio waves, light, and sound in the atmosphere.

In the years that have elapsed since the publication of the first edition, there has been a great deal of experimental research on turbulence in clouds and in clear air in the USSR and several other countries. In view of this, the present edition has had to be considerably revised, new material has been added to almost all the chapters, and the format of the book has been modified.

In Chapter 2, which deals with experimental methods of investigating turbulence in the free atmosphere, the section devoted to radar and laser methods of studying turbulence in the atmosphere has been enlarged, and an aircraft instrument for measuring turbulent fluctuations of temperature is described.

A separate chapter (Chapter 4) is devoted to the structure of turbulence in the thermally stratified atmosphere. This chapter gives the results of experimental investigations in the USSR and abroad and discusses the various theories in which the effect of buoyancy forces on the spectral structure of the velocity field and temperature irregularities is taken into account.

Chapter 5, which deals with the results of research on clear-air turbulence in the lower half of the troposphere, is supplemented with a description of methods of determining the turbulent viscosity coefficient. Values of this coefficient are given and the dissipation of turbulent energy as heat with the concomitant heating of the atmosphere is discussed.

Chapters 6–8, dealing with atmospheric convection, thermal turbulence, turbulence in clouds, and also waves and atmospheric turbulence, have been radically revised and enlarged. For instance, Chapter 6 includes a new section devoted to the statistical characteristics of thermal turbulence. The sections dealing with rotor motions (Section 4, Chapter 7) and vertical motions and turbulence in convective clouds (Sections 2 and 3, Chapter 8) have been rewritten. Chapter 9, dealing with turbulence affecting aircraft flight, has been enlarged by the addition of new experimental data on turbulence in the troposphere and stratosphere; it gives atmospheric parameters, which form the basis of the physical model of clear-air turbulence. An account of the methods of diagnosis and prediction of the turbulence responsible for aircraft buffeting is given.

Chapter 10 is supplemented with data on the statistical distribution of the spectral density of wind-velocity fluctuations for different wave numbers, the variation of the energy of turbulent motions of different scales with altitude, and the energy balance in zones of clear-air turbulence.

At the end of this monograph there is a bibliography containing papers published after 1961. The books cited are not subject to this restriction.

This monograph was written by a team of authors consisting of N. K. Vinnichenko (Section 4, Chapter 2; Chapter 3), N. Z. Pinus (Chapter 1 in collaboration with G. N. Shur, Chapters 5, 9, and 10), S. M. Shmeter (Chapters 6–8), and G. N. Shur (Chapters 2 and 4).

The authors regard it as a pleasant duty to express their deep gratitude to L. S. Gandin, the scientific editor of the first edition and the reviewer of the second edition, for valuable critical comments and advice, and also to V. D. Litvinova, V. I. Silaeva, and other members of the staff of the Laboratory of Atmosphere Dynamics and TsAO, who gave great assistance in the preparation of the second edition of the book.

Contents

1. Some Aspects of the Theory of Atmospheric Turbulence 1
1. Nature of Atmospheric Turbulence 1
2. Turbulent Energy Balance. Coefficient of Turbulent Viscosity . . . 3
3. Richardson Stability Criterion 5
4. Structure and Energetics of Turbulent Flow 8
5. Statistical Description of Turbulence Field 10
6. Energy Spectrum of Turbulence 15
7. Interaction of Mean and Turbulent Fields of Temperature and Wind Velocity 20

2. Methods of Experimental Investigation of Turbulence in the Free Atmosphere 25
1. The Airplane as a Means of Investigating Atmospheric Turbulence 25
2. Method Using Reaction of Airplane to an Atmospheric Gust (Load Method) 26
3. Method Using Measurements of Absolute and Relative Airplane Velocities (Doppler Method) 38
4. Method Using Measurements of Velocity Fluctuations of Incident Air Flow (Hot-Wire Anemometer Method) 46
5. Method Using Pressure Measurements at Various Points on a Body in the Flow 51
6. Method Using Measurements of Angle of Attack (Wind-Vane Method) 53
7. Method Using Measurements of Speed of Sound (Acoustic Method) 54
8. Gyroscopic Instruments for Determination of Airplane Attitude 56
9. Measurement of Temperature Fluctuations on an Airplane 58
10. Table Summarizing Aircraft Methods of Investigating Turbulence 59

11. Airplanes–Flying Laboratories . . . 60
12. Radar and Optical Methods of Investigating Atmospheric Turbulence . . . 61

3. Statistical Analysis of Measurements . . . 65
1. Calculation of Spectra from Continuous Records of Finite Length . . . 65
2. Calculations of Spectra from Discrete Records of Finite Length . . 75
3. The Fast Fourier Transform . . . 87
4. Preliminary Analysis of Nature of Spectrum . . . 89

4. Spectrum of Turbulence in a Thermally Stratified Atmosphere . . . 97
1. Spectrum of Turbulence in a Stably Stratified Atmosphere . . . 97
2. Relation between Shape of Spectral Curve and Thermal Stratification of Atmosphere . . . 105
3. Energy Transfer in Spectrum and Generalized Spectrum of Turbulence in Free Atmosphere . . . 109
4. Some Features of Experimental Spectra of Clear-Air Turbulence . . 116
5. Spectra of Temperature Fluctuations in Thermally Stratified Atmosphere . . . 118

5. Clear-Air Turbulence in Lower and Middle Troposphere . . . 121
1. Some Experimental Data on Turbulence Intensity . . . 121
2. Structure of Turbulent Layer of Atmosphere. Intermittency of Turbulence . . . 125
3. Experimental Values of Turbulent Viscosity Coefficient . . . 129
4. Spectral Characteristics of Turbulence . . . 132

6. Atmospheric Convection and Thermal Turbulence . . . 145
1. Disordered Atmospheric Convection . . . 146
2. Cellular Convection . . . 159

7. Waves and Atmospheric Turbulence . . . 165
1. Gravity-Shear Waves . . . 165
2. Criteria of Stability of Internal Gravity-Shear Waves and Relation between Turbulence and Disintegrating Waves . . . 172
3. Mountain Waves . . . 176
4. Rotor Motions and Turbulence in Mountain-Wave Region . . . 186

8. Turbulence and Vertical Flows in Clouds . . . 195
1. Turbulence in Stratiform Clouds . . . 195
2. Vertical Motions in Convective Clouds . . . 200
3. Turbulence in Cumulus and Cumulonimbus Clouds . . . 214

9. Turbulence Affecting Flight of Aircraft **225**
1. Methods of Assessing Intensity of Turbulence Affecting Flight of Aircraft 225
2. Characteristics of Turbulence in Troposphere and Lower Stratosphere 237
3. Turbulence in Jet Streams and in Region of Tropopause 250
4. Relation between Turbulence Causing Aircraft Buffeting and Thermal and Dynamic Conditions in Atmosphere 259
5. Prediction of Clear-Air Turbulence 271

10. Mesostructure of Wind Field **279**
1. Mesoinhomogeneities of Wind Field 279
2. Energy Spectra of Mesoscale Turbulence 280
3. Vertical Distributions of Wind Gusts and Their Spectra 287
4. Mesometeorological Energy Minimum in Turbulence Spectrum . . . 290
5. Energy of Turbulent Motions of Different Scales in Relation to Altitude 292
6. Energy Balance in Clear-Air Turbulent Zones 295

REFERENCES 301

1

Some Aspects of the Theory of Atmospheric Turbulence

1. Nature of Atmospheric Turbulence

Most of the processes occurring in the atmosphere, such as the transport of water vapor and atmospheric dust, heat transfer, formation of clouds, and precipitation, are very closely connected with the turbulent nature of atmospheric motion. Atmospheric turbulence has a significant effect on the propagation of sound, light, and radio waves, and also on the flight of aircraft.

Turbulent motion of a viscous fluid, as distinguished from laminar motion, is characterized by a variable velocity distribution, the presence of inhomogeneities, or the so-called turbulent eddies, leading to mixing of the streams. By a turbulent eddy we mean an element of turbulent flow of a characteristic size (scale of eddy).

The instantaneous velocity u in a turbulent flow can be represented as the superposition of small fluctuations on the mean motion. If we denote the mean velocity by $\bar{u}$, then $u = \bar{u} + u'$, where u' is the so-called fluctuating, or eddy, velocity. If $u' = 0$ the moton is laminar.

If we denote the characteristic dimension of a viscous flow by L, and the characteristic flow velocity by U, then from L, U, and ν (ν is the kinematic viscosity of the fluid) we can form a dimensionless combination known as the Reynolds number:

$$\mathrm{Re} = \frac{UL}{\nu} \tag{1.1}$$

Laminar motion becomes turbulent only when the Reynolds number exceeds some critical value Re_{cr}, i.e., when

$$\frac{UL}{\nu} > \mathrm{Re}_{cr}. \tag{1.2}$$

The physical meaning of Re_{cr} is as follows. Inertial forces bring together initially well-separated volumes of fluid which have different velocity, and thus produce sharp differences in velocity at close points. Viscosity forces, on the other hand, lead to equalization of the velocities at close points and thus tend to smooth out small flow inhomogeneities. At low values of Re, when viscosity forces predominate over inertial forces, the flow is laminar.

With increase in Re the smoothing effect of viscosity forces is reduced and the flow becomes turbulent, i.e., irregular velocity fluctuations appear in it; Re_{cr} corresponds to conditions where the inertial forces are so large in comparison with viscosity forces that a stable turbulent regime is produced.

Experimental investigations in tubes have shown that laminar flow becomes turbulent at Re = 2500–5000.

In application to the atmosphere, the Reynolds number becomes largely indeterminate owing to the difficulty of determining the scale of the flow as a whole. The indeterminate geometry of atmospheric flows necessitates the introduction of new concepts; for instance, as a characteristic dimension we may choose the distance from the earth's surface to the level of the maximum wind velocity, or the height of the tropopause.

If we formally calculate the values of Re for the atmosphere, we find that it is many times greater than the critical value and, hence, motion in the atmosphere is always turbulent. In fact, if a sufficiently sensitive and fast instrument is used to measure the wind velocity, irregular fluctuations of velocity about a stable mean value can always be found. In some cases, however, these fluctuations are so small that motion in the atmosphere can be regarded as quasi-laminar.

The main reasons for the turbulence of air flows in the atmosphere are the sharp differences, of diverse origin, in the temperature and wind-velocity fields. The atmospheric processes responsible for these sharp differences are:

(a) the friction of the air flow at the Earth's surface and the formation of a wind-velocity profile with large vertical gradients in the lower part of the atmosphere;

(b) unequal heating of different regions of the underlying surface of the Earth and the concomitant development of thermal convection;

(c) cloud-forming processes involving the release of heat of condensation and crystallization and changes in the nature of the temperature and wind-velocity fields;

(d) convergence and interaction of air masses with different characteristics near atmospheric fronts and high-altitude frontal zones, where horizontal differences in temperature and wind velocity are large;

(e) loss of stability by waves formed in inversion layers, at the tropopause, and near other atmospheric interfaces;

(f) deformation of air flows by mountain barriers that produce wavelike disturbances and rotor motions on their lee side.

These processes can act simultaneously, strengthening or weakening one another, and thus the turbulence of the air flow.

The viscosity of the air causes continuous conversion of the kinetic energy of turbulent air flows to heat. It is obvious that if there is no external energy source for continuous excitation of turbulent motion, then this motion will die out sooner or later.

The main characteristic of turbulent motion is the random nature of the velocity field in time and space and, hence, the classical hydrodynamic problem of finding the position of all fluid particles at any instant t from the given position and velocities of the particles at an initial instant t_0 and the boundary conditions becomes meaningless in the case of turbulent motion.

In turbulent flow, even when the external force field and boundary conditions are constant, the velocity at a given point undergoes great changes in time. This necessitates the application of statistical methods in the investigation of turbulence.

As already mentioned, the development of atmospheric turbulence is due not only to dynamic but also to thermal factors, and, hence, the Re is inadequate for determining its onset even in those cases where the scale of the flow can be estimated.

This problem has been considered by Richardson, who, on the basis of the equation for the turbulent energy balance, obtained a criterion for the increase and decrease of the kinetic energy of turbulent motions in a thermally stratified medium.

2. Turbulent Energy Balance. Coefficient of Turbulent Viscosity

The quantity of turbulent energy at a given point of an air flow can be altered by conversion of some of the energy of the mean motion to turbulent energy (or the reverse conversion of some of the turbulent energy to energy of mean motion), by fluctuations of external forces, by diffusion of turbulent energy, and by dissipation of turbulent energy due to viscosity. The role of each of these factors becomes obvious when the turbulent-energy balance equation is analyzed.

The balance equation for the case where the flow is horizontal, not stratified in density, and the turbulence is homogeneous in a horizontal plane, i.e., depends only on the height, can be written as follows [75]:

$$\left.\begin{aligned}
&\frac{\partial E_x}{\partial t}+\frac{\partial}{\partial z}\left(\frac{1}{2}\rho\overline{u'^2w'}-\nu\frac{\partial E_x}{\partial z}\right)=-\overline{p'\frac{\partial w'}{\partial z}}-\overline{p'\frac{\partial v'}{\partial y}}-\rho\bar{\varepsilon}_x-\rho\overline{u'w'}\frac{\partial\bar{u}}{\partial z},\\
&\frac{\partial E_y}{\partial t}+\frac{\partial}{\partial z}\left(\frac{1}{2}\rho\overline{v'^2w'}-\nu\frac{\partial E_y}{\partial z}\right)=\overline{p'\frac{\partial v'}{\partial y}}-\rho\bar{\varepsilon}_y,\\
&\frac{\partial E_z}{\partial t}+\frac{\partial}{\partial z}\left(\frac{1}{2}\rho\overline{w'^3}+\overline{p'w'}-\nu\frac{\partial E_z}{\partial z}\right)=\overline{p'\frac{\partial w'}{\partial z}}-\rho\bar{\varepsilon}_z.
\end{aligned}\right\}\quad(1.3)$$

In (1.3) $E_x = \frac{1}{2}\overline{u'^2}$; $E_y = \frac{1}{2}\overline{v'^2}$; $E_z = \frac{1}{2}\overline{w'^2}$; p' are the fluctuations of pressure forces, $\epsilon_x, \epsilon_y, \epsilon_z$ are the components of the rate of dissipation of turbulent energy, ρ is the density of the air, and v' and w' are the transverse and vertical velocity components.

An analysis of these equations leads to the following:

1. Turbulent energy is produced from the energy of the mean motion due to the action of Reynolds stresses against the gradient of the mean velocity.

2. The energy of the mean motion is transmitted directly only to longitudinal fluctuations and is given by

$$A = -\rho\overline{u'w'}\frac{\partial\bar{u}}{\partial z}$$

3. The transverse velocity fluctuations receive energy from the longitudinal fluctuations due to the action of pressure fluctuations and are given by the quantities

$$\overline{p'\frac{\partial v'}{\partial y}} \text{ and } \overline{p'\frac{\partial w'}{\partial z}}.$$

4. In a turbulent flow the energy is redistributed among the components, but the total energy of all three components is dissipated.

For the stationary case Eqs. (1.3) are simplified:

$$\left.\begin{aligned}
\rho\overline{u'w'}\frac{\partial\bar{u}}{\partial z} &= -\left(\overline{p'\frac{\partial v'}{\partial y}} + \overline{p'\frac{\partial w'}{\partial z}}\right) - \frac{\partial}{\partial z}\left(\frac{1}{2}\rho\overline{u'^2w'} - \nu\frac{\partial E_x}{\partial z}\right) - \rho\bar{\varepsilon}_x,\\
\overline{p'\frac{\partial v'}{\partial y}} &= \frac{\partial}{\partial z}\left(\frac{1}{2}\rho\overline{v'^2w'} - \nu\frac{\partial E_y}{\partial z}\right) + \rho\bar{\varepsilon}_y,\\
\overline{p'\frac{\partial w'}{\partial z}} &= \frac{\partial}{\partial z}\left(\frac{1}{2}\rho\overline{w'^3} + \overline{p'w'} - \nu\frac{\partial E_z}{\partial z}\right) + \rho\bar{\varepsilon}_z.
\end{aligned}\right\} \tag{1.4}$$

Proceeding from the idea of the spectral nature of turbulent motions and introducing a coefficient of turbulent viscosity K into the formula for the turbulent tangential stress

$$\overline{u'w'} = -K\frac{\partial\bar{u}}{\partial z}, \tag{1.5}$$

we can express the quantity of turbulent energy E_T produced in unit mass of air from the energy of the mean motion in the frequency range $0 < f < \infty$ in the following way:

$$E_T = \int_0^\infty E(f)\,df = -K(f)\left(\frac{\partial\bar{u}}{\partial z}\right)^2, \tag{1.6}$$

where $K(f)$ is the spectral coefficient of turbulent viscosity.

The quantity E_T can also be interpreted as the amount of energy of the main motion dissipated as a result of turbulent viscosity. Thus, a knowledge of the coefficient of turbulent viscosity as a characteristic of the turbulence field is of great theoretical and practical value.

The coefficient of turbulent viscosity in its physical sense corresponds to the coefficient of molecular viscosity (or thermal conductivity) of liquids and gases. In the treatment of turbulent flow, however, we are dealing with the transfer of moles (volumes) of liquids and gases.

3. Richardson Stability Criterion

By stratification of the atmosphere we mean the variation in density of the air in a vertical direction, i.e., the direction of the force of gravity. The vertical displacement of particles in the atmosphere from their equilibrium position involves a buoyancy force $(\rho - \rho_0)g$, where ρ is the particle density, ρ_0 is the density of the surrounding medium, and g is the gravitational acceleration. The work of the buoyancy force gives rise to an exchange between the potential energy of the stratified fluid in the field of gravity and the kinetic energy of the motions. Richardson assumed that the loss of turbulent energy due to work against the stable stratification would be much greater than the loss due to molecular viscosity, which converts the turbulent kinetic energy to heat, and to diffusion in the presence of a turbulent energy gradient. In this case the turbulent-energy balance equation can be put in the form

$$\frac{dE}{dt} = K\left(\frac{d\bar{u}}{dz}\right)^2 - K_T\frac{g}{\theta}\frac{d\theta}{dz}, \tag{1.7}$$

where K is the coefficient of turbulent momentum transfer, K_T is the coefficient of turbulent heat transfer, and θ is the potential temperature.† The quantity

$$\mathrm{Ri} = \frac{g}{\theta}\frac{\dfrac{d\theta}{dz}}{\left(\dfrac{d\bar{u}}{dz}\right)^2} \tag{1.8}$$

is called the Richardson number. Since

$$\frac{1}{\theta}\frac{d\theta}{dz} = \frac{1}{T}(\gamma_a - \gamma),$$

†The potential temperature $\theta = T(1000/p)^{0.288}$ is the temperature which an air particle assumes if it is displaced in dry adiabatic conditions from a level with pressure p to a level where $p = 1000$ mb.

expression (1.8) can be written in the form

$$\mathrm{Ri} = \frac{g}{T}\frac{(\gamma_a - \gamma)}{\left(\frac{d\bar{u}}{dz}\right)^2} \tag{1.9}$$

where γ_a and γ are the adiabatic and actual vertical temperature gradient, respectively. The quantity

$$\mathrm{Rf} = \alpha\frac{g}{T}\frac{(\gamma_a - \gamma)}{\left(\frac{d\bar{u}}{dz}\right)^2} \tag{1.10}$$

is called the dynamic (flux) Richardson number. In (1.10) $\alpha = K_{\mathrm{T}}/K$. It follows from (1.9) and (1.10) that

$$\mathrm{Ri} = \frac{1}{\alpha}\mathrm{Rf}. \tag{1.11}$$

The dynamic Richardson number defines the role of thermal convection in the generation of turbulent energy in comparison with the dynamic factors responsible for the transfer of energy from the mean motion to turbulence, whereas Ri can be interpreted as the ratio of the buoyancy force to inertial forces.

It follows from (1.7) and (1.10) that

$$\frac{dE'}{dt} = K\left(\frac{d\bar{u}}{dz}\right)^2(1 - \alpha\,\mathrm{Ri}) \tag{1.12}$$

and, hence, steady (nondecaying) turbulence is possible only when $\mathrm{Rf} = 1$ and, if the dissipation of turbulence energy is taken into account, when $\mathrm{Rf} < 1$.

The effect of the buoyancy force on the turbulent regime depends on the vertical gradient of potential temperature. When $d\theta/dz$ is positive (stable thermal stratification) the buoyancy force opposes the development of turbulence: Some of the turbulent energy is spent on overcoming the buoyancy force. When $d\theta/dz$ is negative (unstable thermal stratification, favoring the development of convection) the buoyancy force promotes the onset and development of turbulence. Thermal instability is of great importance in clouds and also at low heights in the cloudless troposphere. In the middle and upper troposphere outside the clouds, the main factor in the excitation of turbulence is probably the vertical shear of the mean wind velocity. On cloudless days with strong heating of the underlying surface, superadiabatic temperature gradients are observed from the Earth's surface up to 500–1000 m and sometimes even higher. At high altitudes such temperature gradients can be observed only in very thin layers, a few tens or hundreds of meters thick.

Finally, at values of $d\theta/dz$ close to zero (neutral thermal stratification), the buoyancy force has no effect on the turbulent regime.

The turbulence-inducing effect of the vertical gradient of the mean wind velocity in the atmosphere is largely compensated by the stabilizing effect of thermal stratification.

In the general case, turbulence can develop in a thermally nonuniform atmosphere when Ri is less than some critical value, which, as (1.12) indicates, will be less than $\alpha_1 = K/K_T$. The critical value of Ri characterizes the conditions of development and decay of turbulence in a thermally stratified atmosphere. At $\mathrm{Ri} < \mathrm{Ri}_{cr}$ the turbulence will probably show a tendency to increase more or less sharply, just as in the case of excitation of turbulence in relation to Re in conditions where the buoyancy force has no effect.

The value of the critical Richardson number has been estimated by many investigators mainly on the basis of theoretical considerations. The value of Ri_{cr} depends, as (1.12) indicates, on the value of K/K_T adopted. In the classical theory of turbulence, as is known, the transferred substance is assumed to be passive, i.e., its addition to the air has no effect on the turbulence regime. It follows from this that the turbulence coefficient does not depend on the kind of substance and, hence, $K = K_T$ and, consequently, $\mathrm{Ri}_{cr} = 1$.

In fact, momentum transfer between air masses can be effected by dynamic interaction without significant mixing of these masses, i.e., the indestructibility postulate is violated. Moreover, the nature of the temperature distribution may be such that the passivity postulate is violated. For instance, in an isothermal atmospheric layer, in which a vertical gradient of wind velocity can be observed, $K \neq K_T$ presumably. In a thermally stable atmosphere, momentum transfer is more rapid than heat transfer. Hence, there is no basis for asserting that $\mathrm{Ri}_{cr} = 1$ in the free atmosphere.

In meteorological literature it is often postulated that the turbulence coefficients for different substances (for heat and momentum) are proportional to one another and their ratio is constant. Yet laboratory investigations, and also investigations in the bottom layer of the atmosphere [75], have shown that α depends significantly on thermal and wind stratification. In particular, reduction of Ri increases the value of α, particularly sharply in the region $\mathrm{Ri} < 1$. In convective conditions $\alpha > 1$, and in the case of temperature inversion $\alpha < 1$. When $4 < \mathrm{Ri} < 10$, $\alpha = 0.3$–0.05. The dynamic Richardson number Rf for these conditions is much less than unity. Thus, turbulence can be excited at relatively large Ri. This must be borne in mind in the study of turbulence, particularly in the upper troposphere and in the stratosphere, where stable thermal stratification predominates.

We now consider the role of diffusion of turbulent energy. We can infer that the diffusive transport of turbulent energy (D) depends on the turbulence coefficient K and the eddy energy gradient dE'/dr. Taking into account the dimensions of these quantities, we can determine the diffusive flow of turbulent energy, accurate to a constant factor, from the following expression:

$$D = K^{1/3} \left[\frac{dE'}{dr} \right]^{4/3}. \tag{1.13}$$

Thus, D depends less on the turbulence coefficient K and is mainly determined by the degree of nonuniformity of the turbulence field.

Taking (1.10) into account, and also assuming that $K_T = K$, we obtain

$$\mathrm{Ri}_{cr} = 1 \pm \frac{\left[\frac{dE'}{dr} \right]^{4/3}}{K^{2/3} \left(\frac{d\bar{u}}{dz} \right)^2}. \tag{1.14}$$

It follows from (1.14) that the critical Richardson number, other conditions being equal, depends on the degree of nonuniformity of the turbulence field, i.e., dE'/dr, which depends in turn on the conditions causing turbulence of the flow. It also follows from (1.14) that in the case of a nonuniform turbulence field the critical Ri cannot be a constant.

The Richardson number characterizes the conditions of conversion of the kinetic energy of mean motion and the instability energy to kinetic energy of turbulent motion, but it does not provide a means of estimating such important characteristics as the magnitude of the fluctuating velocities and the thickness of the turbulent layer of the atmosphere.

The intensity of turbulence depends not only on Ri, although it is a function of this number. If $\mathrm{Ri} < \mathrm{Ri}_{cr}$, we merely have qualitative evidence that, first, turbulence develops and, second, that the smaller Ri, the greater the amount of accumulated turbulent energy that can be observed in steady conditions.

4. Structure and Energetics of Turbulent Flow

It is known from hydrodynamics that with increase in Reynolds number, large-scale fluctuations appear first. At very large Reynolds numbers in turbulent flow there are fluctuations with scales comparable with the characteristic lengths determining the flow dimensions as a whole and with scales at which turbulent energy is dissipated to heat due to molecular viscosity. According to Kolmogorov, turbulent motion in the atmosphere can be represented in the following way. Since the Reynolds numbers in the atmosphere are very large, the mean flow is unstable and is subject to disturbances–"turbulent eddies of the first order" characterized by random relative motions of individual fluid volumes with diameters of the order of the characteristic flow dimensions. The velocity of these relative displacements is less than the mean flow velocity. Since the Reynolds numbers for eddies of the first order are still very large and such eddies are also unstable, there arise "eddies of the second order" with smaller characteristic dimensions and velocities.

The successive breakdown of turbulent eddies will continue until for eddies of sufficiently high order n the Reynolds number

$$\mathrm{Re}^{(n)} = \frac{U^{(n)}L^{(n)}}{\nu} \tag{1.15}$$

is small enough to ensure that the effect of viscosity on eddies of order n becomes appreciable and prevents the formation of eddies of order $n + 1$.

The question of the energetic characteristics of turbulent motion merits special attention.

Modern thermodynamics gives the following picture of energy dissipation in a turbulent flow.

From large-scale eddies, which draw their energy from the main flow, the energy is transferred to smaller-scale eddies with hardly any dissipation. Thus, the kinetic energy of turbulence is continuously transferred from large eddies to small eddies, i.e., from low to high frequencies. This energy flux is dissipated as heat in eddies of the smallest scale.

Since energy transfer from large-scale to small-scale eddies proceeds without loss, the amount of energy dissipated as heat by viscosity can be determined (in order of magnitude) from the main-flow characteristics.

Maintenance of the "stationary" state requires external sources that continuously transfer energy to the main large-scale flow. If the external energy sources cease to act, the turbulent motions will die out.

It is obvious that in the atmosphere, along with the continuous or fairly long-acting stationary sources of turbulent energy, there may also be relatively short-lived sources.

In the investigation of the energetic properties of turbulent flow, it is essential first of all to find out how turbulent eddy motions are maintained in an air flow and to determine the sources and sinks of energy of fluctuating motion. It will be shown later that in the free atmosphere an appreciable part of the kinetic energy of the turbulent motion may be converted to potential energy.

As already mentioned, turbulent motions have a very broad spectrum of scales, from primary eddies with diameters of hundreds or even thousands of kilometers to eddies in which molecular viscosity becomes effective.

At the present time the properties of turbulence on scales much less than the main scale and at the same time much larger than the scales where the turbulent energy is dissipated as heat have been fairly well investigated theoretically. In this region of scales, which has been called the inertial subrange, the turbulence is assumed to be isotropic, i.e., the properties of turbulent motion are independent of direction.

The theory of locally isotropic turbulence, developed by Kolmogorov (1941), enables us to obtain valuable information about the local properties of turbulence in the inertial subrange directly from a dimensional argument.

The local properties of turbulence cannot be determined by viscosity, since its effect becomes appreciable only at the smallest scales. For the same reason the characteristic dimension and velocity of the motion as a whole cannot affect the properties of turbulence in the inertial subrange. According to Kolmogorov's hypothesis, there is a continuous flux of energy from the largest (primary) eddies to the smallest. In the inertial subrange, energy is transferred from eddy to eddy without loss and, therefore, the rate of energy transfer must be numerically equal to the dissipation rate ϵ.

The velocity u_l of turbulent motion on a scale of order l must be determined only by ϵ and by the value of l. Using dimensional theory, Kolmogorov obtained the expression

$$u_l \sim (\varepsilon l)^{1/3}, \tag{1.16}$$

i.e., the change in velocity over a distance l is proportional to the cube root of this distance. The same result was obtained simultaneously, but by a different method, by Obukhov (1941).

The quantity u_l can be regarded qualitatively as the velocity of a turbulent eddy with a characteristic scale of order l. Since $\epsilon =$ const in the inertial subrange, it follows directly from (1.16) that turbulent eddies are self-similar. This means that in a developed turbulent flow there are eddies of all scales entering the inertial subrange. Each eddy, in decaying, generates eddies similar to itself, i.e., satisfying (1.16), but having smaller dimensions.

5. Statistical Description of Turbulence Field

In spite of the complexity and disorder of the velocity field in a turbulent flow, turbulent motion must satisfy the basic equations of hydrodynamics:

the Navier–Stokes equation

$$\frac{d\vec{u}}{dt} + \vec{u}\nabla\vec{u} = -\frac{1}{\rho}\nabla p + \nu\Delta\vec{u}, \tag{1.17}$$

and the continuity equation

$$\nabla\vec{u} = 0. \tag{1.18}$$

The continuity equation is written for an incompressible fluid, which is correct for the condition

$$\frac{(\bar{u})^2}{a^2} \ll 1 \tag{1.19}$$

where a is the velocity of sound, since in this case, as is well known, density fluctuations can be neglected.

The system of equations (1.17) and (1.18) is sufficient for determining $\vec{u}$ as a function of $\vec{x}$ and t if initial and boundary conditions are specified. For example, in the case of homogeneous turbulence, the boundary conditions with respect to $\vec{x}$ are determined by the presence of statistical uniformity in space. The initial conditions must take into account that at a certain instant t_0 the velocity is a random function of the point and the probability laws characterizing this function are specified.

It is obvious that if the initial conditions are given in probabilistic form, the velocity field at subsequent times can be determined only in probabilistic form. One of the basic characteristics of a random quantity is its distribution function $P(\vec{u})$. The interpretation of u as a random quantity refers to each point $\vec{x}$, t. In general, however, there must be a statistical relation between random quantities at different space–time points. The joint probability distribution function for the values of $\vec{u}$ at several given space–time points can serve as a characteristic of this statistical relation.

A turbulence field of infinite extent is determined statistically by the complete set of joint probability distributions for values of the velocity vector $\vec{u}(\vec{x}, t)$ at any n space–time points. It must be borne in mind here that the statistical characteristics of the turbulence field for various values of t are uniquely related to one another by the equations of hydrodynamics (1.17) and (1.18).

In describing a turbulence field we usually restrict ourselves to mean quantities that correspond to small n. Of these mean quantities with small n the correlation tensor of the velocity field (a two-point moment of second order) and the tensor obtained from it by a Fourier transformation are particularly important in the theory of turbulence.

The correlation tensor of the velocity field for two points connected by the vector $\vec{r}$ has the form

$$R_{jk}(\vec{r}) = \overline{u_j(\vec{x}) \cdot u_k(\vec{x} + \vec{r})}\,; \tag{1.20}$$

where $u(\vec{x})$ is a continuous function of $\vec{x}$, since discontinuities due to viscous forces cannot be conserved in time. It follows from this that $R_{jk}(\vec{r})$ is continuous for all values of $\vec{r}$. For $j = k$, we have

$$R_{jj}(\vec{r}) \leqslant R_{jj}(0). \tag{1.21}$$

For the whole class of random processes called stationary there is a theory developed by Khinchin. A random process determined by the set of variables $Y_k(t)$ is called stationary in the case where the probability distribution laws for two groups of values of these variables

$$[Y(t_1), Y(t_2), \ldots, Y(t_n)] \text{ and } [Y(t_1+\tau), Y(t_2+\tau), \ldots, Y(t_n+\tau)]$$

are identical with one another, and n, t, and τ can be selected quite arbitrarily.

Proceeding from Khinchin's theory, Cramer came to the following conclusion: If the tensor $R_{jk}(\vec{r})$ is to be a correlation tensor of a continuous stationary random process, it is necessary and sufficient that this tensor can be put in the form

$$R_{jk}(\vec{r}) = \int_{-\infty}^{\infty} S_{jk}(\vec{\Omega})\, e^{i\vec{\Omega}\vec{r}}\, d\vec{\Omega} \tag{1.22}$$

($\vec{\Omega}$ is the wave vector, and $i = \sqrt{-1}$) provided that

$$\int_{-\infty}^{\infty} S_{jk}(\vec{\Omega})\, d\vec{\Omega} < \infty\,. \tag{1.23}$$

From the Fourier transformation inverse to (1.22) it follows also that

$$S_{jk}(\vec{\Omega}) = \frac{1}{8\pi^3} \int_{-\infty}^{\infty} R_{jk}(\vec{r})\, e^{-i\vec{\Omega}\vec{r}}\, d\vec{r}\,. \tag{1.24}$$

When $\vec{r} = 0$, Eq. (1.23) takes the form

$$R_{jk}(0) = \overline{u_j(\vec{x}) \cdot u_k(\vec{x})} = \int_{-\infty}^{\infty} S_{jk}(\vec{\Omega})\, d\vec{\Omega}\,. \tag{1.25}$$

It is clear from Eq. (1.25) that $S_{jk}(\vec{\Omega})$ is the density of $u_j u_k$ in wave space. The tensor $u_j(\vec{x}) \cdot u_k(\vec{x})$ defines the energy per unit mass of fluid and is called the energy tensor.

The tensor $S_{jk}(\vec{\Omega})$ represents the energy distribution over different wave numbers in an expansion of the velocity field in terms of harmonic components and is called the spectral energy tensor.

Because of its explicit physical meaning, the tensor $R_{jk}(\vec{r})$ and also the tensor $S_{jk}(\vec{\Omega})$ related to it by the Fourier transformation are the most important quantities characterizing a turbulence field.

The spectral and correlation tensors are functions of vector arguments. In a real experiment only one of the velocity components is usually measured. The correlation and spectral functions, which depend on one scalar argument, can be obtained by averaging $R_{jk}(\vec{r})$ and $S_{jk}(\vec{\Omega})$ over all directions of the vector arguments $\vec{r}$ and $\vec{\Omega}$. In the case of isotropic turbulence these tensor functions of the vector modulus play an important role.

The structure functions introduced by Kolmogorov

$$D_j(r) = \overline{[u_j(x_1 + r_1, x_2 + r_2, x_3 + r_3, t) - u_j(x_1, x_2, x_3, t)]^2}, \tag{1.26}$$

where u_j is the velocity component in direction j, are often used, in addition to the correlation and spectral functions, to characterize the internal properties of locally isotropic and homogeneous turbulent flow.

Thus, $D(r)$ is the mean square difference of the values of u_j at two points separated by a distance r:

$$r = \sqrt{r_1^2 + r_2^2 + r_3^2}.$$

It was in terms of the structure functions that the relation between the velocity of fluctuating motion and the fluctuation scales became widely known as the Kolmogorov-Obukhov "two-thirds" law:

$$D_j(r) = A(\delta_j)\,\varepsilon^{2/3} r^{2/3}, \tag{1.27}$$

where $A(\delta_j)$ is a dimensionless coefficient that depends on the angle between x_j and $\vec{r}$.

If the investigated random process satisfies the stationarity condition, the statistical characteristics of this process are uniquely related to one another:

$$D(r) = 2[R(0) - R(r)], \tag{1.28}$$

and since $R(\infty) = 0$ and, hence $D(\infty) = 2R(0)$, then

$$R(r) = \frac{1}{2}[D(\infty) - D(r)]. \tag{1.29}$$

Proceeding from (1.22), we can express the correlation function in the following way:

$$R(r) = \int_{-\infty}^{\infty} S(\Omega)\, e^{i\Omega r}\, d\Omega. \tag{1.30}$$

From the definition of homogeneity $R(r) = R(-r)$ and, hence,

$$R(r) = \int_{-\infty}^{\infty} S(\Omega) \cos(\Omega r)\, d\Omega. \tag{1.31}$$

From (1.30) we can obtain an expression for the spectral function

$$S(\Omega) = \frac{1}{2\pi} \int_{-\infty}^{\infty} R(r)\, e^{-i\Omega r}\, dr. \tag{1.32}$$

The spectral function $S(\Omega)$ is the spectral energy density per unit mass. In the literature on turbulence $S(\Omega)$ is often called the turbulence energy spectrum. The structure function and energy spectrum are connected with one another by the relation

$$D(r) = 2 \int_{-\infty}^{\infty} (1 - e^{i\Omega r}) S(\Omega) \, d\Omega . \tag{1.33}$$

This relation is valid also when the random process has a linearly varying mean value [107].

We have already mentioned that in a real experiment only one velocity component is usually measured and, hence, the statistical characteristics obtained by analysis of the experimental data describe a set of projections of the velocity vector on the selected direction.

It is often difficult to obtain the turbulence energy spectrum from experimental data owing to instrumental problems and because it is impossible to carry out sufficiently long series of independent measurements. In this case attention should be confined to the simpler statistical characteristics of turbulence, such as the variance of the velocity $(\overline{u'^2})$ and the standard deviation $\sigma_u = (\overline{u'^2})^{1/2}$. The latter quantity is often used as quantitative estimate of the average intensity of turbulence. For a quantitative description of the relative intensity of turbulence the ratio

$$\psi = \frac{\sigma_u}{\bar{u}},$$

is frequently used in the literature and is sometimes called the gustiness. The quantities $\overline{u'^2}$ and σ_u can be obtained, in particular, from the energy spectrum, in which case the range of scales (frequencies) for which the corresponding characteristics are calculated must be specified. For instance, the variance of the velocity is defined as

$$\overline{u'^2_{\Omega_{1-2}}} = \int_{\Omega_1}^{\Omega_2} S(\Omega') \, d\Omega' .$$

The result of such a calculation characterizes the intensity of fluctuations of specific scales. In order to compare the data obtained by different investigators, it is necessary to know the scale interval for which the integral characteristics of the turbulence have been determined. As will be shown in Chapter 2, each experimental method of investigation permits measurement of only a limited portion of the turbulence spectrum. To compare the turbulence intensity in the region of different scales, it is more convenient to use the rate of energy transfer in the turbulence spectrum, and not the integral characteristics. This problem will be considered below.

An essential feature of experimental investigations of the structure of atmospheric turbulence is either that the measurements are made at one point in space over a long period, or the measuring instrument moves through space at a relatively high velocity. If a single measuring instrument is used, the result of the experiment represents the random process measured in a particular time interval, but this time interval has a quite different meaning for each case. In the first case it is the time in which the investigated process occurs, and in the second case it is the time of "instantaneous" measurement in a particular direction.

In the first case the instrument moves relative to the flow with the mean flow velocity. For conversion from a temporal to a spatial representation, we can use Taylor's frozen turbulence field hypothesis, i.e., that the shape of the disturbances is not significantly altered as the disturbances move along with the air flow at the mean flow velocity. This hypothesis, based on the assumption that the mean flow velocity is fairly high in comparison with the velocity fluctuations, has been experimentally confirmed for turbulent disturbances of comparatively small scales of several kilometers. It has not been confirmed for disturbances of greater scale. We can postulate that Taylor's frozen turbulence hypothesis may not be applicable to scales close to synoptic. In particular, significant errors may be introduced when this hypothesis is applied to turbulent motion in which the energy of the turbulent fluctuations is commensurable with that of the mean motion.

Thus, numerical conversion from a temporal to spatial representation is not possible in every case. In Monin's opinion, however, a conversion on the basis of the frozen turbulence hypothesis can be made for fluctuations with a period of up to four days.

We shall return later, in the appropriate sections of the book, to a more detailed examination of these questions.

If the measuring instrument moves at a velocity many times greater than the mean flow velocity, the frozen turbulence hypothesis is not required for conversion from temporal to spatial characteristics, and it is sufficient to know only the velocity of the instrument. It should be noted that if we are to regard the representation obtained in this way as a stationary random process and apply the above relations, it is essential that the turbulence field in the direction of motion of the instrument should be homogeneous.

6. Energy Spectrum of Turbulence

As was stated in Section 5, the energy spectrum is the decomposition of the kinetic energy into a Fourier integral of the wave numbers $\Omega = 2\pi/L$. In other words, $S(\Omega)d\Omega$ is the amount of kinetic energy per unit mass contained in, or carried by, eddies with dimensions characterized by wave numbers from Ω to $\Omega + d\Omega$.

Thus,

$$S(\Omega) = \frac{1}{2}\frac{dV_\Omega^2}{d\Omega}. \tag{1.34}$$

It is very easy to obtain an expression for the energy spectrum of turbulence in the wave-number region where the turbulence satisfies the conditions of homogeneity and local isotropy. According to Kolmogorov's definition, the velocity V_Ω in the wave-number region called the inertial subrange depends on the wave number Ω and the dissipation rate ϵ. An analytical expression for the spectrum of the velocity vector in the inertial subrange can be obtained from dimensional considerations:

$$S(\Omega) = c\varepsilon^{2/3}\Omega^{-5/3}. \tag{1.35}$$

In the inertial subrange the three-dimensional turbulence spectrum and, accordingly, the one-dimensional spectra of the components u (along the main flow), v, and w (across the main flow) are given by the formulas

$$\begin{aligned} S_u(\Omega) &= c_1\varepsilon^{2/3}\Omega^{-5/3}, \\ S_v(\Omega) &= S_w(\Omega) = c_1'\varepsilon^{2/3}\Omega^{-5/3}. \end{aligned} \tag{1.36}$$

For the coefficients c, c_1, and c_1', which have the nature of universal constants, we have the following relations:

$$\begin{aligned} c &= \frac{55}{18}c_1, \\ c_1' &= \frac{4}{3}c_1. \end{aligned} \tag{1.37}$$

The universal constants A for the structure functions of the fluctuations of the wind-velocity components, u, v, and w in expression (1.27) and the universal constants for the spectrum are connected by the relations $A_1 = 4c_1$ and $A_1' = \frac{4}{3}A_1$. According to recent data [75], $A_1 \approx 1.9$ and $c_1 \approx 0.48$ if Ω is measured in radians per unit length, and $c_1 = 0.48/(2\pi)^{2/3} = 0.14$ if Ω is measured in reciprocal length units.

Relation (1.35) can also be obtained from the "two-thirds" law for the structure function. It has been called the Kolmogorov spectrum or the "minus five-thirds" spectral law. The "minus five-thirds" law is valid only for the inertial subrange, i.e., for the region of wave numbers greater than those at which the energy of the main flow enters the turbulence spectrum due to instability, and less than those where the "consumption" of turbulent energy—dissipation as heat due to molecular viscosity—occurs. Within the inertial subrange the only mechanism of energy transfer is inertial transfer from eddies with low Ω to eddies with high Ω. It was on this basis that Kolmogorov and Obukhov equated the rate of energy transfer through a point Ω in the spectrum to the dissipation rate ϵ.

Thus, the spectrum of the wind-velocity field in the inertial subrange is uniquely determined by the quantity ϵ and the single universal constant c. The value of ϵ can be calculated from (1.36) if the spectral density for any wave number belonging to the inertial subrange is known.

On the assumption that the process of energy transfer from large to small eddies is similar to conversion of mechanical energy to heat by molecular motion, Heisenberg introduced a term representing the conversion of energy from large-scale motions to small-scale motions (in addition to the term representing molecular dissipation) into the energy equation:

$$-\frac{\partial}{\partial t}\int_0^{\Omega} S(\Omega)' \, d\Omega' = 2\nu \int_0^{\Omega} S(\Omega')\Omega'^2 \, d\Omega' + \qquad (1.38)$$

$$+ 2a \int_{\Omega}^{\infty} [S(\Omega')]^{1/2}\Omega'^{-3/2} \, d\Omega' \int_0^{\Omega} S(\Omega'')\Omega''^2 \, d\Omega''.$$

The left side of the equation represents the complete dissipation of kinetic energy in unit mass per unit time in motions with scales greater than $L = 2\pi/\Omega$, while the terms on the right side represent the various mechanisms of dissipation and conversion of energy.

The term

$$2\nu \int_0^{\Omega} S(\Omega')\Omega'^2 \, d\Omega'$$

gives the dissipation of kinetic energy directly to thermal energy due to molecular viscosity. The term

$$2a \int_{\Omega}^{\infty} [S(\Omega')]^{1/2}\Omega'^{-3/2} \, d\Omega' \int_0^{\Omega} S(\Omega'')\Omega''^2 \, d\Omega''$$

(a is an absolute constant of order unity) represents the conversion of kinetic energy of large-scale motions to that of small-scale motions. This conversion is a characteristic feature of turbulence.

Since the viscosity ν, accurate to a numerical factor, is equal to the product of the mean free path and the velocity of the molecules, then the turbulent viscosity, by analogy, will be the product of the mean free path $L = 2\pi/\Omega$ of small eddies and their velocity:

$$v_{\Omega} \backsim \sqrt{S(\Omega)\, d\Omega} \, .$$

Energy transfer from motions with scales larger than L to motions with small scales is effected, as Heisenberg postulated, by all eddies with scales less than L. Hence, the turbulent viscosity is given by the expression

$$\nu_{\text{turb}} = a \int_{\Omega}^{\infty} [S(\Omega')]^{1/2} \Omega'^{-3/2} \, d\Omega'. \tag{1.39}$$

The solution of Heisenberg's equation (see [137]) has the form

$$S(\Omega) = \left(\frac{8\varepsilon}{9a}\right)^{2/3} \Omega^{-5/3} \left[1 + \frac{8\nu^3}{3a^2\varepsilon} \Omega^4\right]^{-4/3} \tag{1.40}$$

For $\Omega \ll (3a^2 \epsilon/8\nu^3)^{1/4}$, i.e., for relatively small Ω, where viscosity still has no effect, expression (1.40) becomes the "minus five-thirds" spectral law. For $\Omega > (3a^2 \epsilon/8\nu^3)^{1/4}$, where molecular dissipation plays an appreciable role, the expression for the energy spectrum at the limit takes the form

$$S(\Omega) \backsim \Omega^{-7}. \tag{1.41}$$

The physical idea on which Heisenberg's formula is based is more applicable to the case of energy transfer between widely differing wave numbers than for the more important case of energy transfer between wave numbers of the same order of magnitude. Experimental data do not confirm relation (1.41), but some qualitative inferences can still be drawn from this relation. In particular, the introduction of additional energy "consumption" (dissipation as heat) at large Ω leads to an increase in the absolute value of the exponent of Ω if the energy spectrum is expressed as a power function of Ω of the form $S(\Omega) \propto \Omega^{-n}$.

It follows from the expounded ideas that a turbulent flow is a superposition of eddies of different sizes–from some minimum dimension l_0 to a dimension L_0 comparable with the scale of the flow as a whole. Because of the effect of viscosity the size of eddies cannot be reduced indefinitely. Generally speaking, the smaller the eddy, the larger the velocity gradient across the eddy, i.e., the greater the viscous shear stresses that oppose the eddy motion. Thus, in any turbulent flow there must be a statistical lower limit to the size of the smallest eddy. In the region of turbulence scales where $l \leqslant l_0$ the kinetic energy of turbulence is dissipated directly as heat.

Figure 1.1 shows the curve of spectral density of turbulent energy (energy spectrum) for a flow in which the effect of gravity can be ignored. In the actual atmosphere this corresponds to the case of neutral thermal stratification. Region a in Fig. 1.1 is the region of large (anisotropic) eddies, where turbulent energy is supplied from the energy of the main flow and there is also energy transfer from large eddies to small eddies. Region b is the region of dissipation of turbulent energy due to viscosity; this is the viscous range. The intermediate region c is the inertial subrange. In this region production and dissipation of turbulent energy

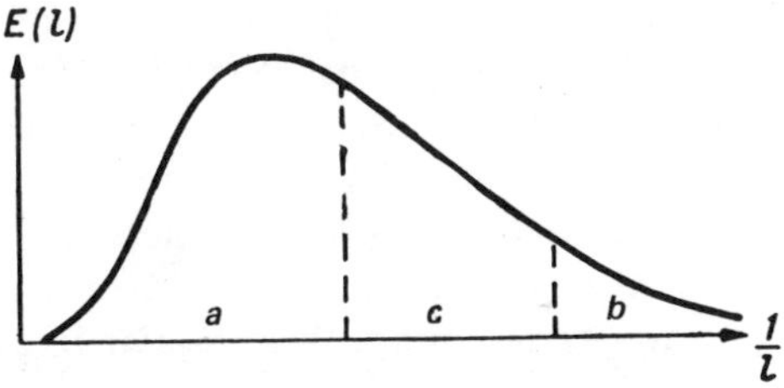

Figure 1.1. Spectral density of turbulent energy.

are small in comparison with the influx and consumption of energy due to transfer of energy over the spectrum by the "breakdown" of eddies. The lower high-frequency boundary of the inertial subrange is the region where the spectrum is determined by the rate of energy dissipation ϵ and the kinematic viscosity ν. On the basis of dimensional considerations we can compile from these quantities a turbulence length scale in which turbulent energy is dissipated:

$$l_0 = \left(\frac{\nu^3}{\varepsilon}\right)^{1/4}. \tag{1.42}$$

Near scale l_0, called the Kolmogorov microscale, the dissipation is greatest. In the region of scales smaller than l_0 the spectral density curve drops sharply due to viscous destruction of the eddies. The lower boundary of the inertial subrange (i.e., the boundary on the high-frequency side) is taken as l_0', which is approximately an order of magnitude greater than the Kolmogorov microscale l_0. In the region of l_0' the effect of viscosity becomes appreciable. The value of l_0' in the surface layer of the atmosphere, as experimental investigations have shown, is approximately 4 mm. In the scale region $l_0' \ll l \ll L_0$ the main factor is inertial transfer of turbulent energy over the spectrum without loss.

It should be noted that in the free atmosphere, which, as already mentioned, is characterized by large Re, the inertial subrange is fairly large. This is confirmed by experimental data.

The discussion so far has been concerned with the case of fully developed turbulence, where the source of turbulent energy acts continuously, making up for the energy loss due to dissipation as heat or to other factors. If the energy source ceases to act, the turbulence decays. The decay of turbulence can be divided into three stages: initial, transitional, and final.

If thermal stratification is ignored, the variation of kinetic energy in a neutrally stratified atmosphere is given by expression (1.38). In the initial stage of decay of turbulence the energy dissipation due to viscosity is compensated by the influx of energy from smaller wave numbers due to turbulent mixing. When the source of turbulent energy is removed, however, there eventually comes a time when an appreciable flow of energy from low to high wave numbers, comparable in magnitude with the dissipation rate, can no longer be maintained. In this final stage of decay, redistribution of energy over the turbulence spectrum can be ignored in comparison with dissipation processes.

The flow ceases to be turbulent in the ordinary sense of the word: Turbulent mixing ceases and there is only a quiet "laminar" decay of particular large-scale disturbances that have survived since the period of turbulent flow [75].

In the spectral theory of turbulence, developed for flows in tubes [112], the energy spectrum in the final stage of decay is given by the expression

$$S(\Omega,\ t) \backsim \Omega^4 e^{-2\nu\Omega^2 t}. \tag{1.43}$$

The law of decay of turbulence intensity is given by

$$u'^2 \backsim t^{-5/2}. \tag{1.44}$$

Although the applicability of relations (1.43) and (1.44) to the atmosphere requires theoretical and experimental verification, it is obvious that in the decay stage in the atmosphere small-scale eddies will decay rapidly, whereas large eddies, which are incapable of transferring their energy to smaller eddies, may exist for a fairly long time.

7. Interaction of Mean and Turbulent Fields of Temperature and Wind Velocity

As already mentioned, the main source of turbulent energy is the energy of the main motion. The intensity of turbulence and the turbulent transfer of momentum or any conservative impurities in the atmosphere depend not only on the velocity gradients of the main motion (vorticity),† but also on the nature of the interaction of the vorticity ($d\bar{u}/dz$) of the main flow and the vorticity of turbulent motion ($d\bar{u}'/dz$).

Tchen distinguishes the case where the vorticity of the main motion is small in comparison with the turbulence vorticity in the frequency (or scale) range considered and the case where they are comparable with one another. In the first case there should be a barely detectable interaction, and the vorticity of the main motion serves only to transfer the kinetic energy of the mean motion to turbulent energy. If the vorticities of the main motion and the turbulent motion are comparable with one another, however, they can interact very strongly with one another. In certain conditions this interaction, according to Tchen, can lead to strong resonance.

†Strictly speaking, the vorticity rot $\vec{V}$ (its component along the horizontal axis y, perpendicular to the main flow) is given by the formula rot $\vec{V} = du/dz - dw/dx$. Since the second term is much smaller than the first, however, the velocity gradient du/dz (the wind shear) is often called the vorticity.

Strong interaction can occur in the region of the sharp changes in the vertical profile of the wind velocity. In the real atmosphere the interaction (weak or strong) is determined not only by the vertical gradients of the mean wind velocity, but also by the vertical gradients of air temperature.

We consider the spectral characteristics of turbulence, taking into account the degree of interaction of the vorticities of the main and turbulent motions. To determine the spectral characteristics of turbulence we mainly follow Gisina [34] and use the spectral equation obtained from the equations of motion and continuity by averaging and expanding in Fourier integrals.

For the equilibrium region of the spectrum $\Omega \gg L_0^{-1}$, and on the assumption of local uniformity of the turbulence this equation has the form

$$\varepsilon = 2\nu \int_0^{\Omega} \Omega'^2 S(\Omega')\, d\Omega' - \frac{d\bar{u}}{dz} \int_{\Omega}^{\infty} \tau(\Omega')\, d\Omega' + F(\Omega) + \beta \int_{\Omega}^{\infty} H(\Omega')\, d\Omega', \tag{1.45}$$

where $\tau(\Omega')$ is the spectrum of the turbulent friction stress, $F(\Omega)$ is a function characterizing the transfer of turbulent energy in the wave-number range from zero to Ω, i.e., to fluctuations with higher wave numbers, $H(\Omega')$ is the spectrum of the vertical heat flux, and $\beta = g/T$ is the buoyancy parameter.

Following Heisenberg, we can put

$$F(\Omega) = K(\Omega) \cdot 2 \int_0^{\Omega} \Omega'^2 S(\Omega')\, d\Omega', \tag{1.46}$$

where

$$K(\Omega) = a \int_{\Omega}^{\infty} [S(\Omega')]^{1/2} \Omega'^{-3/2}\, d\Omega' \tag{1.47}$$

is the coefficient of turbulent viscosity and a is a numerical coefficient.

In the case of weak interaction of the main and turbulent motions in the wave-number region $\Omega \gg L_0^{-1}$

$$\int_{\Omega}^{\infty} \tau(\Omega')\, d\Omega' = -K(\Omega) \frac{d\bar{u}}{dz}, \tag{1.48}$$

and

$$\int_{\Omega}^{\infty} H(\Omega')\, d\Omega' = -\alpha K(\Omega) \frac{dT}{dz}, \tag{1.49}$$

where α is the ratio of the coefficients of turbulent mixing for heat and momentum.

In the case of strong interaction of the main and turbulent motions, we have

$$\int_{\Omega}^{\infty} \tau(\Omega')\,d\Omega' = -K(\Omega)\left[2\int_{0}^{\Omega} \Omega'^2 S(\Omega')\,d\Omega'\right]^{1/2}, \tag{1.50}$$

and

$$\int_{\Omega}^{\infty} H(\Omega')\,d\Omega' = -\alpha K(\Omega)\left[\int_{0}^{\Omega} \Omega'^2 S_T(\Omega')\,d\Omega'\right]^{1/2}, \tag{1.51}$$

where $S_T(\Omega)$ is the spectrum of temperature fluctuations.

Substituting expressions (1.46), (1.48), and (1.49) in (1.45) we obtain the spectral equation for the case of weak interaction of the main and turbulent motions:

$$\varepsilon = 2(\nu + K(\Omega))\int_{0}^{\Omega} \Omega'^2 S(\Omega')\,d\Omega' + a_1 K(\Omega), \tag{1.52}$$

where

$$a_1 = \left(\frac{d\bar{u}}{dz}\right)^2 - \alpha\beta\,\frac{dT}{dz}.$$

The solution of (1.52) with expression (1.47) taken into account gives

$$S(\Omega) = \left(\frac{8}{9a}\right)^{2/3} (\varepsilon + a_1\nu)^{2/3}\Omega^{-5/3}\left[1 + \frac{8\nu^3\Omega^4}{3a^2(\varepsilon + a_1\nu)}\right]^{-4/3} \tag{1.53}$$

It is apparent that in the case of weak interaction the energy spectrum is characterized by the "minus five-thirds" law. The effect of nonuniformities of the fields of mean temperature and mean wind velocity, determined by the quantity $a_1\nu$, is manifested in changes in the dissipation of turbulent energy. The gradients of the temperature and mean wind velocity constitute an additional source of energy.

Strong interaction of the main and turbulent motions can be observed only in the region of turbulence scales commensurable with, or greater than, the scales of variation of the mean fields of temperature and wind velocity. Hence, we can confine ourselves to a consideration of the wave-number interval $L_0^{-1} \ll \Omega \ll l_0^{-1}$.

For this wave-number range the effect of viscosity can be neglected. Substituting expressions (1.46), (1.50), and (1.51) in (1.45), we obtain

$$\varepsilon = K(\Omega)\cdot 2\int_0^{\Omega} \Omega'^2 S(\Omega')\,d\Omega' + \frac{d\bar{u}}{dz}K(\Omega)\left[2\int_0^{\Omega} \Omega'^2 S(\Omega')\,d\Omega'\right]^{1/2} -$$

$$-\alpha\beta K(\Omega)\left[\int_0^{\Omega} \Omega'^2 S_T(\Omega')\,d\Omega'\right]^{1/2} \tag{1.54}$$

Assuming that in the above-indicated wave-number region the interaction between the main and turbulent motions plays a greater role than interaction between eddies of different scales, we can neglect the function $F(\Omega)$. The solution of (1.54) when these assumptions are made and (1.47) is taken into account gives the following expression for the energy spectrum:

$$S(\Omega) = \frac{\varepsilon}{a\dfrac{d\bar{u}}{dz}}\left(1 + \frac{\beta N}{\varepsilon\dfrac{dT}{dz}}\right)\Omega^{-1}, \tag{1.55}$$

where N is the rate of smoothing-out of temperature nonuniformities. The assumption made is valid for wave numbers

$$\Omega \ll \frac{d\bar{u}}{dz}\left[\frac{\varepsilon}{a\dfrac{d\bar{u}}{dz}}\left(1 + \frac{\beta N}{\varepsilon\dfrac{dT}{dz}}\right)\right]^{-1/2} \tag{1.56}$$

It follows from (1.55) that strong interaction of the main and turbulent fields of temperature and wind velocity leads to a change in shape of the energy spectrum in comparison with the shape of the spectrum for conditions where the "minus five-thirds" law is fulfilled.

The above assumption of local uniformity is applicable in principle also in the case where the vertical gradients of the wind velocity vary with height.

Panchev and Syrakov [197], developing the idea of the role of interaction of the mean and turbulent fields of temperature and wind velocity in the formation of the fluctuation spectra, showed that the shape of the spectrum depends on the sign and magnitude of the dimensionless parameter

$$m = \beta N/\varepsilon\gamma\,. \tag{1.57}$$

In the case of a positive value of the vertical temperature gradient γ, when $m > 0$, the modulus of the exponent in the spectral law for the inertial subrange, which includes the buoyancy subrange, is more than $\frac{5}{3}$ and increases with increase in m, reaching a value of 3 for $m > 10$. The question of the effect of stable thermal stratification on the shape of the spectrum is considered more thoroughly in Chapter 4.

In the case of a negative value of γ, when $m < 0$, the spectrum has a special kind of "step" followed by (for the region of smaller wave numbers) transition to a gradient with the exponent in the spectral law equal to $-\frac{5}{3}$. "Steps" in the spectrum, as Panchev and Syrakov's calculations showed, are particularly prominent in the spectrum of temperature fluctuations.

2

Methods of Experimental Investigation of Turbulence in the Free Atmosphere

1. The Airplane as a Means of Investigating Atmospheric Turbulence

A characteristic feature of methods of investigating turbulence in the bottom layer of the atmosphere is that the measuring instruments are mounted on special structures (towers, masts, etc.) that are stationary relative to the ground.

Air motions measured by such instruments are absolute in a coordinate system associated with the Earth. Air velocities at high altitudes are measured by sounding balloons, which are carried along by the air flow. By measuring the velocity of balloons by radar or other methods from the ground, we can determine the velocity of the air flow itself.

The airplane occupies an intermediate position as a device for measuring motions in the atmosphere: On one hand, it can be used as a "platform" on which measuring instruments are mounted and, on the other, the airplane is carried along by the air flow and in this sense is a measuring instrument itself.

It is obvious that an airplane will be entrained better by atmospheric motions, the greater the scale of these motions. On the other hand, for very small-scale motions an airplane is effectively a fixed platform or, what is virtually the same, a platform moving at constant velocity.

Atmospheric turbulence is characterized by random motions of the most diverse scales, some of which entrain the airplane well, while others have practically no effect on the airplane as a whole.

Most of the experimental data on turbulence in the free atmosphere have been obtained by means of specially equipped airplanes ("flying laboratories").

To measure the mean values of the main meteorological parameters, these flying laboratories are equipped with electrometeorographs—instruments which provide quantitative thermodynamic characteristics of the medium in which the turbulent motions take place.

In the following sections we consider airplane methods of measuring the characteristics of turbulence proper.

The apparatus for measuring the characteristics by these methods is usually specially constructed for the purpose.

In describing airplane methods of investigating turbulence we use the conventional aerodynamic coordinate system in which the origin is at the center of gravity, the x axis coincides with the direction of flight and is called the longitudinal aircraft axis, the y axis for an aircraft flying horizontally is directed vertically upwards, and the z (transverse) axis is perpendicular to the plane $x0y$.

2. Method Using Reaction of Airplane to an Atmospheric Gust (Load Method)

An airplane entering a turbulent region is acted on by the disturbed flow and begins to oscillate, being subject to an additional load (buffeting).† The very fact of buffeting of an airplane in flight led to the idea of using data on the intensity of buffeting to assess the intensity of atmospheric turbulence. The loads in this case are measured by special instruments–accelerometers.

The simplest one-component accelerometer, shown in Fig. 2.1, is an inert mass m suspended by a spring S from the body of the instrument. The accelerometer is equipped with a damper D, which opposes the displacement of the mass m relative to the instrument body with a force proportional to the rate of displacement. The equation of free motion of the accelerometer has the form

$$m\frac{d^2x}{dt^2} + D\frac{dx}{dt} + kx = 0, \tag{2.1}$$

where D is the damping coefficient, k is the rigidity of the spring, and x is the displacement of the lower end of the spring from the position corresponding to $m = 0$. The nature of the decaying oscillations depends on the degree of damping. In a critically damped system with $D_{cr} = 2(mk)^{1/2}$ the decay will be aperiodic, and if $D = 0$ nondecaying oscillations with frequency $\omega_0 = (k/m)^{1/2}$ will take place. For accelerometers in general, $0 < D < D_{cr}$. In the undisturbed state the spring of an accelerometer with a vertical working axis will be acted on by a force $F = mg$ and, hence,

$$x_0 = \frac{m}{k}g. \tag{2.2}$$

†The load of an airplane is the ratio $n = A/G$, where A is the lift force and G is the weight of the airplane. It is obvious that in quiet horizontal flight at constant velocity, $n = 1$. The load is usually measured as a fraction of the gravitational acceleration g.

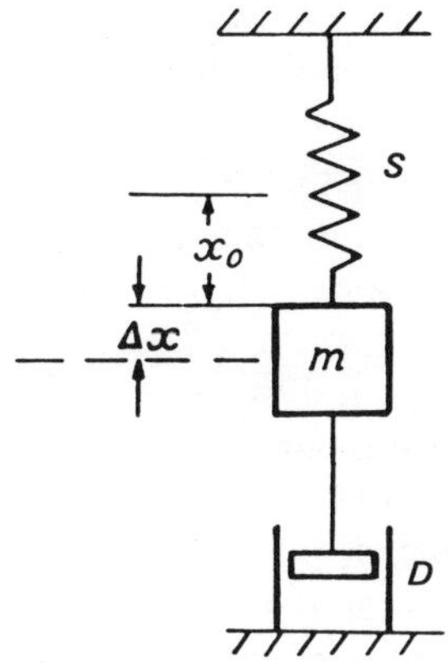

Figure 2.1. Diagram of accelerometer.

When the mass moves vertically with constant acceleration a, we have

$$x = x_0 + \Delta x = \frac{m}{k}(g + a), \tag{2.3}$$

from which

$$\Delta x = \frac{m}{k}a. \tag{2.4}$$

The action of a periodic external force on the accelerometer gives rise to forced oscillations with a frequency equal to the forcing frequency. The frequency characteristic of an accelerometer is uniform at frequencies from zero to ω_0 and then falls off. The accelerometer "piles up" frequencies greater than its natural frequency.

The natural frequency of modern aircraft accelerometers lies in the range 8–14 Hz.

In addition to mechanical accelerometers, there are piezoelectric accelerometers. Piezoelectric accelerometers have a higher natural frequency and better sensitivity.

The investigation of atmospheric turbulence by an airplane can be conducted in various ways. First, the spatial characteristics of regions in which the airplane in flight is subject to buffeting can be investigated, the lifetime of these regions can be determined, their relation with the fields of mean meteorological parameters can be studied, and so on. In this "macroscopic" approach to the study of turbulence, its intensity is assumed to be equal to the intensity of buffeting, and is classified as light, moderate, severe, and very severe. Such estimates are mainly qualitative, of course, and ignore the differences in reaction of the airplane to gusts of different form.

Second, the microstructure of turbulence can be studied. In this case the subject of study is the quantitative characteristics of turbulent fluctuations in the atmosphere, and, hence, it is essential to derive as accurate a relation as possible between the reaction of the airplane to the turbulent gust and the parameters of the gust itself—in other words, the airplane in this case becomes a turbulence-measuring instrument.

Strictly speaking, the turbulence-measuring instrument is the aircraft–accelerometer system. This instrument has a well-defined transmission band as regards scale of disturbances. The airplane responds to disturbances of different scales as disturbances of different frequency. The frequency of the disturbance depends on the scale of the disturbances and the airplane flight velocity.

Turbulence in the atmosphere has a very wide spectrum of scales. Hence, in discussing an airplane as a turbulence-measuring instrument, it must be borne in mind that this instrument can be used to investigate a fairly wide, but nevertheless limited, region of the spectrum.

The loads experienced by an airplane in a disturbed air flow depend significantly on the type of airplane and the flight conditions. To estimate the true velocities of turbulent wind gusts from airplane loads, we must have some method of converting from loads at the center of gravity or other points on the airplane at which the measuring instrument is mounted to gust velocity, i.e., we must obtain, by calculation or experiment, an analytical, graphic, or tabular relationship between these two quantities.

The most rigorous analytical method of converting the reaction of an airplane to a gust to the characteristics of the gust itself is by means of the equations of airplane dynamics, which can be used to relate the readings of a measuring instrument (accelerometer) mounted on the airplane to the turbulent gust acting on the airplane. The system of dynamic equations must, of course, include the accelerometer equation.

Using the method of small perturbations, we can represent the behavior of an airplane by linear (more precisely, linearized) differential equations. We know that if any dynamic system can be represented by a differential equation, and if the factor acting on the system can be represented analytically, then, by solving this equation, whose right-hand side represents the acting factor, we can obtain a complete picture of the behavior of the system. Such an equation, however, cannot always be solved analytically. Even if the system is represented by a linear differential equation with constant coefficients and with a simple right-hand side, the solution of such an equation entails considerable difficulty if its order is greater than 4, since the roots of the characteristic equation cannot be expressed in radicals.

The theory of harmonic analysis provides methods of determining the behavior of a system without solving a differential equation. The function relating the reaction of a linear dynamic system to an acting factor is called the transfer function and can be defined as the ratio of the Laplace transform $X(s)$ of the quantity at the system output to the Laplace transform $F(s)$ of the acting factor

$f(t)$ at its input. When the Laplace operator s is replaced by $j\omega$, the definition still holds. In this case the transfer function $\Phi(j\omega)$ or, as it is otherwise called, the complex gain, acquires a quite definite independent physical meaning. Its modulus is the amplitude gain for each individual harmonic component.

Determination of the transfer functions of systems with a large number of degrees of freedom requires very laborious computations and was impossible prior to the advent of electronic computers. Using the methods of harmonic analysis, we can solve the inverse problem, i.e., knowing the reaction of the system to some acting factor, we can find the acting factor itself.

This is precisely the problem in the investigation of atmospheric turbulence, but in this case the action of turbulent gusts on the airplane and, hence, the airplane response (load) to a gust are random functions of time.

On these random functions we can impose some limiting assumptions. We assume that the investigated process (atmospheric turbulence) is a quasistationary random process. In other words, the form of the probability distribution is independent of the position of the origin on the time axis.

The spectral theory of stationary random processes developed by Khinchin enables us to extend the methods of harmonic analysis to the investigation of random processes. Here the subject of study is not an individual, fully defined, oscillatory process, but the probability distribution of different variants of this process.

The results of experimental investigations of atmospheric turbulence can be represented most conveniently for the solution of many theoretical and practical problems in the form of correlation and structure functions or energy spectra (see Chapter 1).

If a stationary random process $f(t)$ acts on a linear dynamic system, then, knowing the energy spectrum $S_{\text{in}}(\omega)$ of this process and the transfer function $\Phi(j\omega)$ of the system, we can calculate the energy spectrum $S_{\text{out}}(\omega)$ of the response of the system:

$$S_{\text{out}}(\omega) = S_{\text{in}}(\omega)\,|\Phi(j\omega)|^2. \tag{2.5}$$

To solve the inverse problem, i.e., to determine the spectral density (energy spectrum) of the acting factor, we can write

$$S_{\text{in}}(\omega) = S_{\text{out}}(\omega)\,|\Phi(j\omega)|^{-2}. \tag{2.6}$$

This very important deduction of the theory of stationary random processes provides the basis for the method of obtaining the turbulence energy spectrum by means of an airplane. The quantity $|\Phi(j\omega)|^{-2}$ for each specific frequency ω_i is a numerical factor k_{ω_i}; hence,

$$S_{\text{in}}(\omega_i) = S_{\text{out}}(\omega_i)\,k_{\omega_i}. \tag{2.7}$$

In the calculation of the transfer function of such a complex system as an airplane, certain limiting assumptions are made. The first of these is the assumption of smallness of the oscillations of the airplane around stable horizontal flight; hence we can consider the linear problem.

If we ignore the elastic properties of the aircraft, we can calculate the transfer function by using the system of equations derived by Yudin [133] for the oscillations of an airplane in a turbulent atmosphere.

It was assumed in the derivation of the equations that there is no intervention by the pilot and that the undisturbed flight is horizontal.

The transfer function for an elastic airplane was calculated on a computer with the aid of a program compiled by Kuznetsov. The main ideas of this program are as follows. The oscillations of the airplane, simulated by a system of elastic beams, are expanded in a series of orthogonal functions, which are modes of the natural oscillations of the airplane in vacuum. From these modes, using the hypothesis of stationarity, the aerodynamic forces and the coefficients of the equations of motion of the airplane are determined. The movement of the airplane is assumed to be due to a vertical gust of wind, whose intensity varies sinusoidally in space with wavelength L. The frequency ω of the action of such a gust on an airplane flying at velocity v is given by

$$\omega = \frac{2\pi v}{L}. \tag{2.8}$$

The steady motion of the airplane when it is acted on by a gust of prescribed frequency and fixed amplitude is determined; this gives the maximum load amplitudes. Figure 2.2 shows the transfer functions for loads in different sections of the fuselage of an elastic airplane. The respective sections are shown on the outline diagram of the airplane.

It is apparent from Fig. 2.2 that an airplane reacts differently to gusts of different frequencies (scales). An airplane, as is known, is easily entrained by large-scale motions and, hence, does not react to these motions with any appreciable load. In the limiting case of flight in a wind field where the vertical velocity component has zero frequency, the airplane will have a corresponding vertical velocity component and will not experience any load. Hence, the transfer function $\Phi(0)$ will be zero. As the frequency of the acting factor increases, the airplane will respond more and more to the gust with a load; this corresponds to the smooth increase in transfer function in the frequency range 0 to ω_1. In the high-frequency region the transfer function of a rigid airplane is practically constant.

Figure 2.2 shows the transfer function for the instantaneous center of gravity (ICG)—an imaginary point, not rigidly related to the design of the airplane, at which its elasticity is not significant (curve 3). Hence, we can regard the transfer function of the instantaneous center of gravity as the transfer function of an equivalent rigid airplane.

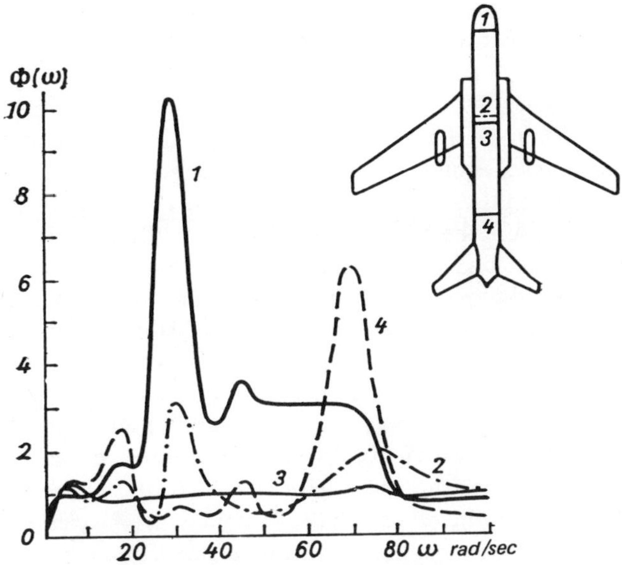

Figure 2.2. Transfer functions in different sections of fuselage of elastic airplane.

As distinct from a rigid airplane, an elastic airplane has a whole series of natural resonant frequencies, and the resonance properties are different in different sections of the airplane. As Fig. 2.2 shows, this leads to transfer functions of very complex shape, which differ appreciably from $\Phi(j\omega)$ for an equivalent rigid airplane. In the high-frequency range, however, the airplane will not react rigidly to gusts as a whole, since its linear dimensions will have some effect. Another factor operating will be the inertia of the measuring instrument (the accelerometer), which will react to the load by the displacement of some mass. This displacement, of course, cannot take place instantaneously and the measuring instrument will "pile up" high frequencies.

Summing up, we can say that the airplane–accelerometer system reacts with an appreciable load only to gusts in a particular frequency (scale) range.

Using the transfer function we can "reconstruct" the actual frequency distribution of gusts in the low-frequency region, down to some frequency ω_{min} for which the transfer function can still be evaluated correctly. In the high-frequency region the limit is the frequency ω_{max} corresponding to the horizontal length of the airplane itself

$$\omega_{max} = \frac{2\pi v}{L_a}, \tag{2.9}$$

where v is the velocity and L_a is the horizontal length of the airplane.

Transfer functions can be obtained experimentally. For electrical systems the problem of experimentally obtaining the transfer functions can be solved relatively simply: A sinusoidal input signal is applied to the system input, and after the transient process has decayed, the amplitude and phase of the output voltage are measured. Such measurements are made for the whole range of investigated frequencies.

The experimental determination of the transfer function of an airplane requires special complex equipment and the construction of a dynamic model of the airplane for which these functions have to be determined. The model is fitted with sensors to measure the parameters: the load at different points of the structure, bending and torsional moments, and so on. The model is mounted in the test section of a wind tunnel and subjected to an air flow in the usual way. In addition, a special device imposes transverse sinusoidal disturbances on the main flow, and the frequency of these disturbances can be varied. Just as in the case of electrical systems, the amplitudes of the fluctuations of the parameters are measured after the transient process has decayed. If we take into account the elasticity of the airplane, we cannot restrict ourselves to the investigation of the motion at one point—the center of gravity of the aircraft, whose motion determines the motion of all the other points of a rigid body. An elastic system can be regarded as a multiterminal network with one input and numerous outputs, i.e., different points behave differently in response to the same acting factor. In obtaining the transfer functions experimentally, the phase shift is not usually measured, i.e., the experiment gives the modulus of the transfer function.

As was shown above, for the investigation of the structure of turbulence we need to know the modulus or, more precisely, the square of the modulus of the transfer function.

A model can be used to "solve" the problem of the behavior of an airplane subjected to any acting factor. There is only one restriction here: the natural oscillations of the system must decay in time when the factor acts. The behavior of the airplane itself will be the "solution of the problem" in this case.

There are other experimental methods of determining airplane transfer functions, which do not require the construction of a dynamic model. If the energy spectrum of the velocities of turbulent gusts and the energy spectrum of loads of the airplane center of gravity are obtained by independent measurements, then their ratio, as (2.6) indicates, will be equal to the square of the modulus of the airplane transfer function. Koprov [58] made flights on an airplane fitted with instruments for determination of the vertical load spectrum at the same level as, and close to, masts on which anemometers for measurement of the vertical velocity component were mounted. Assuming the turbulence to be "frozen" and allowing for the mean wind velocity, he calculated the spatial energy spectrum of the vertical component of the turbulent gusts.

Dividing the gust spectrum by the load spectrum, Koprov obtained the square of the modulus of the transfer funtion.

Shur devised an automatic method of obtaining the energy spectrum of the vertical component of atmospheric turbulence. The method was based on the following limiting assumptions:

(1) Turbulence in the atmosphere is a stationary random process.
(2) The turbulence is "frozen" and, hence, on the assumption of constancy of the airplane velocity during the measurements, the time scale can be converted to the length scale (conversion from frequency ω rad/sec to wave number Ω rad/m $= \omega/V$, where V is in m/sec);
(3) The airplane with the accelerometer mounted on it is a linear system and, hence, the relation

$$S_w(\Omega) = |\Phi(j\Omega)|^{-2} S_n(\Omega), \tag{2.10}$$

which is easily obtained from (2.6), is valid.

The apparatus for this method consists of two units: an on-board unit and a ground unit. All the data processing is done on the ground. Figure 2.3 shows a block diagram of the method.

The essence of the method is as follows. The loads to which the airplane is subjected are measured by a device with an electrical output, are encoded by pulse-frequency modulation, and are recorded on magnetic tape. In processing the data (on the ground) the record is played back through a pulse decoder and fed to a spectrum analyzer. The circuit of the spectral device includes an electronic analog of the inverse squared modulus of the transfer function of the airplane–accelerometer system. The voltage at the output of each filter is multiplied by the coefficient

$$k_{\omega_i} = |\Phi(j\omega_i)|^{-2}. \tag{2.11}$$

The outputs of all the channels are then commutated in a specific sequence to an electronic display unit. With the magnetic memory system the time scale

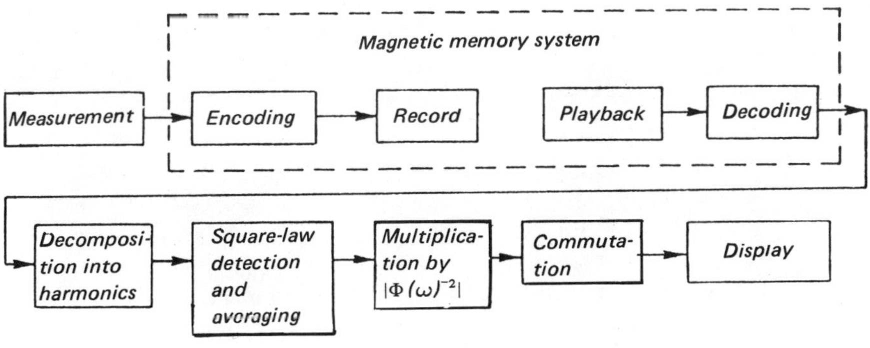

Figure 2.3. Block diagram of method of obtaining energy spectrum of atmospheric turbulence.

can be varied and the frequency bands of the investigated process can be superimposed on those of the spectral instrument. This method is statistical in nature and, hence, its application requires a statistically reliable series of measurements, i.e., of sufficient duration under conditions in which stationarity can be assumed. Neither individual nor, in particular, extreme velocities of the gusts can be determined from the final results.

The above description of the method indicates that the signal containing information about the load is subject to numerous transformations and spectral processing; in addition, the computation procedure includes multiplication by the inverse squared modulus of the transfer function, which is determined analytically by the adoption of certain assumptions. Hence, the main sources of error of this method are:

first, all the measuring, converting, and processing equipment;
second, the conversion from load to gust, i.e., the transfer function of the airplane-accelerometer system.

In addition, since the transfer function is calculated for horizontal flight without pilot intervention, any pilot action creates an additional source of error.

The root-mean-square relative error of measurement of the amplitude of the vertical load is ±0.01.

Frequency distortions introduced by the load-measuring device are of a systematic nature and are taken into account in the transfer function of the airplane–accelerometer system.

The magnetic memory system, consisting of a pulse-frequency modulator, recorder, playback unit, and a demodulator introduces a relative root-mean-square error of ±0.06; this error is determined mainly by the nonuniform motion of the magnetic tape in the recorder ("flutter").

The spectral equipment used for statistical processing is an infrasonic spectrometer, which consists of units that perform a sequence of operations on the analog signal: amplification, decomposition into harmonics, square-law detection, and averaging. The averaging time is the same for all channels, but if the amplitude of each oscillation selected by the filter is regarded as an independent measurement, averaging over a number of these measurements depends on the mean transmission frequency of the respective filter. Hence, the root-mean-square error for the spectrum, including the errors introduced by the load-measuring device and the magnetic memory system, is a function of frequency. For the lowest frequency it is ±0.08, and for the highest frequency it is ±0.01.

The errors in determination of the transfer function are due to the fact that the transfer function for an airplane is calculated theoretically with the aid of certain simplifying assumptions. The transfer function of a real airplane will differ, of course, from the theoretically calculated function. To test the correctness of determination of the transfer function and the estimation of the obtained error, measurements are made at two points on the airplane. The points chosen are such that the calculated analytical transfer functions differ significantly. The experimentally obtained load spectra, as is to be expected, differ considerably from one another, but after the introduction of the transfer functions of the respective points of the airplane into the calculation the obtained spectra of the vertical gusts are in good agreement with one another. The relative difference of the obtained values does not exceed ±(0.2–0.3).

Errors due to pilot intervention in the control of the airplane cannot be determined analytically and, hence, the measurements are usually made with the controls locked. If in the course of the flight in the turbulent flow the pilot has to intervene in control of the airplane, the parts of the record during which the deviation of the elevator, according to the instruments on board, exceeded 1° are not included in the processing of the data.

Airplane measurements can be used to find the velocities of individual air gusts, as well as the statistical characteristics of turbulence. The calculations are based on the relations between the gust velocity and the reaction of the airplane to it, which are found from the equations of flight dynamics. For instance, Dubov [41], using Yudin's system of equations, obtained the following formula for determination of the vertical gust velocity w (m/sec):

$$w = \frac{2G/S\,\Delta n}{c_y^\alpha \rho_H v_i \sqrt{\dfrac{\rho_0}{\rho_H}}} + g\int_{t_0}^{t} \Delta n\,dt - \frac{v_i\sqrt{\dfrac{\rho_0}{\rho_H}}}{205.2}\Delta\theta. \tag{2.12}$$

Here G is the aircraft flight weight (kg), S is the wing area (m^2), $\Delta n = n - 1$ is the load increment in g's, $c_y^\alpha = \partial c_y/\partial\alpha$ is the derivative of the lift force c_y with respect to the angle of attack α, ρ_0 and ρ_H are the air densities at a level of 760 mm Hg and the flight altitude H, respectively, v_i is the equivalent (i.e., reduced to a pressure of 760 mm Hg) airspeed (km/h), and $\Delta\theta$ is the deviation of the pitch angle (deg). The second term in (2.12) is equal to the vertical velocity acquired by the airplane due to its entrainment by the gust.

As Dubrov showed, Eq. (2.12) can be used also in cases where the pilot intervenes in control of the airplane. This makes it very convenient, since controls-fixed flight is impossible in a region of severe buffeting.

In the derivation of Eq. (2.12) several simplifying assumptions were made, viz., it was assumed that horizontal wind gusts are absent, the aircraft is absolutely rigid, and, finally, the aerodynamic coefficients in (2.12) are constant during flight in a zone where vertical flows occur.

The greatest errors are introduced by neglect of horizontal wind gusts.

We will estimate these errors approximately, assuming that the airplane is rigid.

The airplane lift is

$$A = c_y \frac{\rho v^2}{2} S, \tag{2.13}$$

and the aircraft load is

$$n = \frac{A}{G} = c_y \frac{S}{G} \frac{\rho v^2}{2}. \tag{2.14}$$

Since n is a function only of α and v, the change in load is given by the expression

$$\Delta n = \frac{1}{2} \rho \frac{S}{G} (v^2 c_y^\alpha \Delta\alpha + 2c_y v \Delta v). \tag{2.15}$$

It is obvious that Eq. (2.15) is correct only if the airplane within the air gust does not acquire an appreciable velocity in the gust direction. Since in this case $w/v = \Delta\alpha$, then

$$\Delta n = \frac{1}{2} \rho \frac{Sv}{G} (wc_y^\alpha + 2c_y \Delta v). \tag{2.16}$$

The first term in (2.16) corresponds to the load due to the vertical gust w, while the second corresponds to the load due to change in the air velocity Δv, which can be caused, for instance, by a horizontal air gust. The ratio of the load due to a horizontal gust with velocity Δv (Δn_h) to the load in a vertical gust with velocity w (Δn_v) is

$$\frac{\Delta n_h}{\Delta n_v} = \frac{2c_y \Delta v}{c_y^\alpha w} \tag{2.17}$$

The value of $\Delta n_h/\Delta n_v$ is usually 5 to 25%.

The apparatus used for calculation of turbulent gust velocities from the Dubov formula is much simpler than the apparatus for obtaining the turbulence spectrum and is mounted directly on the airplane. As Eq. (2.17) shows, to obtain the vertical air velocity we must measure Δn_y and take into account the entrainment of the airplane by the gust, i.e., integrate the vertical load with respect to time:

$$v_y = g \int_0^t \Delta n \, dt + c. \tag{2.18}$$

We will assume that $c = 0$, i.e., that the airplane is flying horizontally before it encounters the gust. This corresponds to assignment of zero initial conditions.

Devices for integration of electrical signals are widely known and are based on the integrating properties of certain circuits. An electronic load integrator in conjunction with an amplifier with inverse derivative feedback was used by the authors to investigate turbulence in mountainous regions. The special device, simulating the Dubov formula, was used for the first time to study turbulence in frontal zones. A block diagram of this instrument is shown in Fig. 2.4. The load at the airplane center of gravity is measured and converted to an electrical signal by an electric accelerometer (dashed rectangle on diagram). A voltage proportional to the load is then delivered to the integrator and the unit for multiplication by b. The coefficient b is adjusted manually to correspond with the readings of the indicator instruments measuring the airplane altitude and speed and the flying weight (fuel consumption).

Since, according to Eq. (2.12), fluctuations of the pitch angle must be taken into account in the calculation of w, the airplane load-measuring system included a gyro sensor (vertical gyro) to detect changes in angles of pitch and roll. The coefficient $(-c)$ in the "multiplication by $(-c)$" unit is determined by manually introduced values of V and H.

In airplane strength calculations a simple formula, called the strength norm formula, is widely used:

$$w_{\text{eff}} = \frac{2\Delta n \frac{G}{S} \Delta}{k\rho_0 c_y^\alpha v_i}, \tag{2.19}$$

where $\Delta = \rho_H/\rho_0$ is the relative reduction of density at the flight altitude, and k is a coefficient characterizing the smoothness of the increase in velocity in the gust.

Formula (2.19) is used to estimate the velocities of "effective" gusts; the gust is endowed *a priori* with a particular configuration, usually a trapezium. It is assumed that the airplane responds to the atmospheric gust as to an external force, but does not succeed in acquiring an appreciable velocity in the direction of the force during the time in which it acts. Equation (2.19) gives underestimated velocities in those cases where the airplane is entrained by the gust. This must be borne in mind when the abundant data on effective gusts are used for the study of atmospheric turbulence.

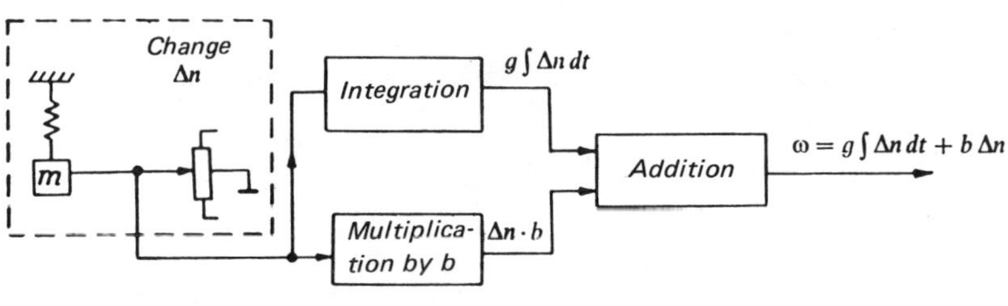

Figure 2.4. Block diagram of device for obtaining w from the Dubov formula.

3. Method Using Measurements of Absolute and Relative Airplane Velocities (Doppler Method)

An airplane in flight moves in an air flow which itself moves relative to the Earth. The data required for determination of the wind velocity and variations of the horizontal component of the turbulent fluctuations in the study of atmospheric turbulence are the instantaneous values of the velocity and direction of motion of the airplane relative to the Earth, and the velocity of the airplane relative to the surrounding air.

Instruments employing the Doppler effect are used to measure the airplane velocity vector in a coordinate system associated with the Earth. The Doppler effect is the difference observed in the frequency of oscillations emitted by a source and detected by a receiver when the source and receiver move relative to one another. The Doppler effect is also observed in the case where the source and receiver are located at the same point, but the receiver receives the reflected, and not the direct, signal. In this case the Doppler effect is due to the relative motion of the transmitter–receiver and the reflecting object. Airplane Doppler devices investigate the oscillations in a narrow beam directed toward the ground, and the velocity of the aircraft relative to the ground is determined from the change in frequency of the reflected signal.

In a coordinate system associated with the Earth, this velocity is the absolute velocity of the airplane. If the direction of the reflecting object (ground) coincides with the direction of flight, the frequency f_a of the oscillations detected by the receiver is given by the relation

$$f_a = f_0\left(1 + \frac{2v_a}{c}\right) = f_0 + F_D. \tag{2.20}$$

Here f_0 is the frequency of the investigated oscillations, v_a is the (absolute) airplane flight velocity, c is the velocity of light, and F_D is the Doppler frequency, equal to

$$F_D = \frac{2v_a}{c} f_0 = \frac{2v_a}{\lambda}, \tag{2.21}$$

where λ is the wavelength of the transmitter.

The direction of the beam actually makes an angle γ with the direction of flight and, hence,

$$F_D = \frac{2v_a}{\lambda}\cos\gamma. \tag{2.22}$$

Equation (2.22) ignores the fact that the air in which the airplane moves itself moves relative to the ground. In the general case the direction of flight does not coincide with the direction of motion of the air mass. The result of this is that the direction of absolute motion of the airplane does not coincide with its longitudinal axis and differs by some angle β, called the drift angle. To determine β it must be possible to change the direction of the beam or, to be more precise, the direction of its projection on the horizontal plane. When the horizontal projection of the beam coincides with the direction of absolute motion of the airplane, the Doppler frequency F_D will be a maximum. Thus, (2.22) should be written in the form

$$F_{D\,\max} = \frac{2v_a}{\lambda}\cos\gamma \tag{2.23}$$

The angle between the horizontal projection of the beam and the longitudinal axis of the airplane will be equal to the drift angle β. If the direction of the longitudinal axis of the airplane is known in some, for example, geographic, coordinate system, the vector of the absolute velocity of the airplane in this coordinate system can be completely determined.

Figure 2.5 shows a block diagram of a single-beam Doppler system for measuring ground speed and drift angle. The transmitter generates sinusoidal oscillations of frequency f_0, which are fed through an antenna switch to an antenna and emitted in the direction of the ground. The Doppler-shifted oscillations reflected from surface inhomogeneities are received by the antenna and delivered through the antenna switch to the receiver mixer. Attenuated direct oscillations with frequency f_0 from the transmitter also arrive at the mixer through the

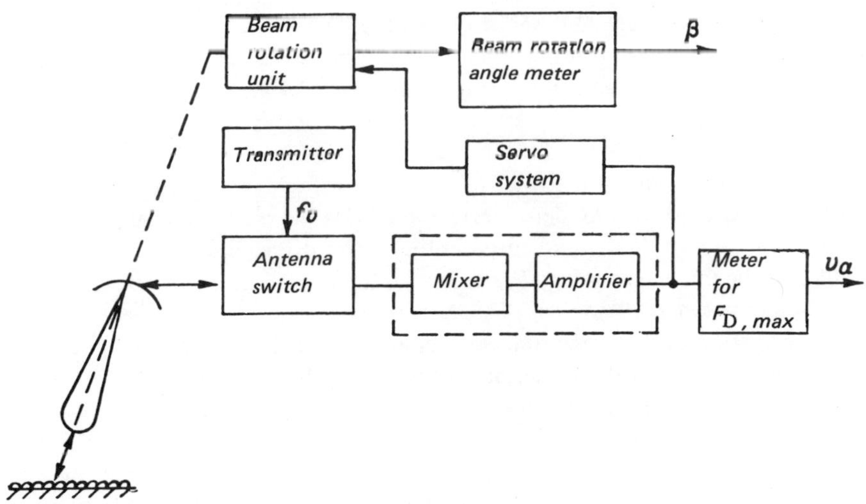

Figure 2.5. Block diagram of single-beam Doppler system for measuring ground speed and drift angle (DSDM).

antenna switch. The interaction of these oscillations leads to a difference- (i.e., Doppler-) frequency (F_D) voltage at the mixer output, and this voltage is then amplified by the amplifier.

The amplified Doppler-frequency signal is delivered to a special servo system, which acts on the antenna rotation system and turns the beam in a horizontal plane in the direction of the ground speed vector. The Doppler frequency will then become a maximum; the frequency meter converts $F_{D,\max}$ to a signal proportional to the ground speed v_a, and the beam rotation meter gives a signal proportional to the angle between the horizontal projection of the beam and the longitudinal axis of the airplane—the drift angle β.

The above-described principle of measurement of v_a and β by means of a single rotating beam is not used in practice owing to the low accuracy of measurement of the ground speed and drift angle. Multibeam Doppler speed and drift meters (DSDM) are used at present. The operation of multibeam DSDM is based on determination of the components of the ground-speed vector in a linear coordinate system attached to the airplane, and from them the modulus of the ground-speed vector v_a and the drift angle β are calculated automatically. A detailed description of the operating principle and the design of Doppler speed and drift meters is given in [43].

In aviation, Doppler speed and drift meters are usually used in conjunction with navigation computers and compass systems, forming Doppler navigation systems (DNS). The main function of DNS is dead reckoning.

In recent years inertial navigation systems (INS), which are more accurate than DNS, have become more and more widely used for navigation and measurements.

There are now a large number of types of DSDM and DNS, differing from one another in number of beams, their shape, transmission conditions, methods of measuring Doppler frequencies, and so on. Table 2.1 gives some basic characteristics of several foreign Doppler meters and Doppler navigation systems.

The tabulated figures for the accuracy of determination of airplane ground speed (the components of the total velocity vector for some systems) and drift angle are of interest. For modern, perfected DSDM the probable relative error of determination of the ground speed in flight over dry land can be as low as ± 0.002 of the maximum value. The errors of determination of the airplane drift angle lie in the range $\pm 0.15°$.

Investigation of the horizontal component of atmospheric turbulence necessitates measurement of the instantaneous values of the wind vector. This requires not only measurement of the ground speed and drift angle of the airplane, but also the airspeed v of the airplane and the direction of the longitudinal axis of the airplane in a geographical coordinate system. The airplane airspeed is measured with pressure-measuring velocity sensors, which use the relation between the speed of the airplane relative to the air and the associated dynamic velocity head:

$$\Delta p = \rho \frac{v^2}{2}, \tag{2.24}$$

where Δp is the velocity head, i.e., the difference between the dynamic and static pressures, ρ is the air density at the airplane flight level, and v is the relative speed (airspeed) of the aircraft.

Modern airspeed meters compute the relation

$$v = \sqrt{\frac{2\Delta p}{\rho}} = \sqrt{\frac{2\Delta pRT}{p_{st}}}. \tag{2.25}$$

For this purpose the airspeed meter receives not only information about the velocity head Δp, but also information about the static pressure p_{st} and external air temperature T. The airspeed v indicated by the meter (gauge) is numerically equal to the velocity of the airplane relative to the air in the direction of the longitudinal axis of the airplane.

Figure 2.6 shows the velocity vector triangle. The elements of this triangle, which is sometimes called the navigation triangle, are: v, the aircraft airspeed; ψ, the relative bearing; u, the wind velocity; δ, the wind direction; δ_ψ, course wind angle; v_a, airplane ground speed; and β, drift angle.

The basic relations between the elements of the navigation triangle are:

$$(v_a \cos\beta - v)^2 + v_a^2 \sin^2\beta = u^2, \tag{2.26}$$

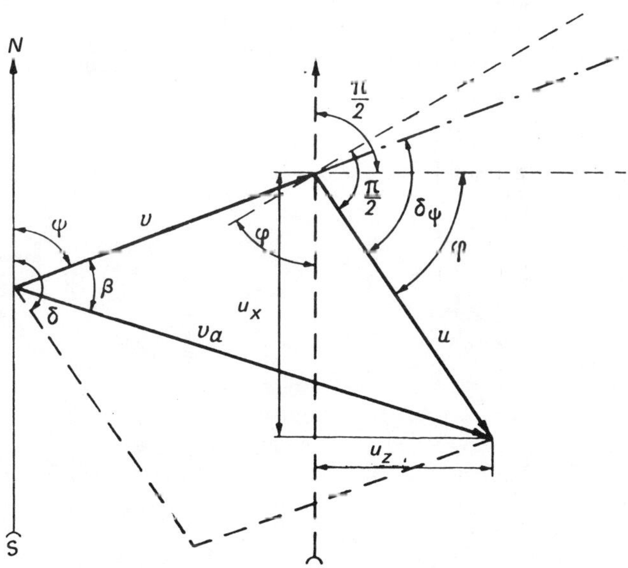

Figure 2.6. Velocity vector triangle.

TABLE 2.1. Basic Data of Foreign DSDM and DNS

Characteristics	AN/APN-66, AN/APN-81	AN/APN-78	AN/APN-79	AN/APN-96	AN/APN-67	AN/APN-105	AD-2000
Number of beams of antenna system	4	3	3	4	2	3	4
Mounting of antenna system	Rotating, mounted on stabilized platform	Fixed	Fixed	Rotating, mounted on stabilized platform	Fixed	Fixed	Rotating, not stabilized
Transmitter frequency (MHz)	8800	9799	13,500	8800	13,500	9799	8800
Transmission conditions	Pulse	Quasi-pulse	Continuous	Quasi-pulse	Continuous	Quasi-pulse	Pulse
Pulse length (μsec)	0.9	–	–	1.25–3	–	–	0.45
Pulse repetition frequency (kHz)	50 with modulation ±2.4	15.6–125	–	83–200	–	4–125	50
Range of speed measurement (km/h)	130–1300, ground speed	92–555, longitudinal velocity component	185–2200, longitudinal velocity component	140–1400, ground speed	148–1480, ground speed	185–2780, longitudinal velocity component	185–1300, ground speed
Range of measurement of vertical velocity component (m/sec)	Not measured	±15	Not measured	Not measured	Not measured	±250	Not measured
Error of speed determination (as % of maximum scale reading; results averaged over a stage of 18.5 km)	±0.2, ground speed	±0.7, longitudinal velocity component	±0.7, longitudinal velocity component	±0.1, ground speed	+0.25, ground speed	±0.7, longitudinal velocity component	+0.1, averaged over a stage of 185 km, ground speed

Range of measurement of drift angle (deg)	±50	Transverse velocity component is measured	Transverse velocity component is measured	±30	±40	Transverse velocity component is measured	±20
Accuracy of measurement of drift angle (deg)	±0.15	Accuracy of measurement of transverse velocity component ±0.7%	Accuracy of measurement of transverse velocity component ±0.7%	±0.15	–	–	0.1
Error of determination of airplane position (as % of distance traversed)	1–gyromagnetic compass, 0.5–astrocompass (for AN/APN-66)	0.4 for prescribed course accuracy 0.25°	±0.5 for prescribed course accuracy 0.25°		±0.5 for prescribed course accuracy 0.1°	±0.5 for prescribed course accuracy 0.25°	1
Altitude limits (m):							
maximum	21,000	7500	21,000	21,000	21,000	24,000	21,000
minimum	150	0	0	60	0	0	120
Power consumption (W)	AN/APN-66: 91–205 W d.c., 2620–2950 VA a.c.; AN/APN-81: 67–95 W d.c., 1660–1760 VA, a.c.	600	850	700	900	850	24
Weight (kg)	AN/APN-66: 333; AN/APN-81: 173	43	76.5	78.8	82	29.3	108.5

$$\delta = \psi + \delta_\psi , \tag{2.27}$$

$$\delta_\psi = \arctan \frac{v_a \sin \beta}{v_a \cos \beta - v}, \tag{2.28}$$

$$\delta = \frac{\pi}{2} + \varphi , \tag{2.29}$$

where

$$\varphi = \arctan \frac{v_a \cos (\psi + \beta) - v \cos \psi}{v_a \sin (\psi + \beta) - v \sin \psi}, \tag{2.30}$$

$$u = [v_a \sin (\psi + \beta) - v \sin \psi] \cos \varphi + [v_a \cos (\psi + \beta) - v \cos \psi] \sin \varphi . \tag{2.31}$$

As already mentioned, a Doppler speed and drift meter (DSDM) and an air-speed indicator (ASI) in conjunction with a gyro compass system allows continuous determination of v_a, β, v, and ψ. A knowledge of these quantities allows solution of the navigation triangle for the wind vector parameters u and δ. In the design of apparatus for automatic solution of this problem, it was found more convenient to convert expressions (2.26)-(2.31) to a form suitable for computation by automatic devices:

$$u = v_a \sqrt{1 + \left(\frac{v}{v_a}\right)^2 - 2 \frac{v}{v_a} \cos \beta}\ , \tag{2.32}$$

$$\delta = \psi + \arctan \frac{v_a \sin \beta}{v_a \cos \beta - v} \tag{2.33}$$

$$u_z \sin \varphi - u_x \cos \varphi = 0 , \tag{2.34}$$

$$u_z \cos \varphi + u_x \sin \varphi = u , \tag{2.35}$$

where

$$u_x = v_a \cos (\psi + \beta) - v \cos \psi , \tag{2.36}$$

$$u_z = v_a \sin (\psi + \beta) - v \sin \psi . \tag{2.37}$$

Figure 2.7 shows the block diagram of one version of a device for continuous measurement of the instantaneous values of the wind vector. The operation of the device is based on Eqs. (2.36), (2.37), (2.34), (2.35), and (2.29). The solution of Eqs. (2.26) and (2.27) requires no explanation, since all the quantities involved are given by the respective gauges. These equations are independent, and the solution of each of them consists of successive performance of the operations contained in them. As the block diagram shows, these operations are

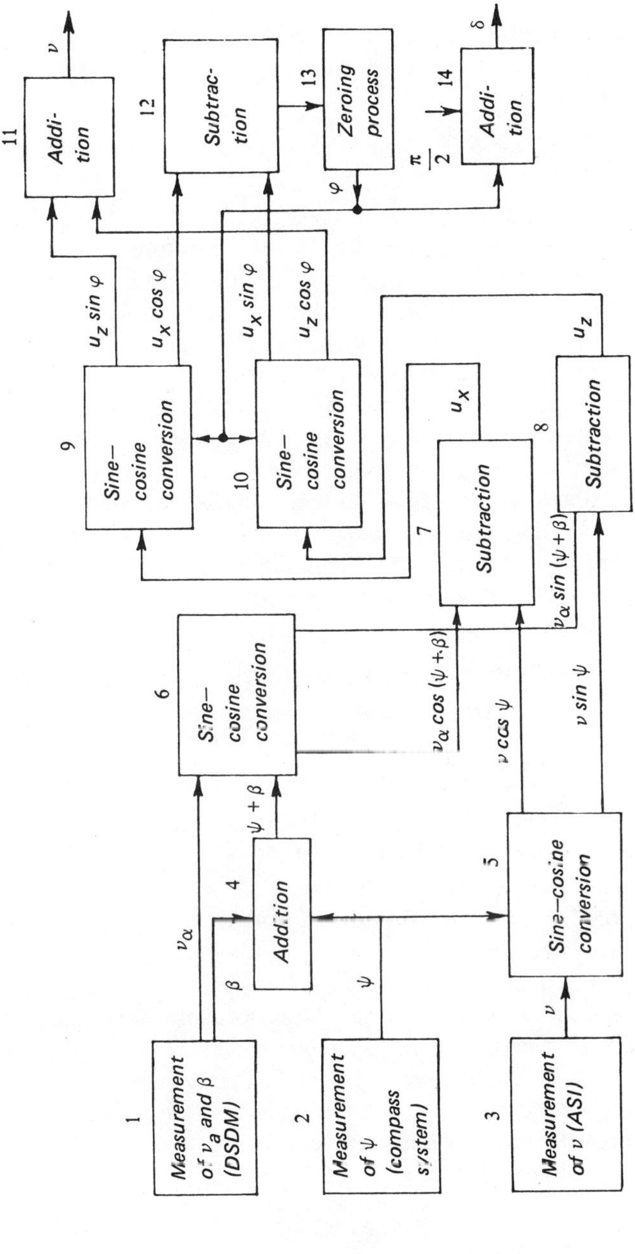

Figure 2.7. Block diagram of wind vector measurement.

effected by functional units that perform addition (subtraction) and sine–cosine conversion. Functional units of this kind are of standard type and are used in analog computers.

Equations (2.34) and (2.35) are interrelated and must be solved together. The determination of u and δ requires assignment of the value of φ. This value is produced by unit 13.

To explain the operation of the circuit, we assume that φ is given an arbitrary value. In this case

$$u_z \sin \varphi - u_x \cos \varphi \neq 0, \tag{2.38}$$

i.e., a quantity different from zero is obtained at the output of unit 12. The function of unit 13 is to alter φ until the signal at the output of unit 12 becomes zero, i.e., until Eq. (2.34) is solved. The solution of Eqs. (2.35) and (2.29) is clear from the block diagram without further explanation.

Thus, the above-described method can be used to measure the wind velocity vector and its fluctuations by using measurements of the vectors of the absolute and relative velocities of the airplane. We must also add that several more complicated systems which, in addition to the wind vector, give its components in the airplane coordinate system are used in practice. This is essential for comparison with data on the fluctuations of the horizontal component of the airplane airspeed measured by methods which, in principle, give the longitudinal (in the airplane coordinate system) component.

4. Method Using Measurements of Velocity Fluctuations of Incident Air Flow (Hot-Wire Anemometer Method)

The fundamental difference betwee this method and that described above is that the airplane does not act as a sensor, but is merely a "platform" for the apparatus. In this case of course, the behavior of the airplane in the turbulent atmosphere must still be given full consideration, but it is no longer the object of investigation and is now an interfering factor.

One of the methods in which the aircraft is used as a measuring platform is the hot-wire anemometer method. The hot-wire anemometer has for long been the most widely used instrument for the conduction of investigation in wind tunnels [112] and in the surface layer of the atmosphere. Zubkovskii [46] described a method of using hot-wire anemometers to investigate turbulence in the surface layer of the atmosphere and gave the main results obtained. Quite recently this method has been used as an airplane method [18, 19].

The operation of a hot-wire anemometer is based on the relation between the heat loss of a hot wire, immersed in a flow of liquid or gas perpendicular to the flow, and the flow velocity (King's law):

$$Q = l(T_w - T_0)(K + \sqrt{2\pi K \rho d c_p v}), \tag{2.39}$$

where Q is the heat lost by the wire in unit time, l is the length of the wire, d is its diameter, T_w is the wire temperature, T_0 is the flow temperature, K is the thermal conductivity of the fluid, ρ is the density of the fluid, c_p is the heat capacity of the fluid, and v is the flow velocity. In all practical hot-wire anemometer devices the wire is heated by an electric current and, hence, the heat loss Q of the wire can be related to the current i or the voltage U applied to the wire; then King's law can be rewritten in the form

$$U^2 = A + B\sqrt{v} \ , \tag{2.40}$$

where the coefficients A and B are determined by the parameters of the wire (its length, diameter, material) and are, in addition, functions of the temperature T_0 and density ρ of the air.

In investigations of atmospheric turbulence the airborne hot-wire anemometer is used to measure small-scale fluctuations of the horizontal wind-velocity component (longitudinal relative to the flight directions). As distinct from experiments in wind tunnels, where the wire is usually stationary, the airborne hot-wire anemometer moves, generally speaking, in an arbitrary direction relative to the wind at the particular flight altitude, and the mean velocity of the flow impinging on the wire is determined entirely by the flight conditions and is not connected in any way with the mean wind velocity. Hence, an airborne hot-wire anemometer cannot, in principle, measure the mean wind velocity (this problem is solved by the methods described above), but measures only the mean airplane airspeed and its fluctuations (the wire is assumed to be perpendicular to the longitudinal axis of the airplane).

It can be shown, however, that in a certain range of scales the fluctuations of the airplane airspeed are due entirely to fluctuations of the horizontal component of the wind velocity. In fact, an analysis of the equations representing the longitudinal motion of the aircraft [133] indicates that large-scale wind fluctuations entrain the airplane, thus altering its ground speed (speed relative to the ground), whereas the airspeed remains constant. On the other hand, small-scale wind fluctuations cannot entrain the airplane owing to its inertia and, hence, cause fluctuations of the airspeed. Thus, the airplane is characterized by some function which represents the reaction of the airplane to horizontal wind fluctuations in relation to their scale. An example of such a function, called the longitudinal transfer function of the airplane, is shown in Fig. 2.8, where Δv are the airspeed fluctuations, Δu are the longitudinal wind-velocity fluctuations, and L is the spatial scale of the fluctuations. It is obvious that an airborne hot-wire anemometer can provide undistorted measurements of horizontal wind-velocity fluctuations with scales from zero to a certain value L_{lim} (for the TU-104, for instance, $L_{\text{lim}} \approx$ 1-2 km).

In a turbulent atmosphere there are fluctuations of the wind velocity in all directions. We will show that an airborne hot-wire anemometer measures only

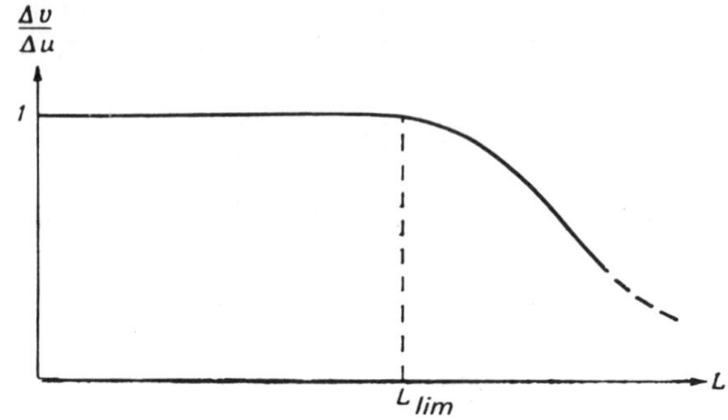

Figure 2.8. Longitudinal transfer function of airplane.

fluctuations that are longitudinal relative to the flight direction. We associate with the airplane center of gravity a coordinate system xyz such that the x axis coincides with the flight direction, the y axis coincides with the direction of gravity, and the z axis is perpendicular to the plane $x0y$. Let the wire of the hot-wire anemometer be situated parallel to the z axis. The hot-wire anemometer will then measure the magnitude of any velocity vector $\vec{v}$ situated in the plane $x0y$, i.e., perpendicular to the wire. It is obvious that the component v_x of the vector $\vec{v}$ is the instantaneous airspeed of the airplane, and v_y corresponds to the vertical wind gusts. (It is assumed here that the airplane is flying steadily in a horizontal plane). Hence $\overline{v}_x$ is equal to the mean airspeed, while $\overline{v}_y$, generally speaking, is zero.

We assume that a longitudinal (along the x axis) gust of amplitude $+\Delta v_x$ acts on the airplane. The instrument then measures a velocity equal to $\overline{v}_x + \Delta v_x$. If the airplane encounters a vertical gust of magnitude $\Delta v_y = \Delta v_x$, the instrument will measure a velocity equal to

$$[(\overline{v}_x)^2 + (\Delta v_y)^2]^{1/2}. \tag{2.41}$$

The measured velocity fluctuation in the first case is obviously equal to Δv_x, whereas in the second case it is equal to

$$\{[(v_x)^2 + (\Delta v_y)^2]^{1/2} - \overline{v}_x\} \tag{2.42}$$

and tends to zero when $\overline{v}_x/\Delta v_x \to \infty$. Thus, if the mean airspeed $\overline{v}_x$ is sufficiently high, the hot-wire anemometer measures, in effect, only the longitudinal fluctuations Δv_x and hardly responds at all to the vertical gusts Δv_y. For instance, when $\overline{v}_x = 200$ m/sec, which correspond to the speeds of airplanes of the TU-104 type, a vertical wind gust with amplitude 5 m/sec will be detected as a fluctuation of 0.05 m/sec. We also note that if the argument presented is valid for a wire parallel to the y axis, the wire should in practice be placed along the

z axis, since in this case the intrinsic motions of the airplane in the pitch plane (xOy) would have no effect on the quality of the measurements.

Practical hot-wire anemometer schemes can be divided into two main types: the constant-current scheme and the constant-temperature scheme. In the first case the wire is heated by a current whose strength is independent of the velocity, and the temperature (more accurately, the resistance) of the wire is measured. In the second case it is the temperature (or resistance) which is kept constant, while the heating current, which keeps the wire temperature constant, varies according to the velocity. The constant-temperature hot-wire anemometer has several advantages over the constant-current instrument. The chief one is the much better frequency characteristic, which is particularly important in the investigation of high-speed flows. Since a hot-wire anemometer mounted on a modern airplane will always operate in a high-speed flow ($\overline{v} \geqslant 100$ m/sec), it is better to use the constant-temperature scheme. Figure 2.9 shows the block diagram of such an instrument.

The operating principle of the constant-temperature hot-wire anemometer is as follows. Let us assume that for some constant flow velocity the measuring bridge M (consisting of the resistance R of the hot wire, the variable resistance R_v, and the constant resistances R_1 and R_2) is balanced. This means that the voltage U applied to the bridge M ensures a current through the wire at which its resistance $R = R_1 R_2 / R_v$ and between points a and b of the bridge there is no voltage difference. When the velocity v changes, the wire resistance R changes, and at the input of the regulating dc amplifier (DCA) there appears an "error" voltage, since the bridge M is unbalanced and there is a voltage difference between points a and b. This leads to a change in the output voltage U of the amplifier DCA and, hence, in the current flowing through the wire. The temperature of the wire, and hence its resistance, is altered until the bridge is balanced again. The variable resistance R_v ensures that the wire has the required temperature (resistance), which is automatically maintained by the instrument throughout the experiment.

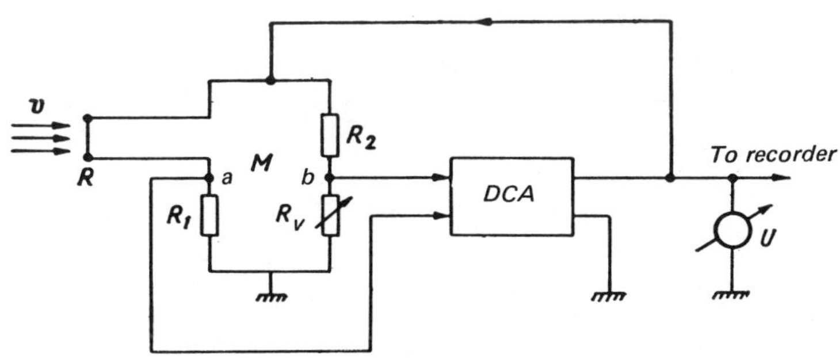

Figure 2.9. Block diagram of constant-temperature hot-wire anemometer.

One of the main features of the use of a hot-wire anemometer on an airplane is the variation of the thermodynamic characteristics of the investigated flow. When the flight altitude varies from 0 to 10–12 km, the air temperature and density change significantly, and this must be taken into account when measurements are made. Generally speaking, if the particular wire is calibrated on the ground at constant temperature T_0 and air density ρ and different values of v, the coefficients A and B in Eq. (2.40) can be determined and then, if the variation of these coefficients with T_0 and ρ is known [18], the corrections to A and B can be calculated from the measured T and ρ at the particular flight altitude. This method is very attractive, since the instrument becomes absolute, and the mean airspeed of the airplane is measured by an independent method. At present, however, this method cannot be used in practice. First, the functional dependence of A and B on T_0 and ρ is fairly complex and, hence, the introduction of corrections without the use of automatic computing devices is very laborious. Second, present airplane methods of measuring T_0 and ρ are not accurate enough for the correct computation of the errors. Third, the wire of the hot-wire anemometer has usually such a short life that the time required for on-ground calibration often exceeds its life. Hence, when a hot-wire anemometer is used on an airplane, the following method must be used. We note first of all that the actual value of the mean airspeed $\overline{v}$ is required only for determination of the "working point" on the anemometer characteristic (2.40) and in this respect is an auxiliary quantity. The only value required for the investigation of fluctuations is that of the instrument sensitivity, given by the expression

$$\alpha = \left(\frac{dU}{dv}\right)_{\overline{v}}, \tag{2.43}$$

which gives the gradient of the instrument characteristic at the working point $\overline{v}$. Differentiating (2.40) with respect to v, we find that

$$\alpha = \frac{dU}{dv} = \frac{B}{4U\sqrt{v}}. \tag{2.44}$$

At the same time, it follows from (2.40) that $B = (U^2 - A)/\sqrt{v}$. We note that, generally speaking, A is equal to a certain U_0^2 if we put $v = 0$ in (2.40). We finally have

$$\alpha = \frac{dU}{dv} = \frac{U^2 - U_0^2}{4Uv}. \tag{2.45}$$

Thus, the determination of α requires a knowledge of the wire voltage U for a given mean flight speed $\overline{v}$, the wire voltage U_0 for zero airspeed, and the value of $\overline{v}$. The quantity U is measured by the hot-wire anemometer, and $\overline{v}$ can be determined sufficiently accurately by standard flight instruments. The situation is more difficult with U_0, since it is impossible, of course, to obtain zero airspeed of the airplane. Hence, when a hot-wire anemometer is mounted on an airplane, the wire must be placed in a special holder [18] which can, at the experimenter's command, move the wire from the working position into a protective case in which the air velocity is zero. In this way the instrument can be calibrated directly in the air at a given altitude (and, hence, at a given T_0 and ρ), which is very convenient.

We must emphasize that an airplane hot-wire anemometer can be used only for investigating clear-air turbulence, since the mechanical strength of the wires used does not allow measurements in clouds. Despite this serious limitation, the hot-wire anemometer can provide valuable information about the microstructure of the wind field owing to its good frequency characteristic and high scale resolution. It greatly extends the spectrum of the investigated horizontal fluctuations of wind velocity. A hot-wire anemometer must be used for measurement of temperature fluctuations on high-speed aircraft.

When temperature fluctuations are present in the flow impinging on the wire, the instrument produces a fluctuating current to keep the wire temperature constant and, hence, "issues" spurious fluctuations of airspeed. At the same time, in measurements of temperature fluctuations themselves, especially on high-speed aircraft, fluctuations of the flow velocity produce fluctuations of the stagnation temperature. The presence of two instruments–a hot-wire anemometer and a fluctuation thermometer–allows mutual corrections to be made to their readings. This question is discussed in more detail below in the description of the method of measuring temperature fluctuations.

The airplane hot-wire anemometer designed by Vinnichenko [18] has the following basic characteristics: The sensor is a platinum wire 20 μ in diameter (or a platinum-coated tungsten wire 18 μ in diameter); the wire is about 2 mm long; the working temperature of the wire is 250–400°C. The gauge operates at flight speeds of up to 200 m/sec at altitudes from zero to 20,000 m in clear air. The scales of the measured fluctuations are 10 m to 1–2 km. The sensitivity (at flight speeds of 100–200 m/sec) is 50–70 mV · m^{-1} · sec. The error of measurement of velocity fluctuations (when standard aircraft instruments are used to measure the mean airspeed) is 10%. The instrument is fitted with an automatic sensor-carrying stand, which can be mounted on any pressurized subsonic airplane.

5. Method Using Pressure Measurements at Various Points on a Body in the Flow

The velocity of an airplane in air (airspeed) is measured by pressure-measuring devices in conjunction with a Pitot tube. The Pitot tube has a hole that is parallel to the longitudinal axis of the airplane and faces the impinging air flow. The other hole (or system of holes) is located on the side of the tube, where the pressure is equal to the static pressure. The difference in dynamic and static pressures is proportional to the square of the incident flow velocity. The Pitot tube measures the projection of the velocity on the flow direction. The velocity vector in wind tunnels is measured with anemoclinometers–spherical probes with a system of dynamic holes in the hemisphere facing the main flow.

Pakhomov [79] used an anemoclinometer to measure the airspeed vector on an airplane in clouds. Burns [147] used a similar apparatus for investigating clear-air turbulence at altitudes up to 300 m over Great Britain and North Africa.

The sensitive element of an airplane anemoclinometer is also a spherical probe with a system of holes on the hemisphere facing the incident flow.

An airplane anemoclinometer has two pairs of holes located in vertical and horizontal planes, symmetric relative to the central hole which, like the dynamic hole of the Pitot tube, perceives the pressure of the total velocity head.

Figure 2.10 shows a schematic section of the probe in the plane xOy (x is the longitudinal axis of the airplane, y is the vertical axis). The pressure p at the entrance to the hole is given by the expression

$$p = p_0 + \frac{1}{2}\rho v^2\left[1 - \frac{9}{4}\sin^2(\varphi_0 + \Delta\varphi)\right], \tag{2.46}$$

where p_0 is the static pressure, ρ is the air density, v is the velocity of the incident flow, and $\varphi_0 + \Delta\varphi$ is the angle between the radius vector connecting the

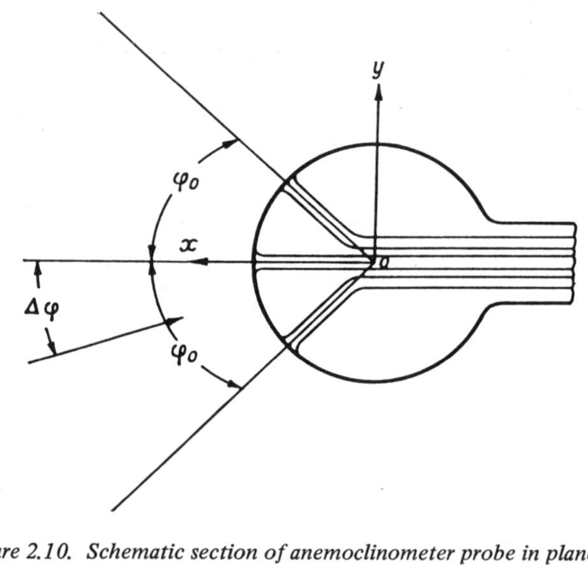

Figure 2.10. Schematic section of anemoclinometer probe in plane xOy.

center of the hole to the center of the sphere and the velocity vector of the incident flow. The longitudinal airspeed component is measured by the usual method as the difference in the total head perceived by the central hole and the static pressure. The substitution of $\varphi_0 = 0$ and $\Delta\varphi = 0$ in Eq. (2.46) gives the well-known relation

$$\Delta p = p_0 + \frac{1}{2}\rho v^2. \tag{2.47}$$

To secure fulfillment of the condition $\Delta\varphi = 0$, the entire probe before the start of measurement is oriented in the direction of the incident flow velocity. It can be shown that when the mean incident flow velocity is 200 m/sec, corresponding to the airspeed of a TU-104 airplane, a gust of magnitude 10 m/sec perpendicular to this velocity deflects the instantaneous airspeed vector through an angle of less than 3°. The error in determination of the longitudinal airspeed component does not exceed 0.5%. As was shown in [79], the difference in pressures at the entrance of the vertical and horizontal pairs of probes is proportional, respectively, to the vertical and transverse components of the airspeed vector. The pressure difference is measured by differential pressure gauges with an electrical output. An airplane anemoclinometer can be fitted with a computer which introduces corrections for air density, change in pitch angle, and entrainment of the aircraft by the air gust. The true values of the airspeed vector components are obtained at the computer output. The results of the measurements are recorded by an optical recorder. The instrument error in measurement of longitudinal and vertical wind velocity fluctuations is about 10% [147]. Transverse wind velocity fluctuations cannot be measured with an anemoclinometer owing to the characteristic transverse oscillations ("Dutch roll") of an airplane.

6. Method Using Measurements of Angle of Attack (Wind-Vane Method)

An angle-of-attack wind vane is sometimes used to measure fluctuations of the vertical wind-velocity component on an airplane. The essence of this method is that vertical wind-velocity fluctuations lead to fluctuations of the airplane angle of attack.

The instantaneous angles of attack are measured by a special vane mounted on a nose boom about 1.5–2.0 m long. Owing to its low inertia the vane reacts to relatively small-scale gusts, whose amplitudes are so small that they do not cause "buffeting" of the airplane.

It is obvious that for calculation of the velocity of vertical gusts from the vane readings α the motion of the airplane itself in the vertical plane must be taken into account. First, the instantaneous position of the airplane, i.e., its pitch angle θ, must be considered. Second, the vane readings have to be corrected for the vertical velocity v_y of the airplane if it is entrained by the gust. Third, the possible rotation of the airplane around the transverse axis, which will lead to a spurious gust velocity, must be taken into account.

Finally, the fluctuations of the vertical component of the wind velocity u_y can be calculated from the formula

$$u_y = (\alpha - \theta)\, v - v_y - l\theta', \tag{2.48}$$

where v is the true mean airspeed, l is the distance from the airplane center of gravity to the vane, and θ' is the rate of change of the pitch angle (angular speed of rotation of the airplane).

The value of v_y is determined by integration of the vertical load of the airplane (see Section 2 of this chapter). An ordinary hodoscope and a differential hodoscope are used to obtain θ and θ'. It is obvious that the above equation also includes all the corrections for any action of the pilot in controlling the airplane. This is of particular importance, since the pilot usually has to take vigorous action to control the behavior of the airplane when it is subjected to intense buffeting.

If the vertical load measured at the point of attachment of the vane is integrated, then it can be assumed that in (2.48) $v_y - l\theta' = v_y^*$.

It should be noted that the method described will be effective in practice only if the computation of u_y is automated, since the treatment otherwise becomes far too laborious.

According to Ridland, the error of measurement of the vertical wind-velocity fluctuations by an angle-of-attack vane is 10%.

7. Method Using Measurements of Speed of Sound (Acoustic Method)

An acoustic method is also used for airplane investigations of the fine structure of atmospheric turbulence. It is known that the velocity of propagation of sound in a liquid or gas flow depends on the flow velocity. This physical relation underlies the operation of the instruments called ultrasonic (or acoustic) anemometers. Acoustic anemometers have provided fundamental information about the nature of turbulence in the atmospheric surface layer [75]. An acoustic anemometer on an airplane was used relatively recently to investigate wind-velocity fluctuations [59].

The sensitive element of an airplane acoustic anemometer consists of an omnidirectional emitter of ultrasonic oscillations and two pairs of microphones mounted symmetrically relative to the emitter along the vertical y and transverse z axes of the airplane. By measuring the fluctuations of the velocity of propagation of sound from the emitter to the microphones, the wind-velocity fluctuations on the respective axes can be determined.

In the xyz coordinate system (the emitter is at the origin of the system, the x axis coincides with the longitudinal axis of the airplane, the y axis is vertical, and the z axis is transverse) the velocity vector of the flow incident on the airplane can be decomposed into components v_x, v_y, and v_z. The quantity v_x is the instantaneous airspeed of the airplane, while v_y and v_z are the vertical and transverse fluctuations, respectively, of the wind velocity. We denote the coordinates of the receiving microphones by $(0, l, 0)$ and $(0, -l, 0)$ for the pair located along the vertical axis, and by $(0, 0, l)$ and $(0, 0, -l)$ for the pair located along the horizontal z axis. The time of propagation of sound from the emitter to the microphone $(0, l, 0)$ is

$$t_1 = \frac{l\,[(c^2 - v_n^2)^{1/2} - v_y]}{c^2 - v^2}, \tag{2.49}$$

where $v_n^2 = v_x^2 - v_z^2$, $v^2 = v_x^2 + v_y^2 + v_z^2$, and c is the velocity of sound.

The time of propagation of the acoustic wave to the symmetric microphone $(0, -l, 0)$ is, similarly,

$$t_2 = \frac{l\,[(c^2 - v_n^2)^{1/2} + v_y]}{c^2 - v^2}. \tag{2.50}$$

The difference $t_1 - t_2$ obviously characterizes the wind-velocity fluctuations (on the y axis), since

$$t_1 - t_2 = -\frac{2lv_y}{c^2 - v^2}. \tag{2.51}$$

from which

$$v_y = (t_2 - t_1)\frac{c^2 - v^2}{2l}. \tag{2.52}$$

A similar relation holds for the pair of microphones situated along the transverse axis, which, consequently, measures the transverse wind-velocity fluctuations. Unfortunately, an acoustic anemometer mounted on an airplane cannot measure longitudinal wind-velocity fluctuations, since the microphone mounted in front of the emitter (on the x axis) "masks" it, while the turbulent noise produced in the wake behind the microphone and emitter interferes with the operation of the second microphone.

The acoustic anemometer is used successfully on LI-2 and IL-14 aircraft for investigation of turbulence at altitudes of 50 to 4000 m. The instrument includes a device that automatically introduces corrections for oscillations of the airplane in the pitch plane ($x0y$) and for entrainment of the airplane by vertical gusts.

At present there is still no experience on the use of acoustic anemometers on high-speed, high-altitude aircraft, and it is not clear what technical and methodological difficulties may be involved.

The instrument has the following characteristics: ultrasonic carrier frequency 100 kHz; upper limit of frequency characteristic 0.5 kHz; scales of measurable fluctuations 100 cm to 1 km (for aircraft of the IL-14 type). The error of measurement of the vertical wind-velocity fluctuations is about 10%. Transverse wind fluctuations cannot be correctly measured by the acoustic method owing to the strong distorting effect of airplane yaw (see Section 5 of this chapter).

8. Gyroscopic Instruments for Determination of Airplane Attitude

All methods of investigating atmospheric turbulence by means of airplanes require careful determination of the attitude of the airplane and its oscillations about this mean position. In some cases, where the airplane is used as a sensor, these data are a useful "signal," carrying information about the state of the atmosphere (load method), and in other cases, where the airplane acts as an instrument "platform," its natural oscillations become an "interfering" factor, distorting the measurements (hot-wire anemometer, anemoclinometer, angle-of-attack vane).

By the attitude of an airplane we mean the relative angular position of the coordinate system associated with the airplane and the coordinate system associated with the Earth, i.e., the relative motion of the origins of the coordinate systems is ignored. The angular position of the airplane coordinate system relative to the Earth coordinate system can be specified by means of three angles—the pitch angle θ, the roll angle φ, and the heading angle ψ; θ is the angle between the longitudinal axis of the airplane and the plane of the true horizon, φ is the angle between the transverse axis of the airplane and the horizon, and ψ is the angle between the vertical plane passing through the longitudinal axis of the airplane and the meridian plane.

These angles are determined on the airplane by gyroscopic sensors. The heading angle ψ is also measured by means of magnetic compasses and astrocompasses. In the investigation of atmospheric turbulence attention is centered mainly on measurement of the pitch angle θ and the roll angle φ and their fluctuations. These angles are measured by a gyroscopic instrument called a gyro vertical (Fig. 2.11). The gyro vertical measures the deviation of the longitudinal x and transverse z airplane axes from the plane of the true horizon. The operation of the instrument is based on the property of a free gyroscope to maintain its axis of rotation (situated vertically relative to the y axis) in a constant position relative to world space. A gyroscope consists of a rapidly rotating rotor 1 suspended in two gimbals—an inner gimbal 2 (pitch gimbal) and an outer gimbal

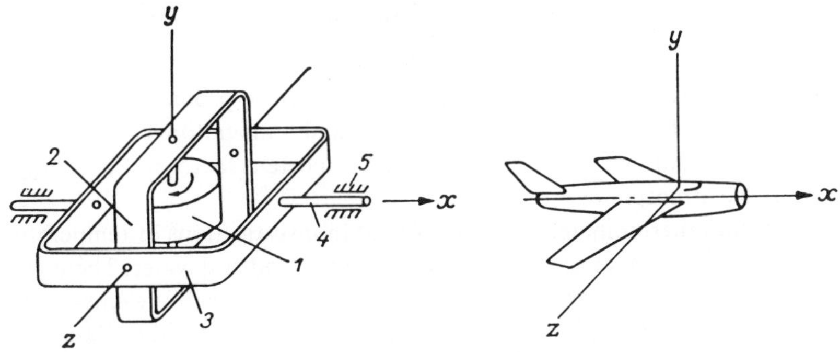

Figure 2.11. Schematic of gyro vertical.

3 (roll gimbal). The spindle 4 connects the gimbal 3 to the body 5 of the gyroscope, which is rigidly mounted in the airplane. Such a gyroscope has three degrees of freedom, since the rotor can rotate around three mutally perpendicular axes $0x$, $0y$, and $0z$. The gimbal suspension 2, 3 allows the axis of the rotor to maintain its vertical position in space for any pitch angle θ and roll angle φ. For transmission of data about the angles θ and φ, potentiometers (or selsyn stator coils) are mounted on the gyroscope case, and potentiometer sliders (or selsyn rotor coils) are mounted on the gimbal suspension 2, 3. An alteration of the pitch and roll angles of the airplane leads to a corresponding change in the relative position of the gyroscope case and the gimbals, which leads in turn to a change in the output voltages on the pitch and roll potentiometers (or selsyns).

For the vast majority of aircraft instruments the most important readings are those of the pitch and roll angles, whereas wind vector computers also require accurate measurements of the heading angle, so that the true direction of the wind can be calculated. Gyroscopic instruments in which the gyroscope axis of rotation is situated horizontally in the meridian plane are called directional gyros.

In cases where the angular displacements of the airplane are an "interfering" factor, it is advantageous to mount the measuring instruments on a special platform, whose position in space is stabilized by gyroscope systems and, thus, does not depend on the airplane attitude. The commonest type are platforms which are stabilized only for pitch and roll, i.e., platforms which remain parallel to the plane of the true horizon. Such platforms are used in some Doppler navigation systems for stabilization of the attitude of the system (see Section 3 of this chapter).

Modern airplane gyro verticals allow measurement of pitch and roll angles to an accuracy of ±0.05°, with a sensitivity of not worse than 0.01°, and the detection of angular displacements of the airplane at rates of up to 360 deg/sec.

9. Measurement of Temperature Fluctuations on an Airplane

A knowledge of the structure of the temperature field of turbulence allows the solution of a whole series of theoretical and practical problems. Simultaneous measurements of fluctuations of wind velocity and temperature can be used to determine turbulent heat fluxes, which are an important factor in the energy balance of the atmosphere.

Fluctuation thermometers are widely used in investigations of temperature fields in the bottom layer of the atmosphere above land and water. There are various types of fluctuation thermometers. In some the sensitive elements are resistance thermometers [5], while others (acoustic types) use the relation between the velocity of sound and the temperature of the medium. A particular feature of measurement of temperature fluctuations on an airplane is that the mean air temperature in relation to which the fluctuations are measured varies with altitude. Hence, before the start of each measurement the instrument has to be set to a new working point [129].

Figure 2.12 shows a block diagram of one type of airplane fluctuation thermometer [5]. A wire resistance thermometer R_T, which is held out in the flow beyond the airplane boundary layer, forms one arm of a measuring bridge. An amplifier A, which amplifies the unbalance voltage, is connected to the bridge diagonal. The system is brought to the working point by a servomechanism consisting of a power amplifier PA, a motor M, reducer R, and a balancing potentiometer R_{reg}. The switch S closes the setting circuit before the start of the measurements. The servo system thus follows the mean temperature of the impinging flow. Before the start of the measurements on a horizontal portion of the flight the switch S opens the feedback circuit, and the output of amplifier A delivers a signal proportional to the deviation of the temperature from the set mean value.

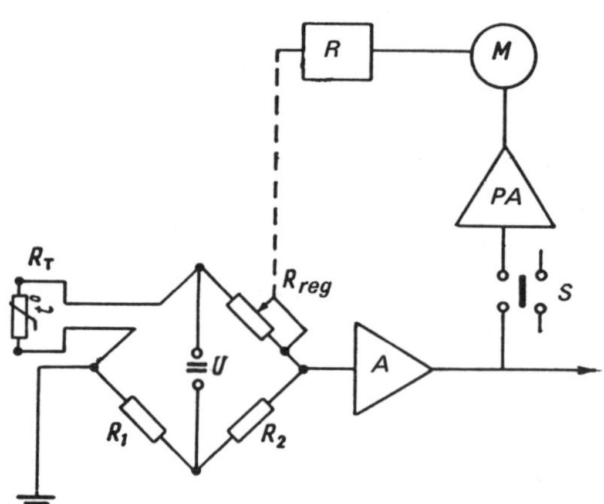

Figure 2.12. Block diagram of airplane fluctuation thermometer.

The resistance thermometer, if its heating by the current is ignored, measures the temperature of the air in immediate contact with the wire. Since the measurements are made at a high relative rate and the sensor impact coefficient is not zero, a dynamic correction for the impact temperature must be made to the instrument readings.

In measurements of temperature fluctuations on high-speed airplanes the errors can be considerable. For instance, in a thermally homogeneous field at a mean flight velocity of 720 km/h a fluctuation of the air velocity of 1 m/sec leads to a spurious temperature fluctuation of 0.2°C. If the airplane carries a hot-wire anemometer, appropriate corrections can be introduced.

Laboratory and flight investigations of a fluctuation thermometer and a hot-wire anemometer by Belyaev and Dmitriev provided true values of the temperature (ΔT_{tr}) and velocity (Δu_{tr}) fluctuations when the readings of both instruments were used simultaneously:

$$\Delta T_{tr} = \frac{\Delta T_{meas} - k_1 \Delta u_{meas}}{1 + k_1 k_2}, \qquad \Delta u_{tr} = \frac{\Delta u_{meas} + k_2 \Delta T_{meas}}{1 + k_1 k_2}, \tag{2.53}$$

where ΔT_{meas} and Δu_{meas} are the measured values of the temperature and velocity fluctuations, respectively, and k_1 and k_2 are coefficients determined from the instrument readings for each measuring "area."

The true values ΔT_{tr} and Δu_{tr} are calculated on a computer by the joint solution of the system of equations (2.53).

10. Table Summarizing Aircraft Methods of Investigating Turbulence

The methods discussed above are used to measure turbulent fluctuations of wind velocity and temperature. The fields of pressure, density, humidity, etc. are also nonuniform in a turbulent flow. If the atmosphere contains a passive contaminant, its distribution is largely determined by the structure of the turbulence field. The motion of cloud particles (both drops and crystals) in developed turbulent flow is also due to the turbulent velocity field. There are various methods of measuring such parameters as fluctuations of the dielectric constant of the air, fluctuations of the density of passive contaminants and suspensions, etc., by means of which the turbulence structure can be investigated.

Aircraft radiorefractometers have been designed and used for turbulence investigations. The data obtained with their aid are of particular value for the solution of practical problems associated with the propagation of electromagnetic waves in a turbulent atmosphere.

In the present book, however, we have confined ourselves to a consideration of the methods of measuring the velocities of turbulent motions, and one

TABLE 2.2. Aircraft Methods of Investigating Turbulence

Method	Measured parameter[a]	Frequency range (Hz)	Scale resolution (m)	Sensitivity (m/sec)	Range of scales of measurable fluctuations (m)	Accuracy (%)	Limitations in use
Load	v_y	0–(~5)	50	0.5	50–5000	15	None
Doppler	v_x, v_z	0–(~0.2)	1000	1.0	1000	5	Calm water surface
Hot-wire anemometer	v_x	0.05–(~1000)	5	0.05	10–1000	10	Clouds, dust
Anemo-clinometer	v_x, v_y, v_z	0–(~20)	10	0.2	10–1000	10	Icing
Angle-of-attack vane	v_y	0–(~20)	10	0.2	10–1000	10	Icing
Acoustic[b]	v_y, v_z	0–(~10000)	0.1		10–1000	10	High-frequency noise
Thermometric	T	0.05–(~1000)	5	0.05° C	10–1000	0.5	Large-droplet clouds, severe icing

[a] The symbols are given in the airplane coordinate system; the x axis is longitudinal, the y axis vertical, and the z axis transverse.
[b] The characteristics relate to airplanes with an airspeed of 50–70 m/sec. The characteristics of the other methods relate to airplanes with an airspeed of 120–200 m/sec.

method of measuring temperature fluctuations. Table 2.2 gives comparative characteristics of the aircraft methods that provided most of the results considered in the later chapters in the book.

11. Airplanes—Flying Laboratories

The development of aircraft methods of investigating the atmosphere has led to the construction of special flying laboratories equipped with measuring, computing, and recording apparatus. Flying-laboratory airplanes can be classed as comprehensive or specialized. The former are designed for investigation of the dynamics and thermodynamics of the atmosphere, its optical and electrical characteristics, clouds and atmospheric dust, etc., while the latter are designed for the investigation of specific atmospheric processes or phenomena.

Considerable difficulties of a technical and organizational nature have to be overcome in conducting airplane investigations of the atmosphere.

First, there is no standard airplane apparatus for investigating the atmosphere. It is a case of either adapting existing air-navigation instruments, or designing models of new measuring instruments or, of course, of constructing systems that include existing standard instruments or models designed for the conduction of a series of particular measurements.

Some of the instruments have sensitive elements that have to be located in the undisturbed flow. For the selection of the optimal position for external sensors, dynamic models of flying laboratories are tested in wind tunnels.

Some of the sensors must be oriented in a particular way in space. Gyro stabilizing systems are required for this purpose.

The apparatus designed for conversion of the signals from external sensors, as well as instruments that do not incorporate external sensors, are usually designed by experimenters themselves. It is often difficult to compare experimental results owing to differences in the characteristics of the instruments.

The problem of recording and processing data is still far from solved. Various on-board information storage devices (oscillographic and magnetic) are now widely used. Processing, particularly statistical analysis, is performed with the aid of computers, but significant differences in the methods of recording and preparation for machine processing, and the absence of standard computer programs of various types, lead to difficulties in the comparison of experimental results.

The organization of a flight experiment, especially one using several airplanes simultaneously, is very complex. The rapid motion of the airplanes and their separation limit the maneuvers of flying laboratories which for the purposes of the experiment have to intersect routes, change altitude and velocity of flight, and perform flights in difficult meteorological conditions. The organization of such flights requires the closely coordinated interaction of aircrew and ground services.

12. Radar and Optical Methods of Investigating Atmospheric Turbulence

Radar methods of tracking ascending or descending objects are used for investigating turbulence in the atmosphere. One of the methods by which the fine structure of the vertical wind velocity profile is investigated is the Jimsphere method of radar measurement used in the U.S.A. A Jimsphere is a constant-volume balloon with special projections on its surface to prevent its rotation. The FPS-16 radar system provides for measurement of the wind velocity at altitude steps of 25 m.

The Jimsphere method allows the detection of layers with large velocity gradients, in which dynamic turbulence may arise. The Jimsphere method, however, is still not a method of measuring turbulent motions in the atmosphere.

Radar devices have relatively recently come into use for measuring motions in the atmosphere, and the results obtained relate mainly to the boundary layer of the atmosphere. Radar methods are extremely promising, however, and in the future these methods will provide data on turbulence structure in the free atmosphere too.

The main difference between radar and aircraft methods is that the airplane, used as a measuring instrument, is located directly in the flow, whereas the radar station is located on the ground, and objects that reflect the radar signal are situated in the turbulent flow. For the investigation of turbulence in clouds and precipitation, their elements are usually used, and for clear-air turbulence, scatters are artificially introduced.

There are several methods of measuring the movements of scatterers, based on the analysis of the characteristics of the signal reflected from a certain volume filled with scatterers that are well entrained by the air flow.

The first method is the pulse analysis method [35].

The pulse transmitter of an incoherent station emits electromagnetic waves in a narrow beam, which penetrates the cloud of scatterers. The radio waves reflected by the scatterers are detected by the antenna of the station, but the radar receiver is on for only a short period of time–in the interval between transmissions of the sounding pulses. The volume subjected to observation is limited by the beam diameter and the duration of the transmission period. In the time between successive sounding pulses the mutual position of the scatterers within the investigated volume changes. This leads to a change in the phase relations in the received signal and, hence, in its magnitude at the receiver output. The fluctuations of the output signal amplitude depend on the relative velocities of the scatterers within the investigated volume. An increase in the relative velocities leads to an increase in the frequency of fluctuations of the output signal.

From the number of maxima n during the time of observation Δt in the envelope of the output signal, it is possible to determine approximately the relative velocity of the scatterers

$$u_S = \frac{n\lambda}{2\Delta t}, \tag{2.54}$$

where λ is the transmitter wavelength. The transmitter frequency has to be fairly stable. If the scatterers are so well entrained by the flow that their falling velocity can be neglected, the spectrum of fluctuations of the radar signal intensity is given by the relation

$$G(F) = \bar{A}^2 P_T\left(\frac{\lambda F}{2}\right), \tag{2.55}$$

where F is the frequency of fluctuations of the radar receiver output signal, $\bar{A}$ is the mean power of the echo from the assembly of scatterers, and $P_T(u_S) = P_T(\lambda F/2)$ is the distribution of the projections of the scatterer relative velocities on the beam direction.

A pulse Doppler radar can measure the mean wind velocity (from the mean Doppler frequency) and its fluctuations (from the nature of the spectrum of the output signal of the phase detector of the Doppler radar station) [68]. This is true, of course, only in the case where the scatterers are well entrained by the flow. The spectrum of Doppler frequencies is connected with the distribution of the projections of the scatterer absolute velocities on the beam direction by the relation

$$G_D(F) = \bar{A}P_T(F), \tag{2.56}$$

where $F = 2u/\lambda$.

A comparison of the spectra of the horizontal turbulence component associated with developed convection, obtained from the data of an aircraft hot-wire anemometer and a coherent-pulse radar station using dipole reflectors, showed good agreement of the results obtained [125].

The further development of radar methods will require an increase in the station potential, which will allow investigation of the wind-field structure in clear skies without the use of balloons or the introduction of artificial reflectors at high altitudes.

A promising future method of measuring clear-air turbulence is optical location by means of lidars (laser radars). Long-range investigation of the atmosphere by a lidar is effected by investigation of the optical signals reflected from various objects. The objects used for investigation of clear-air turbulence are aerosols and minute dust particles suspended in the atmosphere. It is possible in principle to use back-scattering from certain gases present in the atmosphere, particularly carbon dioxide (CO_2) and water vapor. The experience acquired in the development of radar methods can be used in the development of methods of investigating the wind-field structure in clear skies by means of lidars. We know, for instance, that a method of direct measurement of the air velocity gradient from the Doppler shift of a reflected laser signal is being developed.

In addition to methods of investigating the velocity field in the turbulent atmosphere, the development of methods of long-range detection of turbulent zones in the atmosphere by means of radar and optical devices has recently received great attention.

Both incoherent-pulse [35] and Doppler radars [68] can be used to detect turbulent zones in clouds and precipitation, i.e., in a medium containing numerous natural scatterers.

Turbulent mixing in clear skies gives rise to inhomogeneities which can be detected by high-potential incoherent-pulse stations operating with 3.2- and 10.7-cm waves.

Zones and layers of intense clear-air turbulence have been detected by ground lidars [150].

Investigations of clear-air turbulence have shown a close relation between zones of intense turbulence and horizontal inhomogeneties in the temperature field. As was shown in [101], when the horizontal temperature gradients are more than 6°C per 100 km the frequency of aircraft buffeting reaches 85%. Research on an aircraft infrared radiometer for the advance detection of zones of clear-air turbulence is being carried out in several countries.

Methods of measuring the radiation temperature of physical objects are based on the use of the radiation laws.

Since the atmosphere is a mixture of different gases, the test object can be sought among them. It is known that the emission of gases is very small, except in certain spectral regions (absorption bands) where the gas can be regarded as an absolutely black body. In particular, carbon dioxide (CO_2) emits as a black body at wavelength 15 μ. Carbon dioxide is present in the atmosphere at all altitudes; its concentration is about 0.03%. A monochromatic infrared (IR) radiometer at wavelength 15 μ measures the CO_2 radiation temperature in the immediate vicinity of the instrument. A monochromatic IR radiometer tuned to the edge of the CO_2 absorption band measures the mean temperature of the air mass contained within the solid angle of the instrument optical system. The linear extent of this mass is greater, the greater the distance between the instrument filter passband and the center of the absorption band. By measuring the difference in radiation temperatures at the center of the CO_2 absorption band and on its edge, it is possible to determine the horizontal temperature gradient ahead of the airplane. This instrument is of very great importance for experimental investigations of such a complex phenomenon as clear-air turbulence.

3

Statistical Analysis of Measurements

Statistical analysis has been increasingly used in recent years for treatment of the results of observations. The statistical nature of many atmospheric phenomena makes such analysis essential, and the development of diverse computational techniques (analog and digital) has provided the means for conducting the analysis. Hence, we must discuss, albeit briefly, the techniques of statistical analysis and, what is just as important, the accuracy of the results obtained. It should be noted that while the actual methods of statistical analysis are fairly well known, the question of the accuracy of the statistical characteristics obtained has received relatively little consideration in the meteorological literature. This, in turn, makes it difficult (and often simply impossible) to compare heterogeneous experimental data, and in addition, can lead to erroneous conclusions regarding the nature of the phenomenon. The results of statistical analysis are usually variances, structure and correlation functions, and spectral densities, which are simple and also the most important characteristics of meteorological fields.

In this chapter we give an account of methods of calculating the spectra of meteorological parameters, and we also indicate the errors entailed in these calculations. The preference that the authors show for spectra can be attributed to the ease of physical interpretation of spectral characteristics and the relative simplicity of spectral transformations, which is of particular importance for practical applications.

In addition, the size of this book does not allow us to deal with the derivation of most of the formulas cited. Readers can find such information in the special monographs [38, 139].

1. Calculation of Spectra from Continuous Records of Finite Length

1.1. Some Definitions. Although the main concepts of the statistical characteristics of the field of random velocities in a turbulent flow were formulated in Chapter 1, it is useful here to go over some of them again without particular reference to the relation between these characteristics and turbulence.

A function $u(t)$ is called a random function of time if its value at any instant is a random quantity characterized by some probability distribution. In almost all that follows we will assume that the random process generating $u(t)$ is Gaussian, i.e., for any set $t_1, t_2, \ldots, t_n$ the joint probability distribution $u(t_1)$, $u(t_2), \ldots u(t_n)$ is an n-dimensional Gaussian (normal) distribution. Each such distribution is completely determined by the means $\overline{u(t_i)}$ and the covariances

$$R_{ij} = \overline{\{[u(t_i) - \overline{u(t_i)}]\,[u(t_j) - \overline{u(t_j)}]\}}\,.$$

The vinculum indicates averaging over the statistical ensemble, i.e., over an infinite set of functions $u(t)$, each of which is a particular realization of a single random process.

For convenience we assume that all $\overline{u(t_i)} \equiv 0$. We will also assume that the process is stationary (i.e., temporally homogeneous) throughout. Then

$$R_{ij} = \overline{u(t_i)\,u(t_j)} = R(t_i - t_j)\,.$$

Thus, a stationary random process is completely specified by a single function of the single variable $\tau = t_i - t_j$.

It is obvious, however, that in the experimental investigation of random processes in the atmosphere we usually have to deal with a single function of time, and not with an ensemble of functions. This is why it is extremely important that averaging over an ensemble is equivalent to averaging over time (ergodicity) whenever the process is stationary, Gaussian, and has a zero mean value and a continuous power spectrum. In other words, we can write

$$R(\tau) = \overline{u(t)\,u(t+\tau)} = \lim_{T\to\infty} \frac{1}{T} \int_{-T/2}^{+T/2} u(t)\,u(t+\tau)\,dt\,.$$

Khinchin showed that $R(\tau)$ can be represented by a Fourier transform† of some function $S(f)$:

$$R(\tau) = \int_{-\infty}^{+\infty} S(f)\,e^{i2\pi f\tau}\,df,$$

†The term "Fourier transform" means a function obtained by a Fourier transformation.

where

$$S(f) = \lim_{T\to\infty} \frac{1}{T} \left| \int_{-T/2}^{+T/2} u(t)\, e^{-i2\pi f t}\, dt \right|^2 \tag{3.1}$$

Here and henceforth in this chapter $f = \omega/2\pi$. The frequency function $S(f)$ is called the energy spectrum of the stationary random process considered. The quantity $S(f)\,df$ is the contribution of frequencies between f and $(f + df)$ to the variance of the process. In addition to (3.1) there is the inverse Fourier transformation, by which the energy spectrum can be expressed in terms of the autocorrelation function

$$S(f) = \int_{-\infty}^{+\infty} R(\tau)\, e^{-i2\pi f\tau}\, d\tau. \tag{3.2}$$

Since $R(\tau)$ and $S(f)$ are even functions, then (3.1) and (3.2) can be written in the form

$$R(\tau) = \int_{-\infty}^{+\infty} S(f) \cos 2\pi f\tau\, df, \tag{3.3}$$

$$S(f) = \int_{-\infty}^{+\infty} R(\tau) \cos 2\pi f\tau\, d\tau \tag{3.4}$$

or, more simply,

$$R(\tau) = 2 \int_{0}^{\infty} S(f) \cos 2\pi f\tau\, df, \tag{3.5}$$

$$S(f) = 2 \int_{0}^{\infty} R(\tau) \cos 2\pi f\tau\, d\tau. \tag{3.6}$$

1.2. Calculation Techniques. In practice we can obtain only a limited number of realizations of the function $u(t)$, each of which has a finite length. Hence, from the available data we can obtain only estimates (i.e., approximate values) of the autocorrelation function and the energy spectrum.

In fact, when we have a continuous record of finite length it is impossible to estimate the value of $R(\tau)$ for τ greater than the record length. Moreover, as we will show later, it is usually undesirable to use τ greater than 5–10% of the record length.

Thus, instead of

$$R(\tau) = \lim_{T \to \infty} \frac{1}{T} \int_{-T/2}^{+T/2} u(t)u(t+\tau)\,dt,$$

where $R(\tau)$ has a meaning for all τ, we must write

$$R_{00}(\tau) = \frac{1}{T_n - \tau} \int_{-\frac{T_n-\tau}{2}}^{+\frac{T_n-\tau}{2}} u\left(t - \frac{\tau}{2}\right) u\left(t + \frac{\tau}{2}\right) dt, \tag{3.7}$$

where $R_{00}(\tau)$ has a meaning only for $|\tau| \leqslant \tau_m < T_n$ (τ_m is the maximum time difference which we wish to use), and T_n is the record length.

Since the experimental function $R_{00}(\tau)$ is known only for $|\tau| \leqslant \tau_m$, then it is impossible, generally speaking, to use one of the Eqs. (3.2), (3.4), or (3.6) to calculate $S(f)$ since these equations require a knowledge of $R(\tau)$ for all τ. In this case the most obvious approach is to assume that when $|\tau| > \tau_m$, $R_{00}(\tau) = 0$, i.e., to assign some function $R_i(\tau)$ such that

$$R_i(\tau) = \begin{cases} R_{00}(\tau) & \text{when } |\tau| \leqslant \tau_m, \\ 0 & \text{when } |\tau| > \tau_m. \end{cases} \tag{3.8}$$

The subscript i is put in so that the autocorrelation function obtained will not be confused with either the true $R(\tau)$ or with the experimental $R_{00}(\tau)$.

The function $R_i(\tau)$ is called the modified experimental autocorrelation function. Conversion from $R_{00}(\tau)$ to $R_i(\tau)$ is very simple but, as will be shown later, it has a fairly complex effect on the shape of the energy spectrum obtained from $R_i(\tau)$ by means of the Fourier transformation.

In fact, Eq. (3.8) is equivalent to

$$R_i(\tau) = D_i(\tau) R_{00}(\tau), \tag{3.9}$$

where

$$D_i(\tau) = \begin{cases} 1 & \text{when } |\tau| \leqslant \tau_m, \\ 0 & \text{when } |\tau| > \tau_m. \end{cases} \tag{3.10}$$

Since $R_i(\tau)$ is known for all τ, we can perform the Fourier transformation and obtain

$$S_i(f) = Q_i(f) * S_{00}(f), \tag{3.11}$$

where $Q_i(f)$ is the Fourier transform of $D_i(\tau)$, $S_{00}(f)$ is the Fourier transform of $R_{00}(\tau)$, and the asterisk means the mathematical operation of convolution.†

Although $S_{00}(f)$ is not determined [since $R_{00}(\tau)$ is known only for $|\tau| \leqslant \tau_m$], we can write (because of ergodicity)

$$R_i(\tau) = D_i(\tau) R(\tau), \tag{3.12}$$

where $R(\tau)$ is the true autocorrelation function. It follows directly from (3.12) that

$$S_i(f) = Q_i(f) * S(f), \tag{3.13}$$

where $S(f)$ is now the true energy spectrum.

From the definition of convolution we can rewrite (3.13) in the form

$$\overline{S_i(f_1)} = \int_{-\infty}^{+\infty} Q_i(f_1 - f) S(f) df.$$

As (3.13) shows, the mean value of $S_i(f_1)$ is obtained by smoothing (averaging over frequencies) of the true energy spectrum near f_1 with weights proportional to $Q_i(f_1 - f)$. It is therefore natural to call $Q_i(f)$ the spectral window, and the function $D_i(\tau)$ the time window.

It is clear that the quality of the energy spectrum estimates obtained by means of (3.13) will depend largely on the form of the weighting function (or spectral window) $Q_i(f)$ and, hence, on the selection of the time window $D_i(\tau)$. Thus, we must strive not for "simplicity" of $D_i(\tau)$, but to ensure that the Fourier transform of some specially selected $D_i(\tau)$ is concentrated near $f = 0$ and has large side maxima (in comparison with the maximum at frequency $f = 0$).

We will consider some particular examples of the function $D_i(\tau)$ and the $Q_i(f)$ corresponding to them. Let the time window be given by (3.10):

$$D_1(\tau) = \begin{cases} 1, & |\tau| \leqslant \tau_m, \\ 0, & |\tau| > \tau_m. \end{cases} \tag{3.14}$$

The subscript i is replaced by 1 here, to indicate that it is the first window pair to be considered.

†If the functions φ_1 and φ_2 are given, the convolution of these functions is given by

$$\varphi_1 * \varphi_2 = \psi(x) = \int_{-\infty}^{+\infty} \varphi_1(x - y)\varphi_2(y) dy.$$

Then

$$Q_1(f) = \int_{-\infty}^{+\infty} D_1(\tau)\, e^{-i2\pi f\tau}\, d\tau = 2\tau_m \frac{\sin 2\pi f \tau_m}{2\pi f \tau_m}. \tag{3.15}$$

Graphs of the functions $D_1(\tau)$ and $Q_1(f)$ are shown in Fig. 3.1. Figure 3.1 shows that the apparently simple and natural choice of $D_1(\tau)$ leads to a relatively complex function $Q_1(f)$, in which, in addition, the first side extremum is negative and is almost $\frac{1}{5}$ of the height of the main maximum.

Different authors have proposed several types of time windows $D_i(\tau)$ which have a Fourier transform $Q_i(f)$ that is satisfactory for practical purposes.

We will discuss two generally used forms of $D_i(\tau)$. The time window $D_2(\tau)$ proposed by von Hann has the following form:

$$D_2(\tau) = \begin{cases} \frac{1}{2}\left(1 + \cos \frac{\pi\tau}{\tau_m}\right), & |\tau| \leqslant \tau_m, \\ 0 & , \quad |\tau| > \tau_m. \end{cases} \tag{3.16}$$

The corresponding spectral window $Q_2(f)$ can then be written in the form

$$Q_2(f) = \tau_m \frac{\sin 2\pi f \tau_m}{2\pi f \tau_m} \frac{\pi^2}{\pi^2 - (2\pi f \tau_m)^2}. \tag{3.17}$$

Finally, the time window proposed by Hamming,

$$D_3(\tau) = \begin{cases} 0.54 + 0.46 \cos \frac{\pi\tau}{\tau_m}, & |\tau| \leqslant \tau_m, \\ 0 & , \quad |\tau| > \tau_m \end{cases} \tag{3.18}$$

has a spectral window of the form

$$Q_3(f) = 1.08\tau_m \frac{\sin 2\pi f \tau_m}{2\pi f \tau_m} \frac{\pi^2 - 0{,}08\,(2\pi f \tau_m)^2}{\pi^2 - (2\pi f \tau_m)^2}. \tag{3.19}$$

Graphs of the functions D_2, D_3 and Q_2, Q_3 are shown in Fig. 3.2. Comparing the graphs of Q_2 and Q_3 with the graph of Q_1 (see Fig. 3.1), we note first of all that the side maxima of Q_2 and Q_3 are much smaller than those of Q_1. The height of these maxima is not more than 2% of the height of the main maximum, whereas the first side maximum of Q_1 is almost ten times more than in Fig. 3.2.

The differences between Q_2 and Q_3 are less appreciable and not so significant. We merely note that the height of the side maxima of Q_2 decreases more

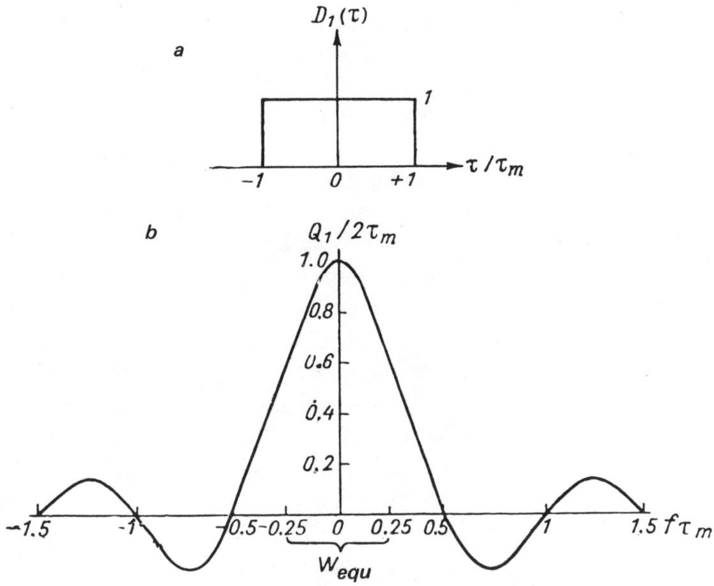

Figure 3.1. (a) Time window $D_1(\tau)$; (b) spectral window $Q_1(f)$.

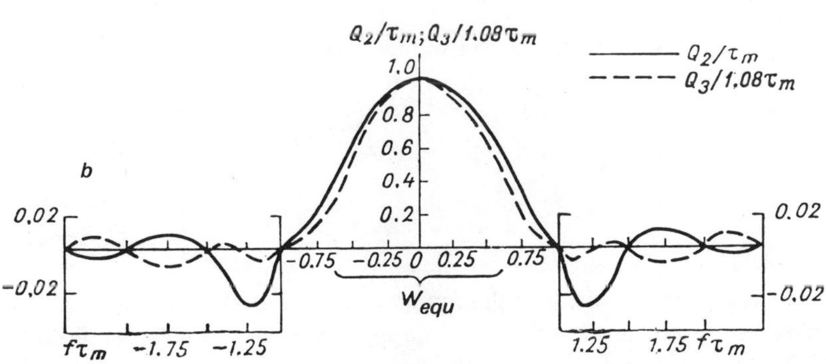

Figure 3.2. (a) Time windows $D_2(\tau)$ and $D_3(\tau)$; (b) spectral window $Q_2(\tau)$ and $Q_3(\tau)$.

rapidly than those of Q_3. Henceforth in calculations of energy spectra the pairs (D_2, Q_2) and (D_3, Q_3) are used. The following point should be noted. Multiplication of the experimental autocorrelation function $R_{00}(\tau)$ by the appropriate even function (say D_2 or D_3) leads to a modified experimental autocorrelation function $R_i(\tau)$, which, of course, is quite unlike the true autocorrelation function. The Fourier transform of this modified function, however, gives a very satisfactory estimate of the smoothed values of the true energy spectrum.

1.3. Stability of Energy Spectrum Estimates. We have discussed the method of calculating energy spectra in general features and have given formulas which can be used to obtain the mean (smoothed) values of the energy spectrum from experimental data. Since the number of experimental data is limited, the obtained estimates will, generally speaking, differ from the values of the energy spectrum by a certain quantity, called the sampling fluctuation. Hence, it is essential to know the size of these fluctuations, i.e., the stability of the experimental spectrum. The accuracy of the results obtained can eventually be estimated if the limits in which the energy spectrum lies are known with a specified probability.

Before proceeding to a quantitative analysis of the stability of spectral estimates, we note that all that has been, or will be, said about the mean values of the energy spectrum is valid if the investigated process is stationary and does not depend on whether it is Gaussian or not. Stability estimates, however, strictly speaking, are valid only for Gaussian processes, although they are approximately true also for processes which are not precisely Gaussian.

The stability of a fluctuating quantity, e.g., $S_i(f_1)$, is most simply defined as the ratio of its variance σ^2 to the square of its mean.

It can be shown that the stability of a spectral estimate is given by the formula

$$\frac{\sigma^2\,[S_i(f_1)]}{[\overline{S_i(f_1)}]^2} = \frac{1}{T'_n W_{\text{equ}}}, \tag{3.20}$$

where $T'_n = T_n - \frac{1}{3}\tau_m$ is the "effective record length," and

$$W_{\text{equ}} = \frac{\left[\int\limits_{-\infty}^{+\infty} Q_i(f_1 - f)S(f)\,df\right]^2}{\int\limits_{-\infty}^{+\infty} [Q_i(f_1 - f)S(f)]^2\,df} \tag{3.21}$$

is called the equivalent width of the function $Q_i(f_1 - f)S(f)$. Equation (3.20) is derived on the assumption that the true spectrum varies slowly in comparison with $1/T'_n$.

The meaning of Eq. (3.21) is simple. We assume that $S(f) = \text{const}$. Then

W_{equ} will obviously be the equivalent width of the spectral window $Q_i(f)$. In particular:

(1) If Q_i is a rectangle of width W, then $W_{equ} = W$;
(2) if Q_i is a triangle with base W, then $W_{equ} = 0.75W$;
(3) if $Q_i = Q_1$ [see (3.10)], then $W_{equ} = 0.5W$, where W is the width of the main maximum;
(4) if $Q_i = Q_2$ [see (3.16)] $W_{equ} = 0.67W$, where W is the width of the main maximum ($f_1 \geqslant 1/\tau_m$);
(5) if $Q_i = Q_3$ [see (3.18)], then $W_{equ} = 0.63W$, where W is the width of the main maximum ($f_1 \geqslant 1/\tau_m$).

The values of W_{equ} for the spectral windows Q_1, Q_2, and Q_3 are given in Figs. 3.1 and 3.2.

Thus, Eq. (3.20) shows that the greater the record length T'_n and equivalent width of the spectral window, the more stable the spectral estimate. We should mention, however, that when W_{equ} is increased the obtained estimate relates to a wide frequency band.

It is obvious that, if the true form of $S(f)$ is unknown, the value of W_{equ} cannot be determined from (3.21). However, assuming that $S(f)$ varies slowly along $Q_i(f)$ we can, without serious error, put

$$W_{equ} \approx \frac{1}{\tau_m} \tag{3.22}$$

Then (3.20) can be rewritten in the form

$$\frac{\sigma^2\,[S_i(f_1)]}{[\overline{S_i(f_1)}]^2} \cong \frac{\tau_m}{T'_n}\,. \tag{3.23}$$

Thus, if, for instance, we require that the standard deviation σ of the energy spectrum value is not more than $\frac{1}{3}$ of its mean value, then τ_m/T'_n must be less than $\frac{1}{9}$.

Hence, to ensure stability of spectral estimates in the range ±(30–40)% of the mean value, we must obtain an experimental autocorrelation function up to values of τ_m that are not more than 10% of the total record length T'_n.

It will be convenient henceforth to treat the stability of spectral estimates on the basis of the assumption that the distribution of the energy spectrum values is a χ^2 distribution. It is known [111] that if $y_1, y_2, \ldots, y_k$ are independent, normally distributed random quantities with zero means and unit variance, the quantity

$$\chi_k^2 = y_1^2 + y_2^2 + \ldots + y_k^2,$$

which is obviously positive, follows, by definition, a χ^2 distribution with k degrees of freedom. The stability of χ_k^2, defined as the ratio of its variance to the square of its mean, is $2/k$. Thus, with increase in the number of degrees of freedom k the quantity χ_k^2 becomes relatively more stable, since its scatter around the mean is reduced. This is also true for any multiple of χ_k^2.

Hence, it is convenient to consider the stability of the estimate of any positive quantity (and the energy spectrum is positive in its physical nature) in terms of the equivalent number of degrees of freedom. By this we mean the number of degrees of freedom of a χ_k^2 distribution such that some multiple of it agrees (in mean value and variance) with the estimate of interest to us. The equivalent number of degrees of freedom is easily found from its definition and Eq. (3.20):

$$k = 2T'_r W_{\text{equ}}, \tag{3.24}$$

or taking (3.23) into account, we obtain

$$k \approx \frac{2T'_n}{\tau_m}. \tag{3.25}$$

From Table 3.1, knowing the number of degrees of freedom, we can determine, with a specified accuracy, the stability of the spectral estimate, i.e., find the limits within which our estimate of the energy spectrum value will lie with a specified probability. It is fairly simple to use Table 3.1. We give some examples.

Example 1. If the mean energy spectrum of wind-velocity fluctuations over a long period is 10 m^3/sec^2, then among estimates with 15 degrees of freedom

TABLE 3.1. Stability (Scatter) of Spectral Estimates in Relation to Equivalent Number of Degrees of Freedom and Accuracy of Determination of Stability Limits[a]

Degrees of freedom	90% accuracy		80% accuracy	
	95% of all values exceed given value	5% of all values exceed given value	90% of all values exceed given value	10% of all values exceed given value
1	0.0039	3.8	0.016	2.7
5	0.23	2.2	0.32	1.85
10	0.39	1.83	0.49	1.60
15	0.48	1.66	0.57	1.49
20	0.55	1.57	0.62	1.42
25	0.59	1.50	0.66	1.38
30	0.62	1.46	0.69	1.34
40	0.66*	1.41*	0.73	1.30
50	0.69*	1.36*	0.75	1.26
70	0.74*	1.29*	0.80	1.20
100	0.76*	1.25*	0.82	1.18

[a]The asterisked values were obtained by extrapolation of standard tabulated data (see [111], for instance).

5% of the values will be less than 4.8 m^3/sec^2, and 5% of the values will exceed 16.6 m^3/sec^2. Thus, to an accuracy of 90% individual values of the energy spectrum will lie in the range 4.8–16.6 m^3/sec^2. If we reduce the accuracy of determination of the stability limits to 80%, the individual values will lie in a narrower range: 5.7–14.9 m^3/sec^2.

Example 2. If an individual value of an energy spectrum with 30 degrees of freedom is 5 m^3/sec^2, then we can say, to an accuracy of 90%, that the true mean will lie within the limits $5/1.46 = 2.43$ m^3/sec^2 and $5/0.62 = 8.07$ m^3/sec^2. If we reduce the accuracy to 80%, the true mean will lie within the limits $5/1.34 =$ 3.74 m^3/sec^2 and $5/0.69 = 7.25$ m^3/sec^2.

Example 3. We assume that we have a continuous record of wind-velocity fluctuations of duration 10 min. We have to calculate the energy spectrum values and we wish to obtain sufficiently stable values which will lie, to an accuracy of 80%, within ±25% of the true value. We find from Table 3.1 that to obtain such stability we must have estimates with at least 50 degrees of freedom. From Eq. (3.25) we can then determine the maximum time interval τ_m for which the autocorrelation function of the wind fluctuations has to be calculated:

$$\tau_m = \frac{2T'_n}{k} \approx \frac{2 \cdot 600 \text{ sec}}{50} = 24 \text{ sec}$$

For simplicity we assume that $T'_n = T_n$.

Thus, in this case τ_m is only 4% of the total record length.

An important property of the χ^2 distribution is that if there is a quantity χ_k^2 with k degrees of freedom and a quantity χ_p^2 with p degrees of freedom, their sum $(\chi_k^2 + \chi_p^2)$ will possess $(k + p)$ degrees of freedom. This property must be borne in mind in calculations of the stability of the mean energy spectrum when the averaging involves several individual estimates with different numbers of degrees of freedom.

2. Calculation of Spectra from Discrete Records of Finite Length

All that has been said about methods of calculating energy spectra from continuous finite records can be put into practice only with the aid of the appropriate analog computing technology (correlometers and spectrum analyzers). In meteorology, however, only very few measurements are processed by means of analog devices. First, in the vast majority of cases standard meteorological measurements provide "point," and not continuous, measurements. Second, even when the measurements are continuous it is usually a case of very low-frequency processes, which allows accurate analog-code conversion and processing of the information in digital form by digital electronic computers. Hence, it is worthwhile giving special attention to methods of calculating spectra from records consisting of a series of equispaced discrete points. Although all the equations given in Section 1 for continuous records have their discrete analogs, the digital calculation of energy spectra from point records has several specific features that require special consideration.

2.1. "Aliasing." Let the random process $u(t)$ be represented not by a continuous function of time, but by a set of discrete equispaced values situated on the time axis at points

$$t = 0, \pm \Delta t, \pm 2\Delta t, \ldots, \pm n\Delta t$$

so that $R(\tau)$ can be calculated only for

$$|\tau| = 0, \quad \Delta t, \ 2\Delta t, \ \ldots, \ n\Delta t .$$

Ignoring for the present the finiteness of the record (i.e., assuming $n = \infty$), we can write the discrete analog of the Fourier transform [compare with (3.2)]:

$$S_A(f) = \sum_{r=-\infty}^{r=+\infty} R(r\Delta t)\, e^{-i2\pi f r\Delta t}\, \Delta t, \tag{3.26}$$

where

$$r = 0, \pm 1, \pm 2.$$

First, if we put $f = 1/\Delta t$ in (3.26), it is quite clear that $S_A(f)$ has period Δt and, since the spectrum is symmetric, most of it is included in the frequency band from $f = 0$ to $f_N = 1/2\Delta t$, and the values of the spectrum subsequently repeat with frequency $f = 1/\Delta t$ (Fig. 3.3). Thus, the highest frequency (the Nyquist frequency) at which an estimate of the energy spectrum can be obtained in the case of sampling at intervals Δt is $f_N = 1/2\Delta t$.

It is extremely important, in addition, to know if there are frequencies higher than f_N in the true spectrum $S_A(f)$ of the real process. If there are, the problem of "aliasing" arises. The essence of this problem is that while frequencies in the $[0, f_N]$ band are clearly distinguishable from one another, any frequency above f_N is indistinguishable from some frequency in the $[0, f_N]$ band. As a result of this the calculated energy spectrum $S_A(f)$ will have greater errors in all its values, the greater the energy of the process at frequencies greater than f_N. The nature of this effect [which, unfortunately, cannot be avoided if $S(f > f_N) \neq 0$] can be explained with the aid of Fig. 3.4.

For instance, Fig. 3.4 shows that when the sampling interval $\Delta t = 0.2$ sec it is impossible to say with certainty from the sine curve of which frequency the sample is actually taken: 1, 4, or 6 Hz. Hence, it is understandable that if the value of S_A (1 Hz) has to be calculated, and the true spectrum of the process extends to, say, 10 Hz, then in sampling at an interval $\Delta t = 0.2$ sec the energy

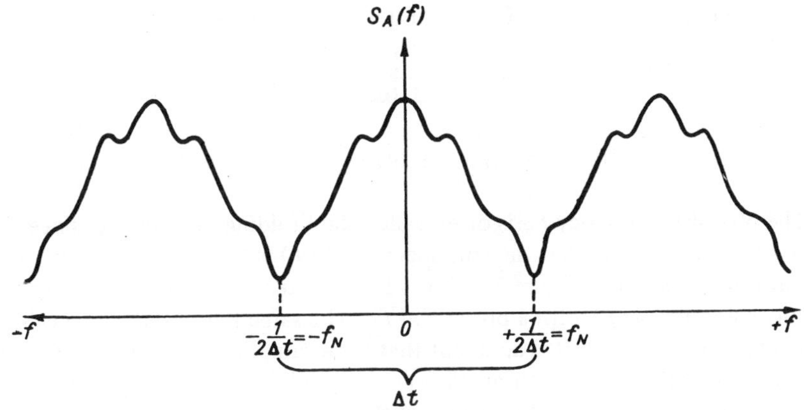

Figure 3.3. Periodicity of spectrum $S_A(f)$.

of the process at frequencies 4, 6, and 9 Hz is added to the true value of S (1 Hz) and a false idea of the shape of the energy spectrum will be obtained. Figure 3.4 indicates the origin of the term "aliasing": Frequencies greater than f_N "impersonate" frequencies in the $[0, f_N]$ band. The aliasing effect can be expressed analytically in the form [139]

$$\begin{aligned} S_A(f) &= S(f) + S(2f_N - f) + S(2f_N + f) \\ &\quad + S(4f_N - f) + S(4f_N + f) + \dots, \end{aligned} \tag{3.27}$$

where $S_A(f)$ is the spectrum calculated from the discrete record, and $S(f)$ is the true spectrum.

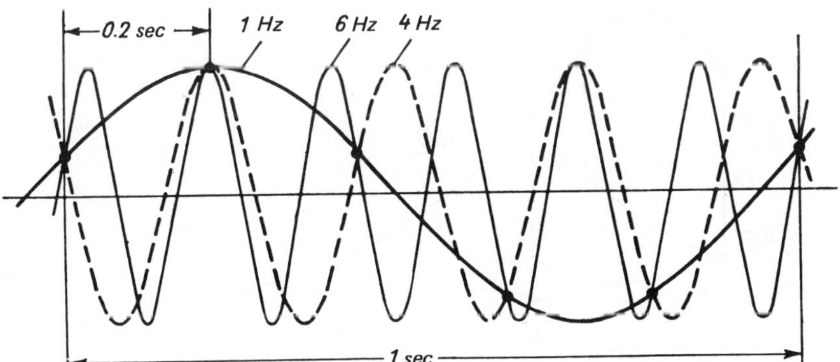

Figure 3.4. Equidistant sampling of sine curves of different frequency.

For the example shown in Fig. 3.4, Eq. (3.27) gives

$$f_N = \frac{1}{2\Delta t} = \frac{1}{2 \cdot 0.2} = 2.5 \text{ Hz},$$

$$S_A(1 \text{ Hz}) = S(1 \text{ Hz}) + S(4 \text{ Hz}) + S(6 \text{ Hz}) + S(9 \text{ Hz}) + S(11 \text{ Hz}) + \ldots$$

The Nyquist frequency f_N is often called the "folding" frequency, since it follows from Eq. (3.27) that the true spectrum $S(f)$ is "accordion-folded" on itself at multiples of f_N. Figure 3.5 shows how the true spectrum is built up on conversion to $S_A(f)$ with a prescribed f_N. The same figure shows the calculated spectrum $S_A(f)$. It is apparent that for a sampling interval $\Delta t = 0.2$ sec and a given $S(f)$, the values obtained for $S_A(f)$ will be 30–50% higher than those of the true spectrum. It is clear that the correct calculation of the spectrum $S_A(f)$ in the case shown in (3.5) would require $f_N = 7.5$ Hz, which corresponds to a sampling interval $\Delta t = 1/2f_N \approx 0.07$ sec. The above account indicates that very serious attention must be given to aliasing in calculations of energy spectra from discrete records. The reduction (or complete elimination) of aliasing requires the filtering out, at the earliest stages of the experiment (measurement, recording), of all the high frequencies that may be in the process and for some reason or other are of no interest to the investigator in subsequent calculations.

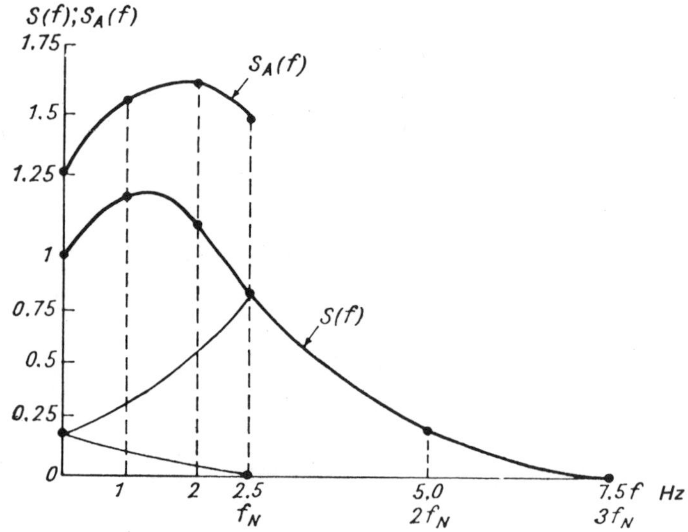

Figure 3.5. "Folding" of the true spectrum $S(f)$ at multiples of f_N.

2.2. Calculation Procedure and Stability of Spectral Estimates. For convenience we denote the obtained values of the random process by $u_0, u_1, \ldots, u_n$, which corresponds to $u(0), u(\Delta t), \ldots, u(n\Delta t)$. Following the procedure described in Section 1, we must calculate the values of the experimental autocorrelation coefficients [and not the continuous experimental function $R_{00}(\tau)$], modify these values, and then perform the Fourier transformation to obtain the spectrum. It is obviously impossible to calculate the experimental autocorrelation coefficients for time intervals other than $0, \Delta t, \ldots, n\Delta t$. Hence, it is convenient to write $R_0, R_1, \ldots, R_m$ instead of $R_{00}(0), R_{00}(\Delta t), \ldots, R_{00}(m\Delta t)$. Here n is the number of available values of the random process corresponding to T_n in Section 1, and m is the number of time intervals Δt for which the autocorrelation coefficients are calculated. This number of intervals corresponds to τ_m in Section 1.

If we now obtain the Fourier transforms of all the autocorrelation coefficients (the number of which is $m + 1$), we can obtain smoothed estimates of the energy spectrum for any frequency between zero and $f_N = 1/2\Delta t$. It can be shown, however, that for extraction of all the information contained in the process it is sufficient to calculate only $m + 1$ values of the energy spectrum, i.e., one value of the spectrum for each R_r ($r = 0, 1, \ldots, m$). The calculated values of the energy spectrum will be uniformly spaced on the frequency axis in the band $0 \leqslant f \leqslant f_N$ at intervals $f_N/m = 1/2m\Delta t$.

In the case of continuous records the experimental autocorrelation function $R_{00}(\tau)$ will have to be modified with the aid of some time window $D_i(\tau)$, and then the Fourier transform of the modified function $R_i(\tau)$ has to be obtained. Thus, estimates of the smoothed values of the energy spectrum are obtained by passage of the true spectrum $S(f)$ of the process through some spectral window $Q_i(f)$ [see (3.14)]. In the case of discrete records it is found to be simpler to perform the Fourier transformation first and then to modify the obtained values of the energy spectrum. In fact, since D_2 and D_3 are finite sums of cosines, their Fourier transforms will simply be the sums of delta functions and, hence, the convolution in Eq. (3.14) will reduce to smoothing with weights 0.25, 0.5, 0.25 for D_2, and 0.23, 0.54, and 0.23 for D_3. We finally obtain the following calculation procedure:

(1) calculation of the experimental autocorrelation coefficients from the formula [compare with (3.7)]

$$R_r = \frac{1}{n-r} \sum_{q=0}^{q=n-r} u_q u_{q+r}, \tag{3.28}$$

where $r = 0, 1, \ldots, m$ ($m \leqslant n$);

(2) calculation of the Fourier transform of R_r from the formula [compare

with (3.6)]

$$P_r = \Delta t \left[R_0 + 2 \sum_{q=1}^{q=m-1} R_q \cos \frac{qr\pi}{m} + R_m \cos r\pi \right]; \tag{3.29}$$

(3) smoothing of the values of P_r to obtain estimates of the energy spectrum from the formulas

$$\left.\begin{aligned} S_0 &= 0.5P_0 + 0.5P_1, \\ S_r &= 0.25P_{r-1} + 0.5P_r + 0.25P_{r+1}, \qquad 1 \leqslant r < m-1, \\ S_m &= 0.5P_{m-1} + 0.5P_m \end{aligned}\right\} \tag{3.30}$$

by using the time window D_2 or

$$\left.\begin{aligned} S_0 &= 0.54P_0 + 0.54P_1, \\ S_r &= 0.23P_{r-1} + 0.54P_r + 0.23P_{r+1}, \qquad 1 \leqslant r \leqslant m-1, \\ S_m &= 0.54P_{m-1} + 0.54P_m \end{aligned}\right\} \tag{3.31}$$

by using the time window D_3. The obtained values relate to frequencies $f_r = r/2m\Delta t; 0 \leqslant f_r \leqslant f_N = 1/2\Delta t$.

The distance (on the frequency axis) between adjacent energy spectrum estimates is $1/2m\Delta t$. The analysis of the stability of the obtained spectral estimates does not differ in any way from that described in Section 1 and is again based on calculation of the equivalent number of degrees of freedom. The analog of Eq. (3.25) for calculation of the equivalent number of degrees of freedom k is written in the form

$$k = \frac{2n'}{m}; \tag{3.32}$$

$n' = n - 1/3m$ is called the equivalent number of points. Knowing k and using Table 3.1, we can determine the stability of the obtained values S_r of the energy spectrum with a specified accuracy.

2.3. "Prewhitening" of True Spectrum. We have described general methods of calculating energy spectra from equidistant records. We must now consider some special problems that do not always arise but, when they do, require the closest attention.

We assume that the true energy spectrum descends rapidly with increase in frequency. We then have the situation illustrated in Fig. 3.6. If the true spectrum $S(f)$ has the form shown in Fig. 3.6a, we find that in estimating the value of the spectrum at frequency f_1 by using the spectral window $Q_2(f)$ (Fig. 3.6b) the effect of the side maxima situated on the left of f_1 becomes impermissibly great (Fig. 3.6c). We finally have the situation where the calculated value of the

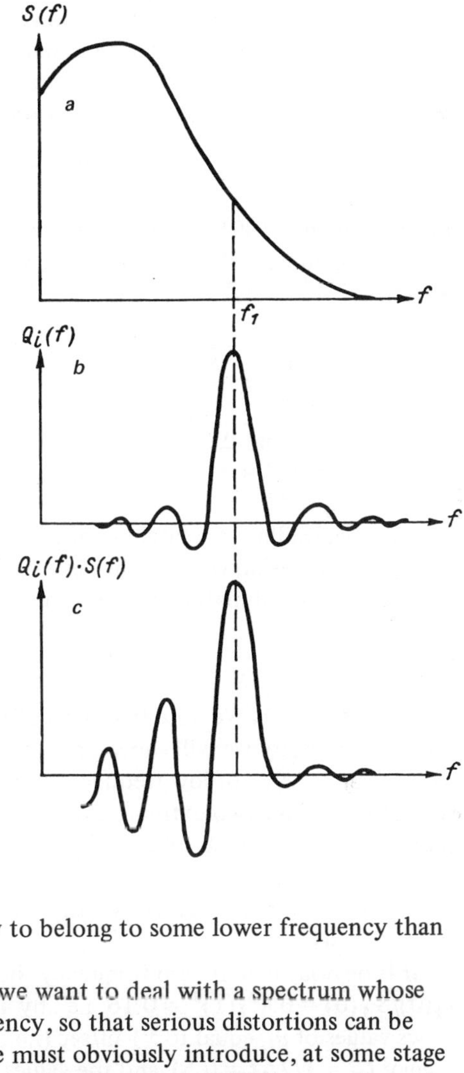

Figure 3.6. Illustration of "prewhitening" of the true spectrum. (a) Steeply descending true spectrum $S(f)$; (b) spectral window $Q_i(f)$; (c) result of passage of $S(f)$ through window $Q_i(f)$.

spectral estimate $S_A(f_1)$ is more likely to belong to some lower frequency than to f_1 proper.

It follows from the foregoing that we want to deal with a spectrum whose value does not vary greatly with frequency, so that serious distortions can be avoided. To obtain such a spectrum we must obviously introduce, at some stage in the experiment, an operation which will emphasize the high frequencies and suppress the lower ones. Such an operation can be called "whitening," since its aim is to obtain the uniform energy spectrum usually called the "white" noise spectrum. It is most convenient to introduce the whitening operation at the earliest stages of the experiment (measurement, recording) by an appropriate choice of frequency characteristics of the apparatus used. However, for various reasons that cannot always be done and, hence, the spectrum has to be whitened after obtaining the experimental data.

One method of whitening the spectrum after obtaining the actual data is to form some moving linear combination whose frequency characteristic is such that the resultant spectrum becomes more uniform than the initial spectrum. In fact, if there is a series of equidistant values of the process $u_i = u_0, u_1, \ldots, u_n$, we can form a moving linear combination of the type

$$\tilde{u}_l = u_l - \alpha u_{l-1} - \beta u_{l-2} - \gamma u_{l-3}. \tag{3.33}$$

It can be shown [139] that the energy spectrum $\tilde{S}_A(f)$ of the obtained process $\tilde{u}_l$ is connected with the energy spectrum $S_A(f)$ of the initial process u_l by the relation

$$\frac{\tilde{S}_A(f)}{S_A(f)} = \frac{\tilde{S}(f)}{S(f)} = |1 - \alpha e^{-i\omega\Delta t} - \beta e^{-i2\omega\Delta t} - \gamma e^{-i3\omega\Delta t}|^2, \tag{3.34}$$

where $\omega = 2\pi f$, and Δt is the interval between successive values of the process.

Expression (3.34) is a nonnegative polynomial of the third degree in $\cos \omega\Delta t$. By an appropriate choice of the coefficients α, β, and γ the required frequency dependence, leading to whitening of the true spectrum, can be obtained.

As an illustration we take the very simple operation of whitening with the aid of a linear combination of the form

$$\tilde{u}_l = u_l - 0.6u_{l-1}, \tag{3.35}$$

and we put $\Delta t = 1$. We note that in (3.35) the values of l begin at unity, and not at zero. For clarity we will show graphically that the operation (3.35) actually leads to suppression of low frequencies and enhancement of high frequencies. To do this we take two extreme cases: zero frequency and the highest frequency, obviously equal to the Nyquist frequency f_N. Figure 3.7a shows the values of u_l, which are all equal to $+1$ m/sec (i.e., we are dealing with zero frequency of the initial process), and the values obtained from (3.35) and equal to $+0.4$ m/sec.

It is obvious that at zero frequency there is suppression equal to $\tilde{S}_A(0)/S_A(0) = 0.4^2/(1)^2 = 0.16$, i.e., by a factor of more than 6. Figure 3.7b shows values of u_l, equal to ± 1 m/sec (i.e., velocity fluctuations with frequency $f_N = 1/2\Delta t = 0.5$), and the values obtained from (3.35) and equal successively to ± 1.6 m/sec.

It is apparent that at frequency f_N there is amplification equal to $\tilde{S}_A(f_N)/S_A(f_N) = (1.6)^2/(1)^2 = 2.56$, i.e., by a factor of more than 2.5. For the linear combination (3.35), Eq. (3.34) gives the relation

$$\frac{\tilde{S}_A(f)}{S_A(f)} = |1 - 0.6e^{-j\omega}|^2 = 1.36 - 1.20 \cos 2\pi f, \tag{3.36}$$

whose graph is shown in Fig. 3.8.

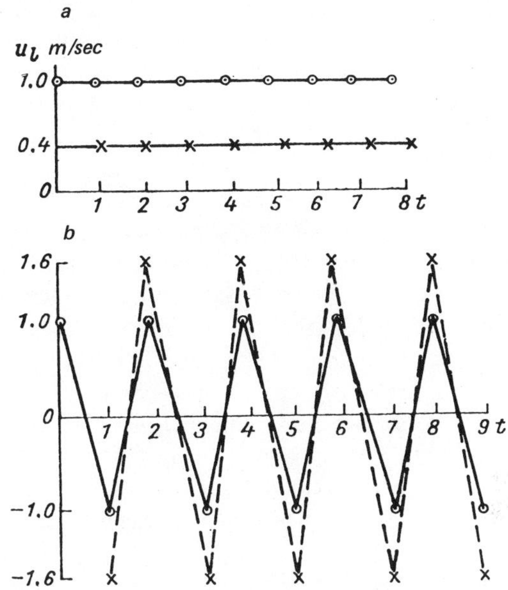

Figure 3.7. Effect of "whitening" operation (3.35) on constant component (a) and high-frequency component (b) of process.

Thus, in the investigation of processes with sharply descending spectra (and it is with this kind of process that meteorologists usually have to deal) we can recommend the application of some procedure for whitening the real spectrum, which leads to much more accurate final results.

2.4. Elimination of Zero Frequencies. Digital processing of equidistant records gives rise to another problem which is not usually encountered in the

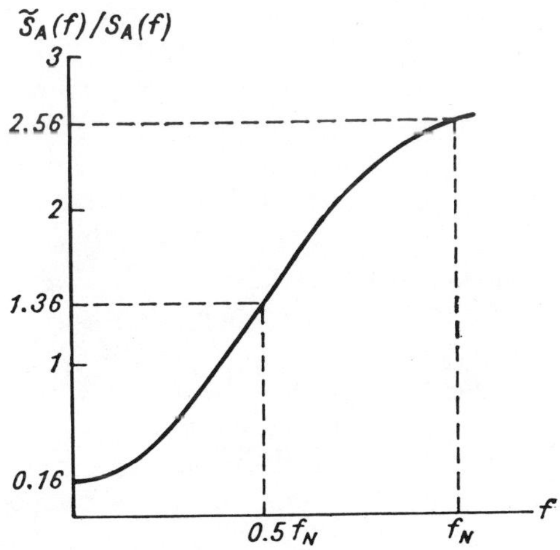

Figure 3.8. Frequency characteristic of "whitening" operation (3.35).

analog processing of continuous records. Analog computers usually filter out the very low frequencies, which allows the investigated process to be regarded as having a zero mean. When discrete data are processed, however, all frequencies, down to zero, have to be included. In this case it is very important to take measures to filter out the zero frequencies to avoid distorted values of the energy spectrum.

The example given by Blackman and Tukey [139] clearly shows how such distortions arise and how to estimate their order of magnitude. Suppose that most of the values (say, 999 out of 1000) of the process in some experiment lie between -100 and $+100$ (arbitrary units). This would be the case when the standard deviation $\sigma = 30$, and the variance $\sigma^2 = 900$. If the mean is 5, or even 10, it would be difficult to detect at a glance that it is not, in fact, equal to zero. Suppose further that these experimental data were obtained in a period $T = 15$ min with a sampling interval $\Delta t = 1$ sec. Thus, there are 900 points altogether. The Nyquist frequency $f_N = 1/2\Delta t = 0.5$ Hz, and the distance (in frequency) between adjacent values of the spectrum is $\Delta f = 1/2T_n = 5 \cdot 10^{-4}$ Hz, i.e., the $[0, f_N]$ band will be divided into 900 elementary frequency bands.

The total power contained in the process is the square of the mean (the so-called power of the constant component) plus the variance. In this example $P_{\text{tot}} = 5^2 + 900 = 925$ or $P_{\text{tot}} = 10^2 + 900 = 1000$. It is clear that the power of the constant component is contained in the lowest-frequency elementary band (i.e., in the first, contiguous with zero). This leads to a value of the energy spectrum in this band equal to $100/\Delta f = 100/(5 \cdot 10^{-4}) = 2 \cdot 10^5$ Hz^{-1}, whereas the mean value of the energy spectrum for all the other 899 elementary bands is equal only to $900/f_N = 900/0.5 = 2 \cdot 10^3$ Hz^{-1}.

Thus, the value of the spectrum at "zero" will be almost 100 times greater than the mean at higher frequencies.

This example clearly shows what happens if zero frequencies are not filtered out. When we speak of zero frequencies we usually have in mind not only processes with a constant component (i.e., with constant, but not zero, means), but also processes in which slow "drifts" of the mean value ("trend") occur. In meteorology the commonest processes are those where rapid fluctuations are superposed on much slower fluctuations, which form the trend. It is usually the fast fluctuations that are of interest, and it is desirable to filter out the trend so as not to affect the statistical properties of the high-frequency component of the process.

This problem can be completely solved, however, only in the case where the low-frequency (LF) and high-frequency (HF) components lie in significantly different, nonoverlapping frequency ranges (Fig. 3.9). In fact, if the LF spectrum lies to the left of some cutoff frequency f_1, and the HF spectrum lies to the right of some cutoff frequency f_2, then by constructing a filter that transmits all frequencies above some f_* and suppresses all frequencies below f_*,

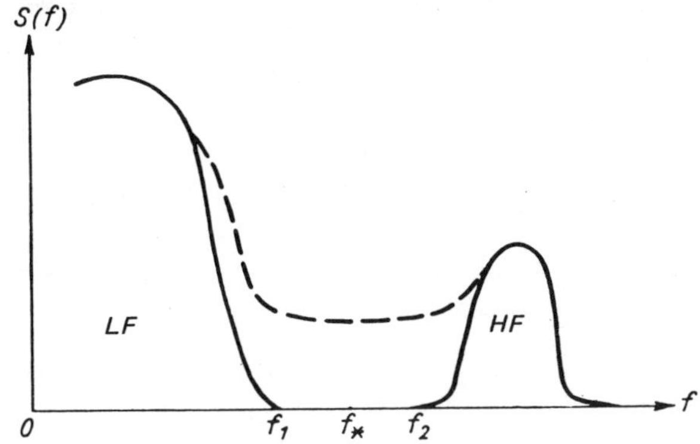

Figure 3.9. Spectrum S(f) with complete or partial (dashed curve) trough between high-frequency (HF) and low-frequency (LF) components.

where $f_1 \leqslant f_* \leqslant f_2$, the trend can be completely eliminated without distortion of the high-frequency component. Such ideal cases, however, are hardly ever encountered in meteorology. The best that we can count on is the presence of a trough of some sort between the LF and HF maxima of the spectra (dashed curve in Fig. 3.9). Even this condition is usually not fulfilled. Hence, in the experimental investigation of an atmospheric phenomenon each particular case will require devising specific ways of dealing with the distorting effect of zero frequencies [62].

Here we will mention only a general (and essential) method of eliminating the constant component. This is most simply accomplished by subtracting the square of the mean of the investigated quantity from each value of the experimental autocorrelation function.

Equation (3.28) for the calculation of R_r is then replaced by

$$R'_r = \frac{1}{n-r} \sum_0^{n-r} u_q u_{q+r} - \left[\frac{1}{n} \sum_0^n u_q \right]^2 \tag{3.37}$$

2.5. Resultant Calculation Procedure. The calculation procedure presented below is not the only one and is not universal. It provides a good illustration of the principles of the method and is useful in many practical cases.

Let there be a set of values of the process u_q $(q = 0, 1, \ldots, n)$ and a sampling interval $\Delta t = 1$. In this case the procedure is:

(1) putting $\Delta t = 1$, ascertain that the true spectrum $S(f)$ of the process is reduced almost to zero when the frequency changes from zero to $f_N = 0.5$. Otherwise Δt must be reduced to avoid the distorting effect of aliasing;

(2) prewhiten the real spectrum by using the operation [see (3.35)]

$$\tilde{u}_q = u_q - 0{,}6u_{q-1} \tag{3.38}$$

(the subscript q varies from unity, and not from zero);

(3) calculate the experimental autocorrelation coefficients of the whitened process and simultaneously filter out the constant component [see (3.37)]

$$R'_r = \frac{1}{n-r}\sum_{1}^{n-r}\tilde{u}_q\tilde{u}_{q+r} - \left[\frac{1}{n-1}\sum_{1}^{n}\tilde{u}_q\right]^2 \tag{3.39}$$

for $r = 0, 1, \ldots, m$ $(m \leqslant n)$. If a trend is present, more complex filters have to be used [139];

(4) obtain the Fourier transform of R'_r from the formula

$$P'_r = \left[R'_0 + 2\sum_{l=1}^{m-1} R'_l \cos\frac{lr\pi}{m} + R'_m \cos r\pi\right]; \tag{3.40}$$

(5) smooth the obtained values of P'_r using the spectral window Q (0.25, 0.5, 0.25) to obtain preliminary estimates of the energy spectrum:

$$\begin{aligned} S'_0 &= 0.5P'_0 + 0{,}5P'_1, \\ S'_r &= 0.25P_{r-1} + 0{,}5P'_r + 0{,}25P_{r+1}. \qquad 1 \leqslant r \leqslant m-1, \\ S'_m &= 0{,}5P'_{m-1} + 0{,}5P'_m; \end{aligned} \tag{3.41}$$

(6) take into account the whitening and filtering out of the constant component to obtain final estimates of the energy spectrum [see (3.36)]:

$$\left.\begin{aligned} S_0 &= \frac{n}{n-m}\,\frac{1}{1{,}36 - 1.20\cos\dfrac{2\pi}{6m}}S'_0, \\ S_r &= \frac{1}{1{,}36 - 1.20\cos\dfrac{2\pi r}{2m}}S'_r, \qquad 1 \leqslant r \leqslant m-1, \\ S_m &= \frac{1}{1{,}36 - 1.20\cos\left(\dfrac{1}{2} - \dfrac{1}{6m}\right)2\pi}S'_m. \end{aligned}\right\} \tag{3.42}$$

The estimate of the energy spectrum value S_0 relates to a frequency a little greater than zero (here f_0 is taken as $1/6m$), the estimates S_r relate to frequencies $r/2m$, and the estimate S_m to a frequency a little lower than $f_N = 0.5$ (here f_m is taken as $f_N - 1/6m$). All the calculated estimates have a stability no greater than that of the χ^2 distribution with $2n/m$ degrees of freedom (see Table 3.1).

3. The Fast Fourier Transform

The fast Fourier transform (FFT) algorithm, introduced by Cooley and Tukey in 1965 [149], has recently found increasing use in the calculation of energy spectra.

It is of interest to note that a similar method was used to investigate x-ray scattering as far back as 1942, and the method was actually rediscovered 23 years later [155].

The fast Fourier transform is an extremely effective method of calculating Fourier coefficients from discrete records of finite length. This method is based on the fact that Fourier coefficients can be calculated by an iteration method, which greatly reduces the computation time. Since it is difficult to envisage this intuitively, it is not surprising that this approach was overlooked for such a long time.

The saving of computation time by the use of the FFT lies in the fact that if the discrete time series consists of $N = 2^p$ values, then calculation of all the Fourier coefficients of this series by the FFT method requires $2pN = 2N \log_2 N$ arithmetic operations, whereas the use of the direct Fourier transform (see Section 2) requires about N^2 operations.

The number of operations required when the direct method and the FFT are used increases extremely rapidly with increase in the length of the series being analyzed. For instance, for $N = 8192 = 2^{13}$ the number of arithmetic operations required for calculation of the Fourier coefficients by the FFT method is less than 213 thousand, whereas the use of the direct Fourier transform requires more than 67 million operations. It is obvious that the use of the FFT not only allows a substantial saving of machine time, but also the spectral analysis of time series of such length that their analysis was fundamentally impossible before the appearance of the FFT.

It is important to note that the use of the FFT not only reduces the required machine time, but also substantially reduces purely computational errors: machine time and the computational errors are reduced by an amount equal to $(\log_2 N)/N$.

Since the FFT algorithms have been fully described [38, 149], we will deal here only with the fundamental relations that reveal the essence of the method.

Let there be a set of N values of u_q, which can be divided into two subsets y_q and z_q, each of which contains $N/2$ values of the process. Given that y_q

contains only even values ($u_0, u_2, u_4, \ldots$), and z_q only odd values ($u_1, u_3, u_5, \ldots$), we can formally write

$$\left.\begin{aligned} y_q &= u_{2q} \\ z_q &= u_{2q+1} \end{aligned}\right\} \text{ when } q = 0, 1, 2, \ldots, \frac{N}{2} - 1. \tag{3.43}$$

It is obvious that y_q and z_q have discrete Fourier transforms, which are written in the form

$$\begin{aligned} B_\nu &= \sum_{q=0}^{\frac{N}{2}-1} y_q \exp(-4\pi i\nu q/N), \\ C_\nu &= \sum_{q=0}^{\frac{N}{2}-1} z_q \exp(-4\pi i\nu q/N) \end{aligned} \tag{3.44}$$

where $\nu = 0, 1, 2, \ldots, N/2 - 1$.

Since we are interested in obtaining the Fourier transform A_ν of the process u_q so that, using (3.1), we can find its energy spectrum, we must show that A_ν can be expressed in terms of B_ν and C_ν, viz.,

$$A_\nu = \sum_{q=0}^{\frac{N}{2}-1} \left\{ y_q \exp(-4\pi i\nu q/N) + z_q \exp\left(-\frac{2\pi i\nu}{N}[2q+1]\right)\right\} \tag{3.45}$$

where $\nu = 0, 1, 2, \ldots, N - 1$ or

$$A_\nu = \sum_{q=0}^{\frac{N}{2}-1} y_q \exp(-4\pi i\nu q/N) + \exp(-2\pi i\nu/N) \sum_{q=0}^{\frac{N}{2}-1} z_q \exp(-4\pi i\nu q/N). \tag{3.46}$$

Using (3.44) we can rewrite (3.46) in the form

$$A_\nu = B_\nu + \exp(-2\pi i\nu/N) C_\nu \tag{3.47}$$

where $0 \leqslant \nu < N/2$.

For values of ν greater than $N/2$, B_ν and C_ν periodically take the values taken at $\nu = N/2$. Thus, substituting $\nu + N/2$ for ν in Eq. (3.47), we obtain

$$A_{\nu \pm N/2} = B_\nu + \exp\left(-2\pi i\left[\nu + \frac{N}{2}\right]/N\right) C_\nu = B_\nu - \exp(-2\pi i\nu/N) C_\nu \tag{3.48}$$

where $0 \leqslant \nu < N/2$.

Hence, from (3.47) and (3.48) we can obtain the Fourier transform of the initial series u_q, and for this we are required to perform $2(N/2)^2$ operations in correspondence with (3.44), and also N addition operations according to (3.47) and (3.48).

If the sets of values of y_q and z_q can be divided, in turn, into subsets, each containing $N/4$ values, the whole argument is repeated, and the required number of operations becomes $4(N/4)^2 + N$.

It is obvious that, having $N = 2^p$, we can continue the operation of successive division until the subsets contain only one value of the initial set u_q. In this case the Fourier transform of such a "one-point" subset coincides with itself.

Since the use of the FFT allows direct calculation of the energy spectrum of the process, bypassing the stage of obtaining the autocorrelation function, then for the calculation of $R(\tau)$ the FFT must be applied again to the calculated energy spectrum $S(f)$.

4. Preliminary Analysis of Nature of Spectrum

The prewhitening of the true spectrum, described in Section 2.3, requires at least an approximate knowledge of the nature of the spectrum so that whitening parameters can be chosen in an appropriate way [Eqs. (3.33) and (3.34)].

In this section we will show how a computer can be used to accomplish operations analogous to electronic filtration, and obtain a preliminary estimate of the characteristics of the investigated spectrum.

In general form the filtering operation can be written as follows:

$$v(t) = \int_{-\infty}^{+\infty} h_p(t - t')\, u(t')\, dt', \tag{3.49}$$

where $u(t')$ and $v(t)$ are the processes at the input and output of the filters, and $h_p(t - t')$ is the impulse response, or weighting, function of the filter. The subscript p denotes a specific form of the function $h(t)$.

The relation between the energy spectrum S_u of the initial process and the spectrum S_v at the filter output is given by the expression

$$S_v = |\Phi_p(f)|^2 S_u, \tag{3.50}$$

where $\Phi_p(f)$ is the filter transfer function, which is connected with $h_p(t)$ by the relation

$$\Phi_p(f) = \int_{-\infty}^{+\infty} h_p(t)\, e^{-i2\pi f t}\, dt. \tag{3.51}$$

Let the function $h_p(t)$ be chosen so that $|\Phi_p(f)|^2$ has the form shown in Fig. 3.10. The process $v(t)$ at the output of such a filter will obviously contain

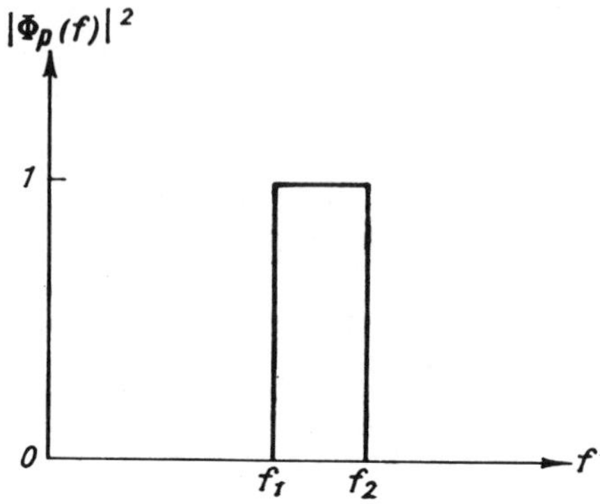

Figure 3.10. Ideal band filter.

only frequencies in the $[f_1, f_2]$ band, and the energy of the process in this band will be

$$\sigma_v^2 = \overline{[v(t)]^2} = \int_{f_1}^{f_2} S_v(f)\,df = \int_{f_1}^{f_2} |\Phi_p(f)|^2 S_u(f)\,df = S_u(f_*)(f_2 - f_1). \tag{3.52}$$

Hence, the spectral density of the energy of the initial process $u(t)$ is

$$S_u(f_*) = \frac{\sigma_v^2}{f_2 - f_1}, \tag{3.53}$$

where f_* lies between f_1 and f_2. If S_u varies slowly in the band $[f_1, f_2]$, we can put $f_* = (f_1 + f_2)/2$.

The filter shown in Fig. 3.10 is called a band filter since it transmits only a certain band of frequencies of the initial process. A similar effect can be achieved by successive use of simpler low- or high-frequency filters. In fact, let there be a filter $|\Phi_p(f_1)|^2$ that transmits all frequencies above some cutoff frequency f_1 and does not transmit lower frequencies (Fig. 3.11a). Such a filter is called a high-frequency filter.

The energy of the process $v_1(t)$ at the output of such a filter is

$$\sigma_1^2 = \overline{|v_1(t)|^2} = \int_{f_1}^{\infty} |\Phi_p(f_1)|^2 S_u(f)\,df. \tag{3.54}$$

If the initial process $u(t)$ is now passed through a similar filter $|\Phi_p(f_2)|^2$ having a cutoff frequency $f_2 > f_1$ (Fig. 3.11b), then for process $v_2(t)$ we will have

$$\sigma_2^2 = \overline{[v_2(t)]^2} = \int_{f_2}^{\infty} |\Phi_p(f_2)|^2 S_u(f)\,df. \tag{3.55}$$

The energy difference $(\sigma_1^2 - \sigma_2^2)$ is obviously the energy of the initial process

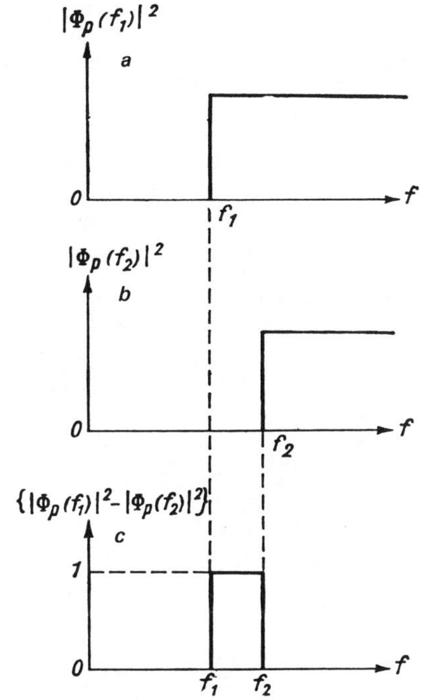

Figure 3.11. Obtaining the effect of band filter (c) as a result of successive application of high-frequency filters with different cutoff frequencies (a, b).

$u(t)$ in the frequency band $[f_1, f_2]$, since it follows from (3.54) and (3.55) that

$$\sigma_1^2 - \sigma_2^2 = \int_{f_1}^{f_2} \{|\Phi_p(f_1)|^2 - |\Phi_p(f_2)|^2\} S_u(f)\, df = S_u(f_*)(f_2 - f_1). \quad (3.56)$$

Whence

$$S_u(f_*) = \frac{\sigma_1^2 - \sigma_2^2}{f_2 - f_1}, \quad (3.57)$$

which is analogous to Eq. (3.53) for a band filter. This agreement is natural, since the difference of the two high-frequency filters with different cutoff frequencies has the same effect as a band filter (Fig. 3.11c). It is obvious that a band filter can also be obtained as a difference of two low-frequency filters. High-frequency filters are usually used, however, since they eliminate the effect of zero frequencies on the results of calculations. The calculation of energy spectra based on the use of successive high-frequency filters can be accomplished by a computer.

4.1. Mathematical High-Frequency Filters. We now consider two pairs of functions $h_p(t)$ and $\Phi_p(f)$, which are used most often in calculations of energy spectra. The simplest impulse response function $h_p(t)$ is the ordinary moving average:

$$h_1(t) = \begin{cases} \dfrac{1}{T} \text{ when } -\dfrac{T}{2} \leqslant t \leqslant +\dfrac{T}{2}, \\ 0 \text{ when } |t| > \dfrac{T}{2}. \end{cases} \tag{3.58}$$

The corresponding transfer function $|\Phi_1(f)|^2$ has the form

$$|\Phi_1(f)|^2 = \left(\frac{\sin \pi f T}{\pi f T}\right)^2. \tag{3.59}$$

The functions $h_1(t)$ and $|\Phi_1(f)|^2$ are shown in Fig. 3.12. The figure shows that $|\Phi_1(f)|^2$ is a low-frequency filter with cutoff frequency $f_c = 1/T$ (curve 2). It is obvious that the transfer function, equal to $\{1 - |\Phi_1(f)|^2\}$, will be a high-frequency filter with the same cutoff frequency (curve 1). The shape of this filter is far from ideal (compare with Fig. 3.11a). First, frequencies above f_c are transmitted nonuniformly and, second, frequencies below f_c are not completely suppressed. It is obvious that the band filter obtained as the difference of the two filters $\{1 - |\Phi_1(f)|^2\}$ with the same frequency f_c will also have a fairly complex form. Filters $\{1 - |\Phi_1(f)|^2\}$ with different f_c correspond simply to different times of averaging in formula (3.58).

The impluse response function $h_2(t)$, called the smoothed moving average [77], has much better properties:

$$h_2(t) = \begin{cases} \dfrac{1 + \cos \dfrac{2\pi}{T} t}{T} & \text{when } -\dfrac{T}{2} \leqslant t \leqslant +\dfrac{T}{2}, \\ 0 & \text{when } |t| > \dfrac{T}{2}. \end{cases} \tag{3.60}$$

The corresponding transfer function $|\Phi_2(f)|^2$ has the form

$$|\Phi_2(f)|^2 = \left(\frac{\sin \pi f T}{\pi f T}\right)^2 \frac{\pi^4}{[\pi^2 - (\pi f T)^2]^2} = |\Phi_1(f)|^2 \frac{\pi^4}{[\pi^2 - (\pi f T)^2]^2}. \tag{3.61}$$

The functions $h_2(t)$, $|\Phi_2(f)|^2$ (curve 2), and $\{1 - |\Phi_2(f)|^2\}$ (curve 1) are shown in Fig. 3.13. The filter $\{1 - |\Phi_2(f)|^2\}$ is a high-frequency filter with cutoff frequency $f_c = 2/T$, and its main difference and advantage in comparison with $\{1 - |\Phi_1(f)|^2\}$ is that it has much smaller side maxima.

In subsequent calculations of the energy spectrum we use band filters consisting of the difference of filters $\{1 - |\Phi_2(f)|^2\}$ with different f_c. Before turning to an account of the calculation procedure we will point out that the

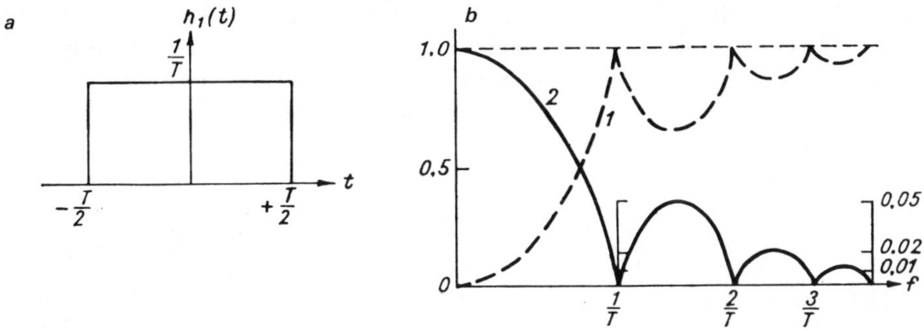

Figure 3.12. "Moving" average (a) and its transfer function (b). (1) Low-frequency filter; (2) high-frequency filter.

difference between the proposed filters and ideal filters necessitates some modification of Eq. (3.57).

The fact is that the cutoff frequencies of filters $|\Phi_1|^2$ and $|\Phi_2|^2$, indicated above and equal to $1/T$ and $2/T$ respectively, are defined formally as the frequencies at which the filters reach the first zero (see Figs. 3.12 and 3.13). Equa-

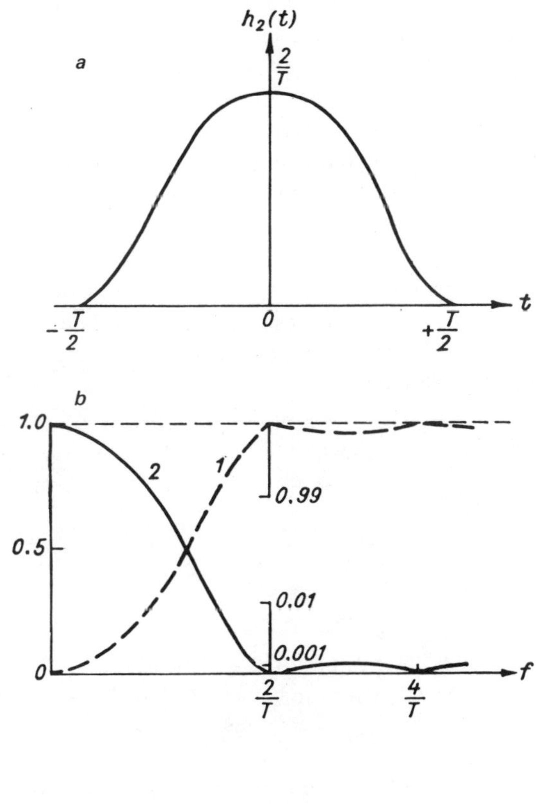

Figure 3.13. Smoothed moving average (a) and its transfer function (b). (1) Low-frequency filter; (2) high-frequency filter.

tion (3.57) is written for the ideal band filter shown in Fig. 3.11c. Hence, when a real filter is used, we must find some equivalent cutoff frequency, which is defined as the cutoff frequency of an equivalent ideal filter whose area is equal to that of the real filter. Thus, to find the equivalent cutoff frequency of the real filter we can write the formula

$$f_c^{\text{equ}} = \int_0^\infty |\Phi_p(f)|^2 \, df, \tag{3.62}$$

which gives $f_c^{\text{equ}} = 1/2T = \frac{1}{2} f_c$ for the filter $|\Phi_1|^2$ and $f_c^{\text{equ}} = 3/4T = \frac{3}{8} f_c$ for the filter $|\Phi_2|^2$.

Formula (3.57) can then be rewritten in the form

$$S_u(f_*) = \frac{\sigma_1^2 - \sigma_2^2}{f_2^{\text{equ}} - f_1^{\text{equ}}}, \tag{3.63}$$

where

$$f_* = \frac{f_2^{\text{equ}} + f_1^{\text{equ}}}{2}.$$

4.2. Computational Scheme. All the formulas given above relate to the case of a continuous record, since it is easier to operate with them in this form. In the computational scheme given below we give discrete analogs of these formulas for the processing of equispaced point records.

Suppose that we have a set of values of the process $u_q, q = 0, 1, \ldots, n$, and the sampling interval is Δt. The record length is $T_n = n\Delta t$.

1. We determine the set of averaging times T_l contained in the formula for h_2^l. The expression for h_2^l means that we will use the transfer function h_2 [see Eq. (3.60)], for which $T \equiv T_l$, $T_l = 2l\Delta t (l = 1, 2, \ldots, m)$. It is obvious that averaging with a prescribed T will involve on each occasion $(2l + 1)$ values of the initial process. The maximum averaging time $T_m = 2m\Delta t$ will not usually exceed 10–15% of the record time T_n to ensure suitable stability of the obtained spectral estimates.

2. The values of T_l uniquely determine the set of frequency bands for which spectral estimates will be obtained:

$$\Delta f_l = f_l^{\text{equ}} - f_{l+1}^{\text{equ}} = \frac{3}{4T_l} - \frac{3}{4T_{l+1}}.$$

There will be m values of Δf_l altogether.

3. We calculate the values of v_s^l at the output of each prescribed filter h_2^l

from the formula

$$v_s^l = \sum_{t=-l}^{t=+l} u_{s-t} \frac{1 + \cos \frac{\pi}{l} t}{2l} . \tag{3.64}$$

Here l is fixed for each h_2^l (or, what is the same, for each T_l), and s takes the values $l, l+1, \ldots n-1$. Thus, for each T_l we obtain $(n-2l)$ values of v_s^l, i.e., with increase in the averaging time the number of values of v_s at the filter output will decrease.

The values of v_s^l are obtained at the output of the low-frequency filter $|\Phi_2^l|^2$. To obtain the values of the process at the output of the high-frequency filter $\{1 - |\Phi_2^l|^2\}$ we must obviously calculate the values of $u_s - v_s^l$.

4. We find the energy of the initial process u_q in the frequency band $[f_l^{\text{equ}}, f_N]$:

$$\sigma_l^2 = \frac{1}{n-2l} \sum_{s=l}^{s=n-l} [u_s - v_s^l]^2 , \tag{3.65}$$

where f_N is the Nyquist frequency.

5. We obtain estimates of the energy spectrum of the initial process u_q from the formula [see Eq. (3.63)]

$$S_l(f_*) = \frac{\sigma_l^2 + \sigma_{l+1}^2}{f_{l+1}^{\text{equ}} - f_l^{\text{equ}}} . \tag{3.66}$$

The calculations of m values of S_l relate to the frequencies

$$f_* = \frac{f_l^{\text{equ}} + f_{l+1}^{\text{equ}}}{2} .$$

Having thus obtained preliminary estimates of the energy spectrum, we wish to know their statistical stability. It can be shown that if $S(f)$ varies slowly in the frequency band $[f_l^{\text{equ}}, f_{l+1}^{\text{equ}}]$, the equivalent width of the filter used is

$$W_{\text{equ}} \approx \frac{1}{T_l} . \tag{3.67}$$

Hence, the equivalent number of degrees of freedom, in correspondence with Eq. (3.24), is

$$k \approx \frac{2T_n}{T_l} , \tag{3.68}$$

and the stability estimate can be obtained with the aid of Table 3.1. We note that as distinct from the method described in Sections 1 and 2, the number of degrees of freedom here is a variable and, hence, the stability of the preliminary estimates of the spectrum depends on the frequency, since in Eq. (3.68) k depends on T_l.

In those cases where the number of initial values n of the process is small (does not exceed a few hundred) and, in particular, if n can be represented in the form 2^p (p is a whole number), a preliminary rough analysis of the shape of the spectrum can be made with an ordinary desk calculator [38, 139].

A preliminary analysis of the nature of the spectrum by the method of mathematical filters is very useful. Experience shows that in a whole series of cases the estimates of energy spectra obtained by this method are quite suitable for further scientific analysis.

4

Spectrum of Turbulence in a Thermally Stratified Atmosphere

1. Spectrum of Turbulence in a Stably Stratified Atmosphere

Experimental spectra of atmospheric turbulence have been obtained in the jet-stream regions in the upper troposphere, in a medium with stable density stratification, but with large wind-velocity gradients [128].

Figure 4.1 shows a typical spectral density curve (energy spectrum) for fluctuations of the vertical wind-velocity component in the jet-stream region. As Fig. 4.1 shows, the curve, plotted in a logarithmic scale, can be approximated by two straight-line segments. Each of these segments corresponds to a power law $S(\Omega) \sim \Omega^{-n}$. The portion BC can be approximated by the power law $S(\Omega) \sim \Omega^{-5/3}$, i.e., it corresponds to the inertial subrange where the Kolmogorov-Obukhov postulate of self-similarity and loss-free energy transfer from scale to scale is valid. Line AD is the line with gradient $-\frac{5}{3}$ drawn through point A. If the spectrum followed AD, it would mean that all the energy obtained by turbulent fluctuations characterized by the wave number Ω_A, being transferred without loss by turbulent formations with large Ω, would eventually be dissipated as heat. The curve ABC lies below AD and, hence, less energy is converted to heat in unit time than the amount arriving at point A. Such cases are characteristic of the upper troposphere. Figure 4.2 shows energy spectra, normalized by variance, obtained in flights in clear air in the weak jet-stream region.

The load method used to obtain these spectra gives reliable results only in those cases where the turbulence intensity exceeds a certain threshold in the region of wave numbers to which the aircraft responds. The energy loss at scales corresponding to the portion AB, however, can be so great that velocity fluctuations with scales corresponding to the portion BC (see Fig. 4.1) will not be detected by the aircraft as an accelerometer. It is obvious that in those cases where

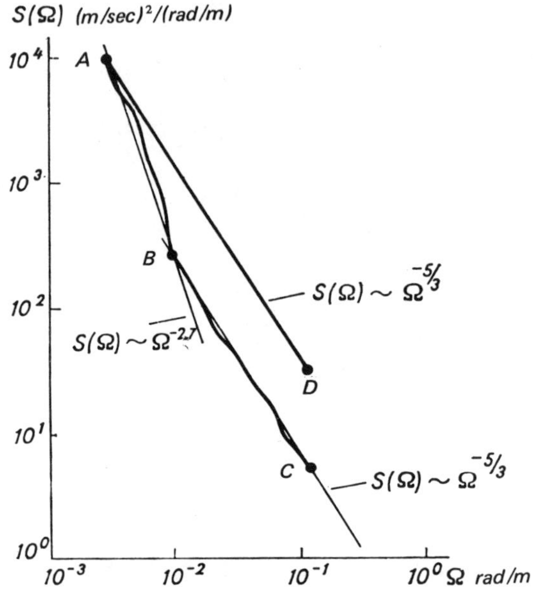

Figure 4.1. Energy spectrum of turbulence (vertical component) in clear air in jet-stream region.

the aircraft as a whole does not respond to these velocity fluctuations it can be used as a platform to carry more sensitive equipment for the measurement of velocity fluctuations. Such measurements, particularly of fluctuations of the horizontal wind-velocity component, have been made with an aircraft hot-wire anemometer. The aircraft wind-vector computer simultaneously recorded large-scale fluctuations of wind velocity. The combination of the two methods provided the spectrum of the horizontal wind-velocity component in a wide range of wave numbers.

Figure 4.3 shows the energy spectrum of fluctuations of the horizontal wind-velocity component obtained in clear air in the jet-stream region [19]. A comparison of Fig. 4.3 with Fig. 4.1 shows that the spectrum of the horizontal component is of the same nature as that of the vertical component. At the same time, the spectral densities in these figures differ by almost two orders of magnitude. In the case corresponding to Fig. 4.3 the aircraft did not experience appreciable loads and only the use of sensitive equipment made it possible to obtain turbulence spectral data. Figure 4.4 shows the spectra of the horizontal component, obtained in flights in clear air [19] and similar in nature to the spectrum in Fig. 4.3. The experimental data indicate that there must be some special mechanism of absorption of the kinetic energy of turbulence in the free atmosphere. This mechanism may depend on special features of the medium, e.g., the "absence of a solid wall" or temperature stratification. It was Obukhov who first drew attention to the special features of turbulent flow in a thermally stratified atmosphere.

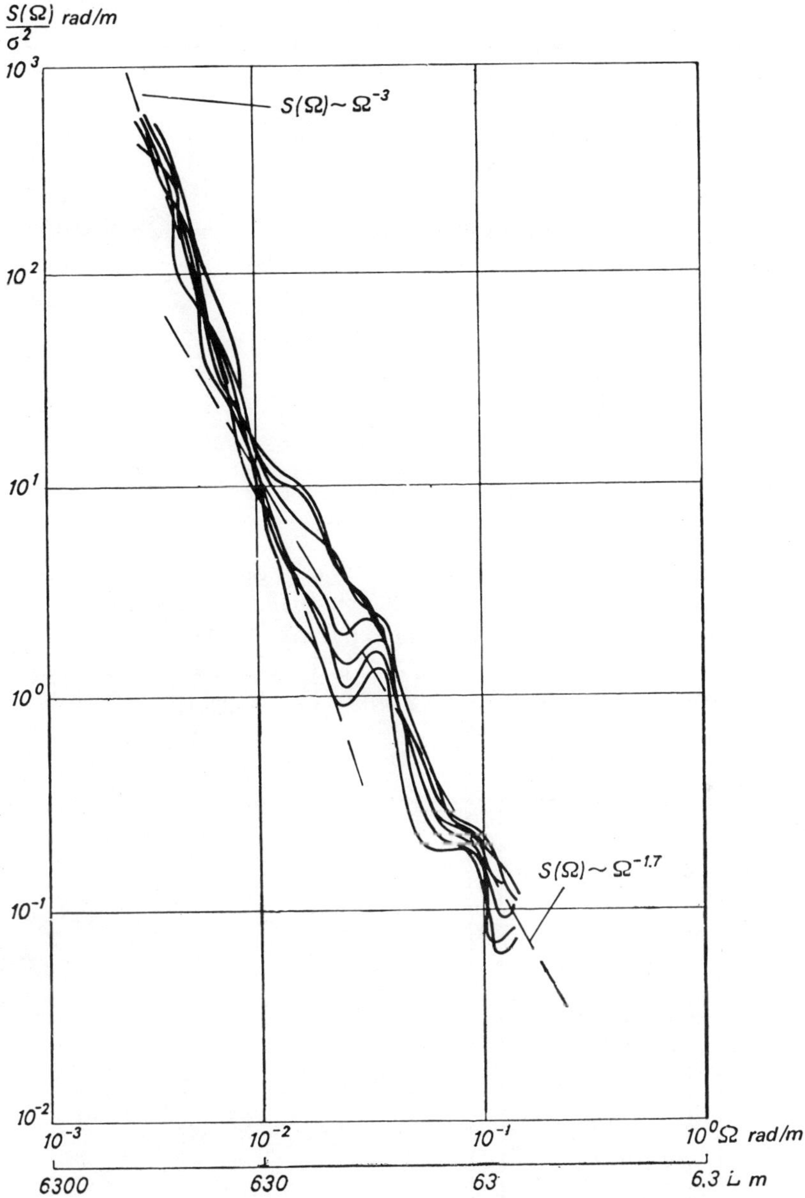

Figure 4.2. Energy spectra of turbulence (vertical component) normalized by variance.

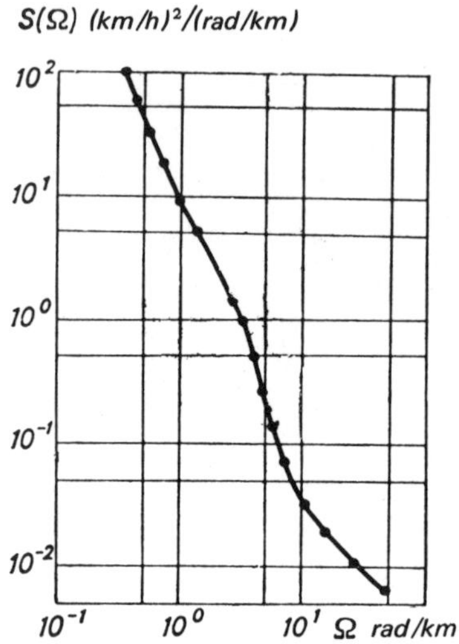

Figure 4.3. Energy spectrum of fluctuations of horizontal component of clear-air turbulence in jet-stream region.

Shur [128] attributed the "leakage" of some of the energy of turbulent fluctuations in a stably stratified atmosphere to the loss of energy due to work against the buoyancy force. The physical hypothesis that he proposed can be expressed as follows. In the free atmosphere with stable density stratification the kinetic energy of an eddy will obviously not be transferred to smaller scale

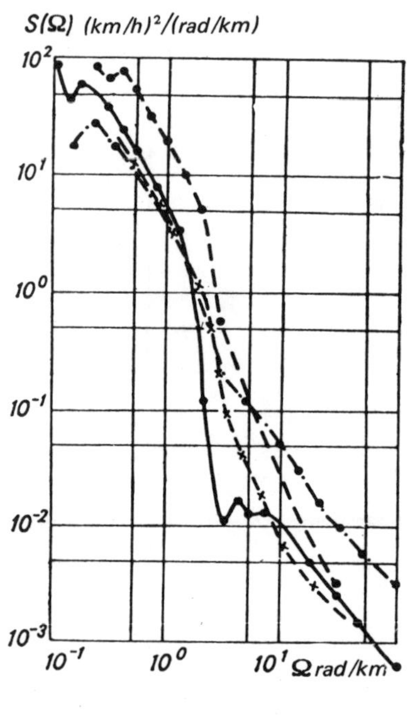

Figure 4.4. Examples of energy spectra of clear-air turbulence (horizontal component).

eddies without loss. During its life the eddy has to perform work against the buoyancy force of stable stratification. Some of the eddy energy will go toward increasing the potential energy of the flow, i.e., to reduction of the stability reserve. It is obvious also that the energy loss of small eddies due to work against the buoyancy force will be negligibly small, whereas large eddies can lose a considerable part of their kinetic energy. The conversion of the kinetic energy of turbulent fluctuations to potential energy will continue as long as stratification remains stable.

In some cases, as shown below, almost all the turbulence energy can be expended on work against the stratification forces, and the turbulence spectrum may be cut off at large wave numbers (small scales).

We consider some quantitative relations on the basis of the generally adopted hypothesis that in a turbulent flow energy is transferred from large-scale eddies to small-scale eddies.

We denote the energy loss per unit time due to work against the buoyancy force by dE'/dt:

$$\frac{dE'}{dt} = gv\,\Delta\rho = gv\frac{\Delta\theta}{\theta}\rho, \tag{4.1}$$

where g is the acceleration of gravity, v is the eddy velocity, ρ is the density, and θ is the potential temperature.

Per unit mass we have

$$\frac{dE'_\rho}{dt} = \frac{1}{\rho}\frac{dE'}{dt} = v\frac{g}{\theta}\frac{\partial\theta}{\partial z}l, \tag{4.2}$$

where $\partial\theta/\partial z \approx (\gamma_a - \gamma)$ is the potential temperature gradient, and l is the characteristic dimension of the eddy.

We assume further that in the region of small l (in the free atmosphere, up to scales of less than a few hundred meters), as has been confirmed by numerous experiments, the Kolmogorov–Obukhov law is fairly well satisfied, and we will investigate the tendency toward alteration of the "five-thirds" law at large l when the effect in question is considered. We have

$$v_l \approx (\varepsilon l)^{1/3}. \tag{4.3}$$

Substituting (4.3) in (4.2) we obtain

$$\frac{dE'_\rho}{dt} \approx \frac{g}{\theta}\frac{\partial\theta}{\partial z}\varepsilon^{1/3}l^{4/3}. \tag{4.4}$$

Expression (4.4) gives the rate of conversion of the kinetic energy of turbulence to potential flow energy; i.e., the rate of loss of kinetic turbulence energy per unit mass characterizes stable stratification of the flow as a turbulent-energy sink. The greater the stability reserve, the greater the amount of turbulent energy that is extracted from the spectrum. In the case of unstable stratification, as (4.4) indicates, dE'_ρ/dt changes sign, and in this case the unstably stratified flow can be regarded as a source of turbulent energy.

The total rate ϵ_l of turbulent energy transfer is made up of the rate of inertial transfer, numerically equal to the dissipation rate ϵ, and the rate of turbulent energy loss dE'_ρ/dt:

$$\varepsilon_l = \varepsilon + \frac{dE'_\rho}{dt} = \varepsilon\left(1 + \frac{g}{\theta}\frac{\partial\theta}{\partial z}\varepsilon^{-2/3}l^{4/3}\right). \tag{4.5}$$

We replace the potential temperature θ by the mean temperature T of the turbulent layer and put

$$b = \frac{g}{\theta}\frac{\partial\theta}{\partial z}\varepsilon^{-2/3} = \frac{g}{T}(\gamma_a - \gamma)\,\varepsilon^{-2/3}. \tag{4.6}$$

Expression (4.3) can now be rewritten in the form

$$v_l \sim \varepsilon^{1/3}l^{1/3}(1 + bl^{4/3})^{1/3}. \tag{4.7}$$

Squaring, we obtain

$$v_l^2 \sim \varepsilon^{2/3}l^{2/3}(1 + bl^{4/3})^{2/3} \tag{4.8}$$

For scales close to those for which (4.3) is valid, we have

$$bl^{4/3} \ll 1\,.$$

Hence

$$v_l^2 \sim \varepsilon^{2/3}l^{2/3}(1 + {}^2\!/\!_3 bl^{4/3}). \tag{4.9}$$

Expressing the scale l in terms of the wave number $\Omega = 2\pi/l$, we obtain

$$v_\Omega^2 \sim \varepsilon^{2/3}\Omega^{-2/3}\left[1 + \frac{2}{3}(2\pi)^{4/3}b\Omega^{-4/3}\right]. \tag{4.10}$$

Differentiating (4.10) with respect to Ω, we obtain after simple algebra

$$S(\Omega) \backsim \varepsilon^{2/3}\Omega^{-5/3}(1 + b_1\Omega^{-4/3}), \tag{4.11}$$

where

$$b_1 = \frac{4}{3}(2\pi)^{4/3}\frac{g}{T}(\gamma_a - \gamma)\,\varepsilon^{-2/3}.$$

The calculations leading to the derivation of expression (4.11) are not rigorous, but the conclusion, derived from (4.11), to the effect that an increase in the modulus of the exponent of Ω in the wave-number region where stable stratification has an effect is in good agreement with the results of experiments [129, 130].

Lumley [188] obtained a more rigorous solution of the problem of the turbulence spectrum in a stably stratified atmosphere. He also proceeded from the assumption that the rate of energy transfer in the spectrum is a function of the wave number Ω and depends on the degree of thermal stability of the atmosphere. For the spectrum of turbulence in a stably stratified atmosphere, Lumley obtained the expression

$$S(\Omega) = \alpha\varepsilon^{2/3}\Omega^{-5/3}\left[1 + \left(\frac{\Omega}{\Omega_A}\right)^{-4/3}\right], \tag{4.12}$$

where Ω_A is the wave number characterizing the region in which the buoyancy force acts.

It is obvious that (4.12) becomes (4.11) if

$$\Omega_A^{4/3} = b_1 = c\frac{g}{T}(\gamma_a - \gamma)\varepsilon^{-2/3}. \tag{4.13}$$

From an analysis of the results of experiments on radio-wave propagation in a turbulent atmosphere, Bolgiano [140] proposed a different model of the energy spectrum for a stably stratified atmosphere. He postulated that in a certain wave-number range the rate of dissipation of fluctuations of the specific buoyancy force has a significant effect on the shape of the spectrum. Bolgiano applied the term "buoyancy subrange" to the wave-number range in which the shape of the spectrum is determined by thermal stratification.

It should be noted that the buoyancy subrange, if it exists, includes scales contained in the inertial subrange in the case of neutral thermal stratification.

Hence, the buoyancy subrange is a portion of the inertial subrange where, in addition to energy transfer of turbulent formations (eddies), some of the energy is converted to potential energy.

For the spectrum in the buoyancy subrange Bolgiano obtained the expression

$$S(\Omega) \backsim \Omega^{-11/5} \tag{4.14}$$

Monin [73] arrived at a similar result. He assumed that in the buoyancy subrange in the case of very strong stability the structure of the turbulent flow ceases to depend on the rate of energy dissipation, and is determined entirely by two parameters: the buoyancy parameter g/T and the mean rate of smoothing-out of temperature inhomogeneities N;

$$N = \nu_T \overline{\sum_{i=1}^{3} \left(\frac{dT}{dx_i} \right)^2}, \tag{4.15}$$

where ν_T is the molecular thermal diffusivity.

From dimensional considerations Monin obtained

$$S(\Omega) \backsim \left(\frac{g}{T} \right)^{4/5} N^{2/5} \Omega^{-11/5}. \tag{4.16}$$

From the three parameters g/T, N, and ϵ and dimensional considerations we can compile the length scale L_*:

$$L_* = \varepsilon^{-5/4} N^{-3/4} \left(\frac{g}{T} \right)^{-3/2}, \tag{4.17}$$

which has been called the Bolgiano–Obukhov scale. For nonuniformities whose scale $L < L_*$, the buoyancy force has no effect on turbulent motions. For $L > L_*$ the regime of turbulent motions in the inertial subrange depends, as (4.17) shows, on g/T and N.

The expression for the spectrum of turbulence in a thermally stratified atmosphere can also be obtained by using Tchen's interaction theory, whose main points are expounded in Chapter 1.

Gisina [34] considered the effect of thermal stratification on the nature of the energy spectrum as a strong interaction of the temperature fields associated with weak interaction of the mean and turbulent wind-velocity fields. From the equations of motion and continuity Gisina used averaging and expansion in Fourier integrals to obtain a spectral equation of the energy balance with buoyancy taken into account. This equation was solved by a fairly widely used method. The parameters contained in the spectral equation were used in conjunction with dimensional considerations to obtain several length scales, for which particular solutions were sought. For scales corresponding to the buoyancy subrange an expression similar to the Bolgiano formula was obtained.

Thus, we can regard it as an experimentally and theoretically established fact that the spectrum of developed turbulence in a stably stratified atmosphere contains a buoyancy subrange in which the gradient of the spectrum exceeds $\frac{5}{3}$. In the region of large wave numbers both the Shur–Lumley and Bolgiano–Monin formulas become, as is to be expected, the "minus five-thirds" spectral law. Different forms of the relations $S(\Omega) = f(\Omega)$ in the buoyancy subrange have been obtained. In the case of the energy spectrum it is difficult, by approximating the experimental results, to give preference to a particular representation, but, as follows from [73, 140, 188], the difference becomes very significant for the spectra of temperature fluctuations. A comparative analysis of the two theories describing the spectrum of turbulence in a stably stratified atmosphere is given by Monin [74] and Phillips [201].

Phillips showed that the two expressions for the spectrum can be obtained from the spectral turbulent energy balance equation. The final form of the expression for the spectrum in the buoyancy subrange depends on certain initial hypotheses. If it is assumed that the turbulence spectrum $S(\Omega_1)$ is determined by the rate of energy transfer in the vicinity of Ω_1, and that the energy density expended on overcoming the buoyancy force is proportional to the mean gradient of the buoyancy force,

$$\frac{\partial}{\partial z}\left[\frac{g}{T}(\gamma_a - \gamma)\right],$$

an expression identical with the Shur–Lumley formula can be obtained.

If it is assumed that the structure of turbulence in a stably stratified flow is described by the rate of dissipation of buoyancy-force fluctuations due to molecular diffusion, an expression identical with the Bolgiano–Monin formula is obtained.

2. Relation between Shape of Spectral Curve and Thermal Stratification of Atmosphere

The analytical expressions (4.11) and (4.16) are suitable for description of the energy spectrum of turbulence in a stably stratified atmosphere. In the case of neutral stratification the buoyancy forces are equal to zero and these expressions asymptotically approach the "minus two-thirds" spectral law. For unstable stratification of the atmosphere expressions (4.11) and (4.16) have no physical sense.

It can be shown on the basis of simple physical considerations, however, that unstable stratification must lead to an increase in turbulence energy.

Thermal instability, like thermal stability, can act in a particular range of wave numbers (scales). If the energy loss due to work against the buoyancy force leads to an increase in the modulus of the exponent [if the spectrum is represented by an exponential function of the form $S(\Omega) \propto \Omega^{-n}$], the energy influx due to thermal instability leads to a reduction of $|n|$ and, in the case of strong instability, to a change in the sign of n.

A good illustration of the relation between the energy spectrum of turbulence and thermal stratification of the atmosphere is provided by the experiment conducted on an IL-18 airplane in the region of Tashkent [22]. This example is typical of the case of thermal stratification, unstable in the lower half of the troposphere and stable at high altitudes, and also for small vertical gradients of mean wind velocity.

During the flight, which took place in September 1964, the weather was cloudless, the wind velocity increased from 20 km/h at an altitude of 1 km to 70 km/h at an altitude of 9 km, and then decreased; up to an altitude of 2.0-2.5 km the atmosphere was unstably stratified, at 3.0 km it was neutral, and above 3 km it was stably stratified. The spectral density curves for fluctuations of the horizontal wind-velocity component for altitudes 1, 3, 6, and 9 km are shown in Fig. 4.5. The low-frequency portions (*AB*) of the curve $S_u(\Omega)$ were plotted from the readings of the aircraft wind-vector meter, and the high-frequency portions (*CD*) were plotted from hot-wire anemometer data. The dashed line joins the portions obtained by the different methods.

Figure 4.5a ($H = 1$ km) shows distinct leveling out of the portion *AB* of the curve in the region of low Ω. The continuation of the portion *AB* by the dashed line into the high-frequency region conforms to the "minus five-thirds" law. The experimental spectral density curve (*CD*) lies above this line with a gradient of about $-\frac{5}{3}$. This position of the portion *CD* of the spectral density curve can be attributed to an influx of energy, in the scale region 1-3 km, due to thermal instability of the atmosphere. This effect, which was distinct at an altitude of 1 km, could be traced up to a height of almost 3 km.

At an altitude of 3 km, where the stratification was close to neutral, the entire curve *ABCD* conformed, as the figure shows, to the "minus five-thirds" law. For an altitude of 6 km, where stratification was stable, there was a distinct increase in the gradient of the $S_u(\Omega)$, which indicates a loss of turbulent energy due to work of the turbulent disturbances against the buoyancy force. Finally, the spectral density of the fluctuations of the horizontal wind-velocity component at an altitude of 9 km (Fig. 4.5d) was of the same nature as $S_u(\Omega)$ for an altitude of 6 km (Fig. 4.5c), but the gradient of the curve $S_u(\Omega)$ at this altitude was even greater, due to the greater thermal stability of the atmosphere at this level.

It was pointed out in [129] that the buoyancy force, although directed along the vertical, will affect both the vertical and horizontal motions, and in a similar way, which is probably due to the three-dimensional nature of the turbulent eddies. The above-described experiment confirms this hypothesis.

Myrup [194] carried out a special experiment on the relation between the shape of the turbulence spectrum and the degree of thermal stability of the atmosphere. Fluctuations of the horizontal and vertical wind-velocity components above a dried-up lake in a desert in North California (U.S.A.) were investigated. The measurements were made on an airplane carrying a device, with a Pitot tube as a sensitive element, for measuring fluctuations of the horizontal velocity, and an angle-of-attack vane for measuring the velocities of the vertical gusts. The flights took place at an altitude of 300 m on the morning of June 25,

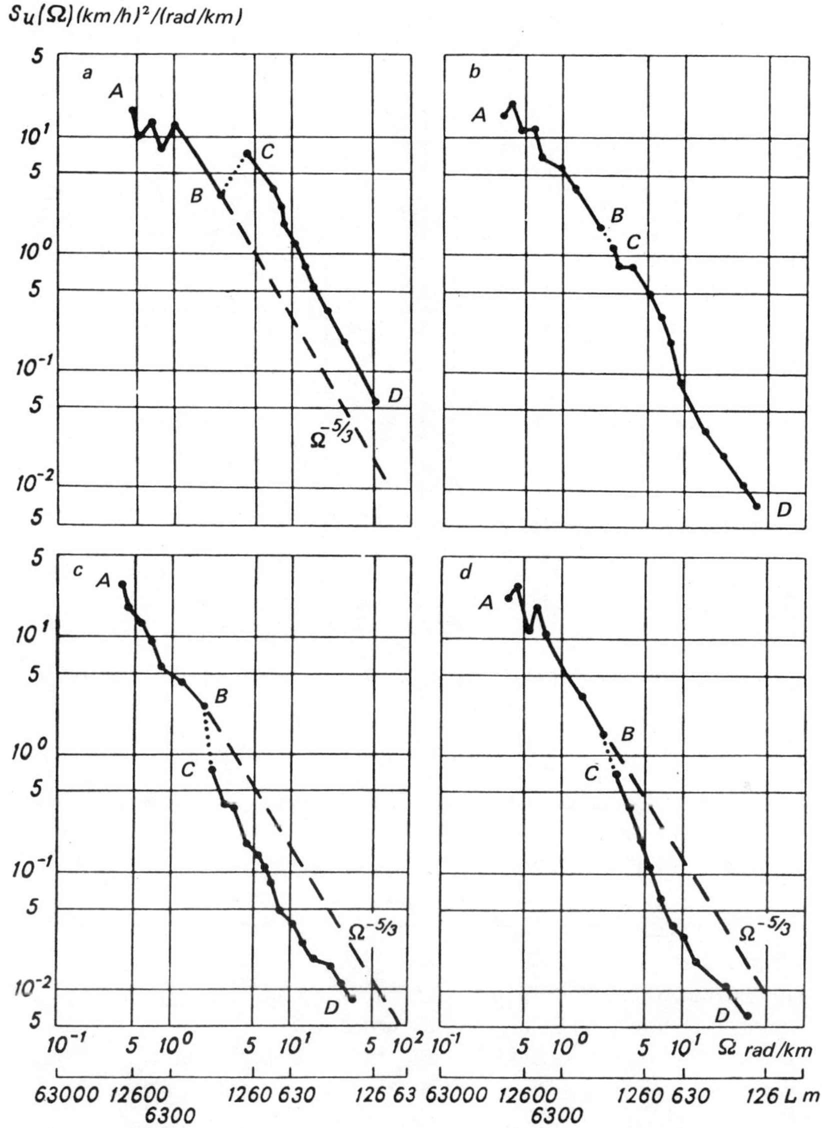

Figure 4.5. Energy spectra of turbulence (horizontal component) at altitudes of 1, 3, 6, and 9 km. Temperature stratification at altitude 1 km (a) unstable, at 3 km (b) neutral, at 6 km (c) and 9 km (d) stable.

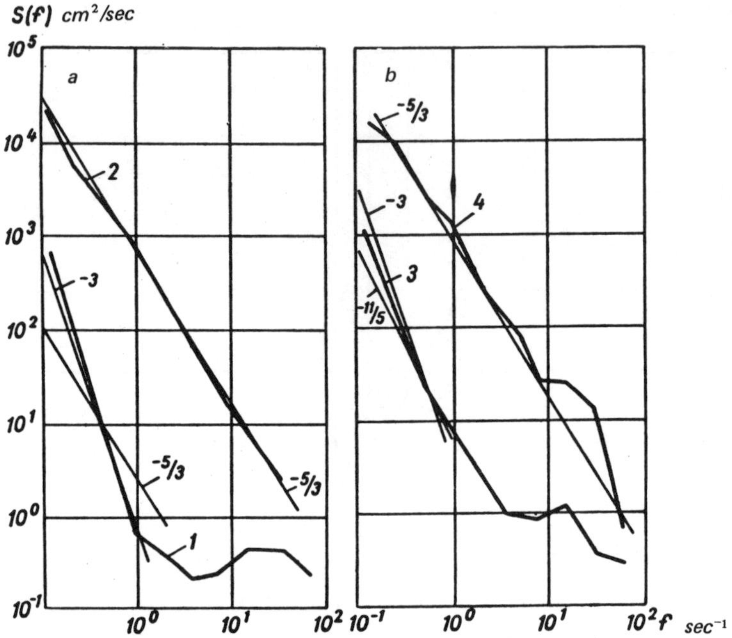

Figure 4.6. Shape of spectral curve in relation to degree of thermal stability (according to Myrup [194]).

1966 in a meteorological situation typical of this region and the season. In early morning above the dried-up lake bed there was a pronounced inversion up to an altitude of about 600 m. Heating due to intense convection converted the stable stratification to neutral stratification. Figure 4.6a shows the energy spectra of fluctuations of the horizontal wind-velocity component, measured at 0655 h Local Time (curve 1) and at 0910 h (curve 2). The gradient of curve 1 is close to −3; the gradient of curve 2 is practically $-\frac{5}{3}$.

Figure 4.6b shows the spectra of the vertical wind-velocity component, measured at 0703 (curve 3) and 0849 h (curve 4). The gradient of curve 3 in the frequency region $<10^{\circ}$ $\sec^{-1}$ lies between −3 and $-\frac{11}{5}$; the gradient of curve 4 is close to $-\frac{5}{3}$. The measurements were accomplished by an increased number of radiosonde and pilot-balloon observations. The mean wind velocity at an altitude of about 300 m above the lake surface did not exceed 3 m/sec. The flight speed of the airplane during the measurements was 69 m/sec.

According to the radiosonde data, conversion of the stratification at the flight altitude from stable to neutral occurred between 0800 and 0830 h, Myrup [194] also investigated a time series of spectral densities at particular frequencies.

3. Energy Transfer in Spectrum and Generalized Spectrum of Turbulence in Free Atmosphere

In the inertial subrange the spectra differ from one another only in the level of turbulent energy, which is determined by the dissipation rate

$$\varepsilon = c^{-2/3} S(\Omega)^{3/2} \Omega^{5/2}. \tag{4.18}$$

We will assume that in a turbulent atmosphere, even if no restrictions are imposed on the spatial structure of the turbulent formations (i.e., as in the case of anisotropic inhomogeneous turbulence), energy is transferred from large-scale disturbances to those of smaller scale. In spectral terms this means that the energy flow is directed toward larger wave numbers. Hence, it is possible to determine the rate of energy transfer ϵ_Ω in the spectrum as a function of the wave number Ω. For the inertial subrange, we have

$$\varepsilon_\Omega = \text{const} = \varepsilon. \tag{4.19}$$

There is no influx of energy into the inertial subrange and no energy extraction within it; there is only energy transfer from some motions (large-scale) to others (small-scale).

Figure 4.7a shows the lines $\epsilon = \text{const}$ ($\epsilon_1 < \epsilon_2 \ldots < \epsilon_n$) in coordinates $S(\Omega)$ and Ω. If the spectral density function $S(\Omega)$ runs parallel to these lines, the rate of energy transfer in the spectrum is constant. In other words, in the wave-number range $\Omega_B - \Omega_C$ there are no sources or sinks of turbulent energy. Segment BC of the spectrum is parallel to the isolines of ϵ. On the segment AB the spectral curve cuts across the isolines as the wave number increases, and ascends to higher and higher energies. In this wave-number range energy enters the spectrum, i.e., in the region $\Omega_A - \Omega_B$ a source of turbulent energy acts. On segment CD the spectral curve intersects the isolines of ϵ in the direction of decreasing ϵ. This means that in the wave-number region $\Omega_C - \Omega_D$ the kinetic energy of turbulent fluctuations is transferred from scale to scale with some loss, i.e., in the region $\Omega_C - \Omega_D$ there is a sink of turbulent energy.

A very convenient way of representing the experimental data in those cases where we wish to investigate energy transfer in the turbulence spectrum is a graph of the function $\epsilon_\Omega = f(\Omega)$, shown in Fig. 4.7b. Such a graph clearly reveals the regions where energy enters the spectrum and the regions where energy is extracted from the spectrum.

The difference between the rate of energy gain ϵ_G in the spectrum and the rate of energy dissipation ϵ may be fairly large, and the sign of this difference depends on the type of stratification. We will assume that the difference $\epsilon_G - \epsilon$ is positive in the case where energy enters the spectrum due to thermal instability, and negative in the case where turbulence energy is expended on overcoming

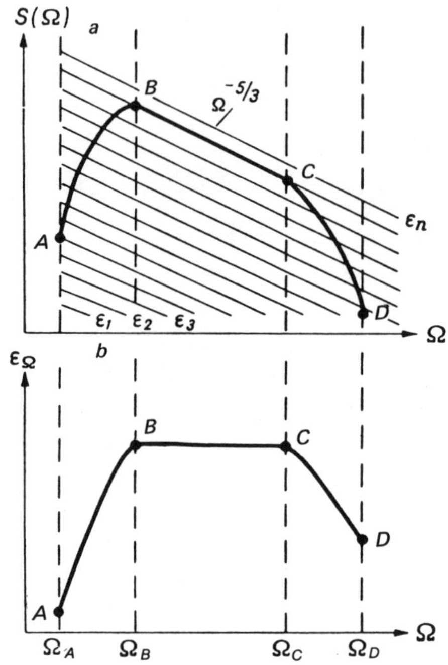

Figure 4.7. Construction of function $\epsilon_\Omega = f(\Omega)$.

negative buoyancy forces. Figure 4.8 shows curves of ϵ_G and ϵ, calculated from the spectra shown in Fig. 4.5. Since the portions *AB* of the spectrum in Fig. 4.5 conform fairly well to the "minus five-thirds" law although we can hardly expect isotropy and local homogeneity at these scales, the value of ϵ_G was calculated from portions *AB* of the spectrum with the aid of Eq. (4.18). The rate of dissipation ϵ was calculated from portions *CD* of the spectrum by means of the same formula.

As Fig. 4.8 shows, ϵ decreases sharply with altitude, whereas ϵ_G, beginning at 3 km, is practically constant. In the zone up to 3 km, $(\epsilon - \epsilon_G) > 0$, which is shown in the figure by the sign +. At greater altitudes, $(\epsilon - \epsilon_G) < 0$, the sign is −. Experimental data for ϵ, given in [35, 48, 68], relate mainly to the lower 500-m layer of the atmosphere and, hence, for comparison with the data given in Fig. 4.8, we show the graph $\epsilon = f(H)$ from Ball [130] in Fig. 4.9, which gives the values of ϵ at altitudes of 1, 3, 6, and 9 km (diamonds) [22]. As the figure shows, the various data are, in general, consistent with one another when one takes into account that during the experiment developed convection was observed at low altitudes.

The reduction of ϵ with altitude does not mean, however, that the intensity of turbulence decreases sharply with altitude. As Figs. 4.5 and 4.8 show, the

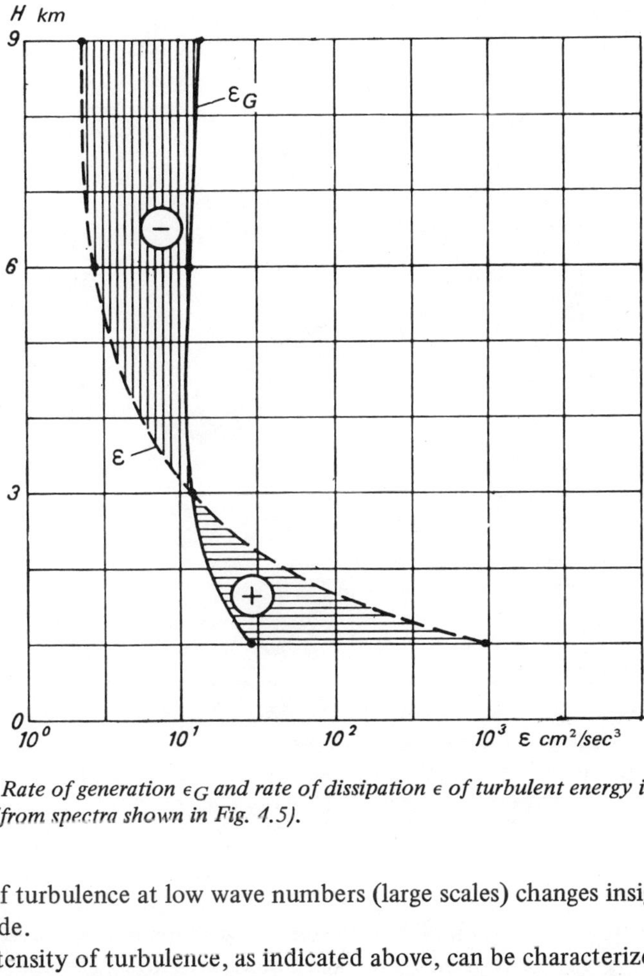

Figure 4.8. Rate of generation ϵ_G and rate of dissipation ϵ of turbulent energy in relation to altitude (from spectra shown in Fig. 4.5).

intensity of turbulence at low wave numbers (large scales) changes insignificantly with altitude.

The intensity of turbulence, as indicated above, can be characterized by ϵ_Ω. This quantity, since it relates to a particular wave number, can also be used as a characteristic of the intensity of turbulence outside the inertial subrange.

Figure 4.10 shows the curves of $\epsilon_\Omega = f(\Omega)$, obtained from the results of airplane experiments in clear air in a stably stratified atmosphere [130]. Curves 1 and 2 are plotted for the horizontal component, and curve 3 for the vertical component. Curve 3 lies well above curves 1 and 2, since it corresponds to intense buffeting of the airplane, i.e., to much higher turbulence energy. All the curves have a distinct region of reduction of ϵ_Ω (buoyancy subrange). It is of interest to note that for large turbulence energies the buoyancy subrange is shifted toward larger wave numbers.

Experimental data on the spectral structure of turbulence in the free atmo-

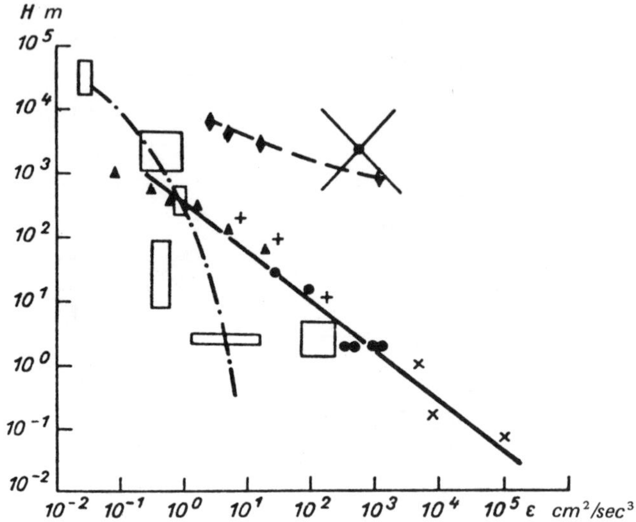

Figure 4.9. Dependence of ϵ on altitude. The large cross corresponds to developed convection. The diamonds give the values of ϵ from Fig. 4.8.

sphere have been obtained by most diverse methods. Each method provides the turbulence spectrum in a particular scale range. It is possible, however, to construct a schematic general spectrum of turbulence in the free atmosphere, which usually has stable thermal stratification, in a wide scale range. For this we use the above-expounded idea of the rate of energy transfer in the spectrum. Figure 4.11 shows a graph of the function $\epsilon_\Omega = f(\Omega)$.

Region I in Fig. 4.11 is the wave-number region in which turbulent energy is generated due to instability of the mean flow. Some of the main-flow energy contributes to the formation of fluctuations. Region I is characterized by an increase in ϵ_Ω. The increase in ϵ_Ω does not mean that in region I there is only an influx of turbulent energy, but we can conclude that the influx of energy is more rapid than the efflux (consumption) of this energy, since at all scales there is energy transfer in the spectrum from smaller to larger wave numbers.

Region II is characterized by quasi-equilibrium between the energy influx into the spectrum and the loss of this energy. By loss we do not mean transfer of energy across the spectrum. The nature of this energy loss requires investigation. The presence of a region of quasi-equilibrium is confirmed by numerous experimental data. The rate of energy transfer in region II is almost constant, although we know that local isotropy conditions are not fulfilled here.

Region III is the wave-number region where a sink for turbulent energy—the buoyancy force of stable stratification—acts in the stably stratified atmosphere. The turbulence energy is converted to potential energy of stratification,

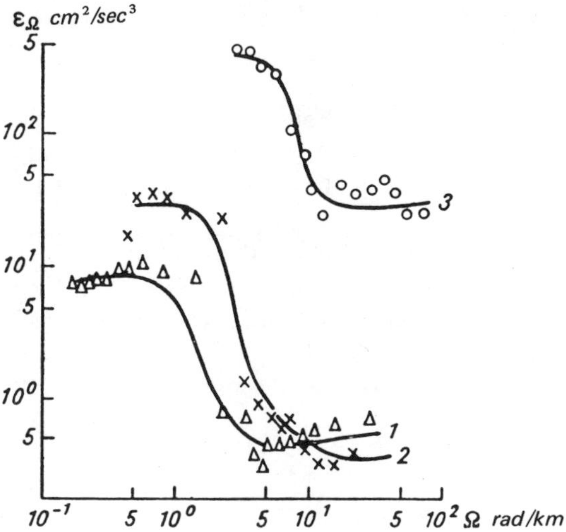

Figure 4.10. Experimental function $\epsilon_\Omega = f(\Omega)$ for clear-air turbulence.

whereas there is practically no influx of energy into the spectrum in the wave-number region $\Omega_3 - \Omega_4$. Owing to loss of energy due to work against negative buoyancy forces, the energy transfer rate ϵ_Ω decreases with increase in Ω.

Region IV is the classical inertial subrange. Here there is neither influx nor loss of energy. The rate of energy transfer over the spectrum in the inertial subrange is constant and equal to the rate at which turbulence energy on the right-hand boundary is converted to heat—the dissipation rate ϵ.

Region V is the viscous subrange, where turbulence energy is converted to heat due to viscosity. In this region ϵ_Ω decreases from ϵ to zero.

The shape of the spectral curve changes in correspondence with the changes

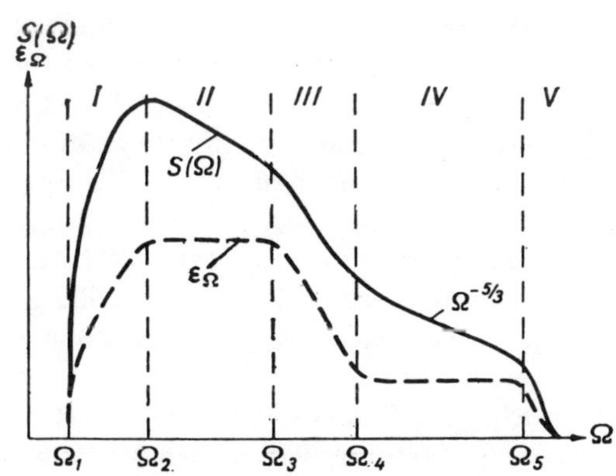

Figure 4.11. Generalized energy spectrum of developed turbulence in free atmosphere and corresponding function $\epsilon_\Omega = f(\Omega)$.

in the rate of energy transfer in the different parts of the spectrum. Figure 4.11 shows the generalized spectrum for a stably stratified atmosphere.

In the inertial subrange the spectrum is represented by the "minus five-thirds" law. The inertial subrange is bounded on the right by the wave number Ω_5, corresponding to the Kolmogorov microscale. Its left-hand boundary Ω_4, as experiments have shown, corresponds in the free atmosphere to scales of about 1000 m.

In the inertial subrange, by definition, there are no sources or sinks of turbulent energy, and there is only inertial energy transfer across the spectrum from large to small scales, whereas in the quasi-equilibrium subrange both sources and sinks of energy act.

Since in this subrange $\epsilon_\Omega \approx$ const, the spectrum will be given by the expression

$$S(\Omega) \backsim \Omega^{-5/3}. \tag{4.20}$$

The expression for the spectrum in the viscous subrange can be theoretically obtained from the well-known Heisenberg equation (see Chapter 1) and has the form

$$S(\Omega) \backsim \Omega^{-7}. \tag{4.21}$$

Experiments in tubes do not confirm this relation. For any approximation, however, the spectrum in the viscous subrange falls more sharply than in the inertial subrange. For the buoyancy subrange, since ϵ_Ω decreases with increase in Ω, the spectrum will be steeper than in the inertial subrange.

The experimental spectra can be approximated by the exponential relation $S(\Omega) \sim \Omega^{-n}$, where $|n|$ varies from 2.2 to 3.1. It should be noted that the gradient of the spectral curve in the buoyancy subrange is a function of the wave number.

The shape of the spectrum in the region of turbulent energy generation, shown in Fig. 4.11, can be derived from general physical considerations and is a direct consequence of increase in ϵ_Ω. It is clear that there are still insufficient experimental data relating to this region of the spectrum for the free atmosphere for the construction even of an empirical formula for the spectrum in this region. The turbulence spectrum shown in Fig. 4.11 is a general scheme for developed turbulence in the free atmosphere. In such developed turbulence there are motions of all scales. The spectrum also includes a buoyancy subrange. In some cases, e.g., when thermal stratification is close to neutral, the buoyancy subrange may be absent from the turbulence spectrum.

Panchev and Syrakov [197] gave the results of a numerical solution of the spectral equations for different values of the dimensionless interaction parameter

m determined from Eq. (1.57): in the case of stable thermal stratification ($m > 0$) the spectrum contains a buoyancy subrange.

The theoretical curves given in [197] have the same shape as the corresponding portion of the spectrum shown in Fig. 4.11. Since the gradient of the spectrum in the buoyancy subrange is variable, it is approximated on the different portions by exponential relations with exponents $-\frac{11}{5}$ and -3. Thus, in the authors' opinion, there is no contradiction between the Bolgiano-Monin and Shur-Lumley theories.

There can also be cases where energy enters the spectrum not only due to instability of the mean flow, but also from additional local sources of turbulent energy. Such a source of turbulent energy in a stably stratified atmosphere could be decaying gravity-shear waves [205]. In this case secondary maxima appear in the turbulence spectra at wave numbers corresponding to the lengths of the breaking gravity-shear waves. Another possible local source of turbulent energy is breaking mountain waves. The mode of formation of such waves and the conditions in which they lose stability are discussed in detail below.

Figure 4.12 shows two spectra of the vertical wind-velocity component obtained in experiments in the region of the Caucasion Range [132]. Curve 1

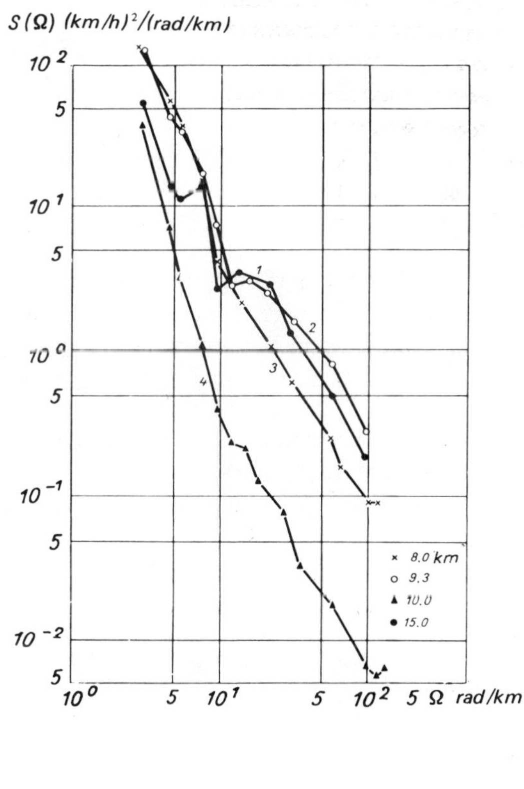

Figure 4.12. Energy spectra of vertical component of clear-air turbulence over mountains and level terrain.

relates to a zone of clear-air turbulence at an altitude of 15 km. The airplane load increments reached ±1.0*g*. Curve 2 represents the spectrum of the vertical wind-velocity component in clear air over mountains at an altitude of 9.3 km in the presence of a jet stream with its axis at an altitude of 13 km. The load increments in this flight reached ±1.2*g*.

For comparison the same figure shows the spectra obtained over level terrain in clear air. Curve 3 corresponds to turbulence in the jet-stream region at an altitude of 8 km with load increments up to ±0.5*g*, and curve 4 corresponds to weak turbulence near the jet axis at an altitude of 10 km with a load increment of about ±0.1*g*.

As Fig. 4.12 shows, the spectral curves above mountains have characteristic bends, as distinct from the more monotonic spectra over level terrain.

Experimental investigations of turbulence in different regions and in different aerosynoptic conditions quite often fail to reveal turbulent formations at the wave numbers usually contained in the inertial subrange. This means that the fluctuations corresponding to these wave numbers either lie below the threshold of sensitivity of the apparatus or are absent altogether.

Such spectra, with a restricted band of wave numbers, are of particular interest and will be discussed separately.

4. Some Features of Experimental Spectra of Clear-Air Turbulence

The experimental results and theories concerning the shape of the generalized spectrum of turbulence in the free atmosphere relate, generally speaking, to cases of developed turbulence with a continuous spectrum containing measurable fluctuations of all scales, including the inertial subrange. A special case that fits into the proposed scheme is the spectrum obtained for a highly stable atmosphere, where practically all the energy entering the spectrum in the generation region is converted to potential energy of stratification in the buoyancy subrange. We consider how the generalized turbulence spectrum may be modified in a stably stratified atmosphere in relation to the degree of thermal stability and the power of the sources producing the turbulence. Thermal stratification is only one of the factors that distort the shape of the spectral curve.

In the case of strong thermal stability, which is usually the situation in the stratosphere, large wind-velocity gradients are required to make the flow turbulent.

The very fact of onset and development of turbulence in such a medium inicates that such conditions prevail.

Although the reserve of thermal stability is not sufficient to prevent the onset of turbulence, buoyancy forces vigorously oppose the cascade transfer of energy across the spectrum. There are three possible cases, corresponding to the three curves in Fig. 4.13.

1. In the case of very strong thermal stability the spectrum of turbulent disturbances may be localized in a narrow band of wave numbers. In this case the spectrum obtained by processing a realization of finite length will be practically the same as that of a purely harmonic oscillation if existing methods of machine statistical processing are used. Such "quasi-wave" disturbances differ from pure wave disturbances in that energy transfer from larger to smaller scales occurs within them, and subsequently all the energy is expended on work against buoyancy forces.

2. When the thermal stability of the atmosphere is a little lower or its dynamic instability is greater, or in those cases where the turbulent energy source is situated in a wave-number region well away from the buoyancy subrange, the turbulence spectrum is much broader and differs significantly from the spectrum of purely harmonic oscillations. In this case, however, buoyancy forces absorb all the kinetic energy of the turbulent fluctuations.

3. This case is frequently encountered where there is a continuous turbulence spectrum in which kinetic energy is transferred from larger to smaller scales. Some of the energy is converted to potential energy, and then, at scales where buoyancy forces become insignificant, inertial transfer takes place in accordance with the "minus five-thirds" law. The kinetic energy ultimately remaining in the spectrum is converted to heat in the viscous subrange.

The different types of spectra shown in Fig. 4.13 are special cases of the general scheme shown in Fig. 4.11 and characterize the energy distribution in the spectrum of developed turbulence.

Finally, when turbulent energy sources acting on particular scales are present, local maxima appear on the turbulence spectrum in the region of the corresponding wave numbers. Examples of continuous spectra with local energy maxima are provided by spectra 1 and 2 in Fig. 4.12.

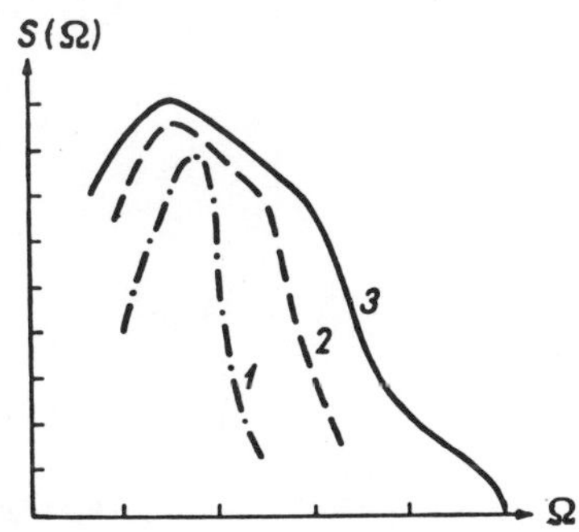

Figure 4.13. Shapes of energy spectra for a stably stratified atmosphere.

5. Spectra of Temperature Fluctuations in Thermally Stratified Atmosphere

Aircraft measurements of temperature fluctuations in clear air, by the method described in Chapter 2, have provided abundant data relating to different types of thermal stratification of the atmosphere.

The measurements were made by airplane (flying laboratories) at altitudes of 200-300 m, where thermal turbulence associated with developed convection was investigated, to 19-20 km, where the temperature field of developed turbulence of dynamic origin was investigated.

It should be noted that in the case of developed convection, owing to strong mixing, the thermal stratification in the lower troposphere is usually slightly unstable or neutral. In this case the temperature fluctuations are frequently small, despite the fact that the airplane is subject to appreciable buffeting. This apparent paradox was pointed out by Monin [74], who predicted amplification of temperature fluctuations on transition from neutral to stable stratification.

If convective jets and bubbles were present, individual spikes of the same sign were superimposed on the weak fluctuations.

Figure 4.14 shows spectral density curves for temperature fluctuations obtained from the results of airplane measurements. The curves in the figure can be divided into three groups.

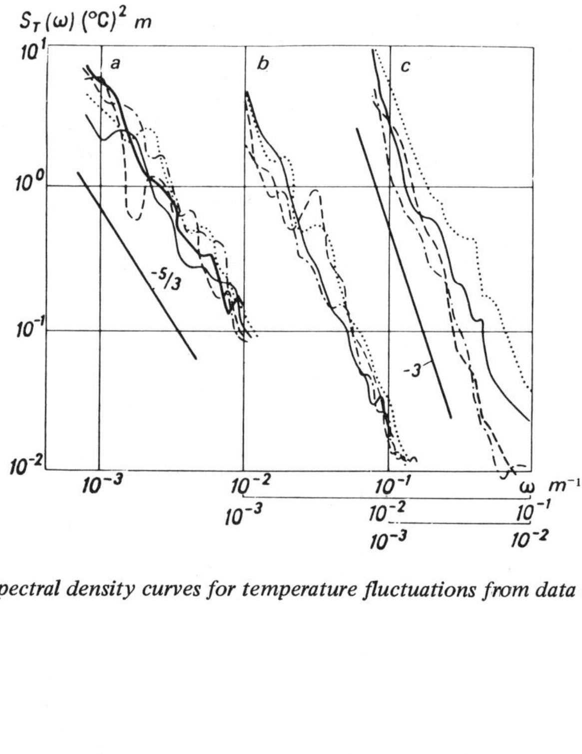

Figure 4.14. Spectral density curves for temperature fluctuations from data of aircraft measurements.

The first group (Fig. 4.14a) contains the spectra of temperature fluctuations in the boundary layer of the atmosphere with developed convection. The gradient of the spectra is close to $-\frac{5}{3}$, which corresponds to the Kolmogorov mechanism of turbulence, where molecular thermal diffusivity plays the role of kinematic viscosity.

The local maxima on the individual low-altitude curves may indicate the presence of ascending jets or bubbles.

The second group of spectra (Fig. 4.14b) relates to developed convection, but outside the boundary layer, at altitudes of 1680 to 3130 m. The nature of the spectral curves depends on the degree of thermal stability; according to the results of aircraft and radiosonde measurements, the temperature gradient varied with altitude, from an approximately dry adiabatic gradient at 1680 m to 0.4°C/100 at 3130 m.

The third group of curves (Fig. 4.14c) relates to the tropopause and stratosphere (altitudes from 10,000 to 17,600 m). The spectra of the temperature fluctuations for the stratosphere are similar in shape to the spectra of the vertical wind-velocity fluctuations.

This similarity of the spectra is not unexpected. If dynamic turbulence arises in a stably stratified flow, the vertical air disturbances will be accompanied by corresponding adiabatic changes in air temperature. The greater the velocity of vertical motion, the greater the vertical distance traversed by the corresponding turbulence element in the thermally stable atmosphere and, hence, the greater the difference between its temperature and that of the surrounding atmosphere.

Investigations of the spectral structure of the temperature field by airplanes (flying laboratories) have led to the following conclusions:

(1) The mean gradient of the spectra curves $S_T(\Omega) = f(\Omega)$ for slightly unstable or neutral stratification is close to $-\frac{5}{3}$, but in those cases where the presence of convective jets or bubbles are to be expected, the spectral curves of the temperature fluctuations are less smooth, with distinct maxima at particular wave numbers.

(2) In a stably stratified atmosphere, significant temperature fluctuations are observed in zones of intense turbulence; these fluctuations are caused by vertical mixing in the presence of a potential temperature gradient.

(3) The spectral density curves of temperature fluctuations in a thermally stable layer of atmosphere have a gradient exceeding $-\frac{5}{3}$, and in some cases close to -3, which agrees well with the shape of the temperature spectrum predicted by Lumley [188].

5

Clear-Air Turbulence in Lower and Middle Troposphere

1. Some Experimental Data on Turbulence Intensity

Experimental investigations of the fluctuations of the longitudinal and vertical wind-velocity components by means of tethered and free balloons, and by aircraft with radar devices, have led to the discovery of a relation between the turbulence intensity and various atmospheric parameters, height above the earth's surface, and so on.

We must first note the relation between the relative turbulence intensity ψ and the mean wind velocity $\overline{u}$. The results of radar measurements of fluctuations of the horizontal wind-velocity components in the scale region from 100 m to 2-3 km have shown that with increase in the mean wind velocity the relative turbulence intensity decreases: Beginning at a velocity of approximately 5-6 m/sec it ceases to depend on the increase in mean wind velocity and is 0.02 on the average.

The relation $\psi = f(\overline{u})$ obtained from the results of aircraft measurements with a Doppler navigation system for scales of a few kilometers to 50-60 km is similar: The values of ψ are greatest at wind velocities less than 50-60 km/h, and at wind velocities greater than 60-70 km/h $\psi \approx 0.1$ on the average.

The relation between ψ and the mean wind velocity can be used for an exploratory estimate of the magnitude of the wind-velocity fluctuations from the results of measurements of the mean wind velocity.

Radar [35] and balloon [27, 96] observations have also revealed a relation between ψ and altitude. Although in each particular case the vertical profile of ψ is of a complex stratified nature due to special features of the vertical profiles of air temperature and mean wind velocity, on the average ψ decreases with altitude.

TABLE 5.1. Relation between Turbulence Intensity and Altitude from Results of Balloon Measurements [96]

	Altitude (m)					
	100–500	500–1000	1000–1500	1500–2000	2000–2500	2500–3500
σ_u	99	53	86	55	63	47
ψ%	9.6	5.9	9.2	4.5	5.5	3.2

Table 5.1 gives some idea of these changes.

For scales of motions close to the scales of radar measurements, the variation of ψ with altitude can be approximated sufficiently accurately by the expression

$$\psi(H) = kH^{-0.5}, \tag{5.1}$$

where H is the altitude in kilometers, and k is a constant that depends on local conditions; for level terrain its value lies in the range 5–15.

Figure 5.1 shows a plot of ψ against H from the results of radar measurements. The reduction of ψ with altitude is sharpest in the bottom 500-m layer of the atmosphere; in the middle troposphere $\psi = 2\%$ on the average; in the upper troposphere ψ increases with altitude. The increase of ψ with altitude in the upper troposphere is also confirmed by the data of aircraft Doppler measurements [84]:

H km	7 ± 0.5	8 ± 0.5	9 ± 0.5	10 ± 0.5	11 ± 0.5
σ_u km/h	11.0	13.5	10.8	14.4	8.2
ψ	0.127	0.190	0.165	0.195	0.110

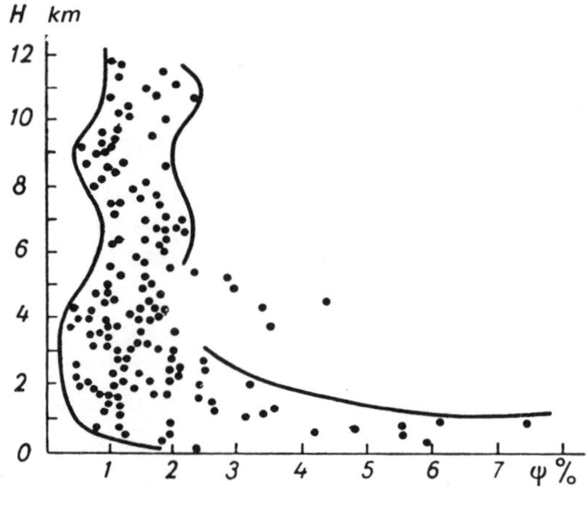

Figure 5.1. Plot of ψ against altitude H from radar data.

Detailed data on turbulence intensity and fluctuations of the wind-velocity components in the layer up to 300 m are given in Vorontsov's monograph [27].

Fluctuations of wind direction are of interest. Experimental investigations [44] showed that in the morning and evening in the atmospheric layer up to an altitude of 1.0-1.5 km the standard deviation σ_φ of the fluctuations of the wind direction relative to the mean direction in a two-hour period is small, varies only slightly with altitude, and is 8-10° on the average. In the afternoon σ_φ increases by a factor of 1.5-2.0, mainly in the bottom 500-m layer of the atmosphere. The value of σ_φ depends on the mean wind velocity: With increase in wind velocity σ_φ decreases, and at velocities exceeding 12-13 m/sec it remains practically constant at approximately 6°. An idea of this relation is given by the following data:

$\bar{u}$ m/sec . . .	1–3	4–6	7–9	9–12	12–15	15–18
σ_φ°	17–14	13–11	10–8	7–6	6	6
$\frac{\sigma_\varphi}{\bar{u}}$ . . .	8.0	2.4	1.1	0.6	0.43	0.36

We now consider the turbulence intensity obtained from measurements of the fluctuations of the vertical wind-velocity component w during flights of free balloons. The velocity averaged over 6-10 sec was taken as a single measurement. Investigations made in clear air or outside clouds showed that the vertical distribution of w was usually of a stratified nature.

In addition, the mean velocities of the ascending motions ($\overline{w}_+$) were not always equal to the mean velocities of the descending motions ($\overline{w}_-$). As an example, Table 5.2 gives the results of measurements made during two flights of free balloons.

The first flight took place in the presence of cumulus cloud. In the layer up to 1000 m ascending motions predominated. Their mean velocity exceeded 1 m/sec, and the maximum velocity reached 2 m/sec. In the atmospheric layer up to 1000 m the mean velocities of the ascending and descending motions were almost the same; the maximum values here were 1-2 m/sec. The second flight took place in cloudless weather in the presence of dense subinversion mist up

TABLE 5.2. Characteristics of Components $\overline{w}_+$ and $\overline{w}_-$ of Vertical Air Motions

Altitude (km)	γ (°C/100 m)	$\overline{w}_+$ (cm/sec)	$\overline{w}_-$ (cm/sec)	Altitude (km)	γ (°C/100 m)	$\overline{w}_+$ (cm/sec)	$\overline{w}_-$ (cm/sec)
	May 12, 1949				May 31, 1949		
0.0	—	—	—	0.0	—	—	—
0.4	0.88	110	37	0.26	−1.96	10	3
1.0	1.24	103	34	0.57	0.71	14	11
1.8	0.68	46	43	1.00	0.78	25	16
2.0	0.70	46	45	1.50	0.70	24	30
2.15	0.13	—	—	2.00	0.64	23	13

to an altitude of 2.0–2.5 km. At the upper boundary of the temperature inversion layer the mean velocities of the ascending and descending motions were less than 10 cm/sec, and even the maximum values did not exceed 18 cm/sec. At greater altitudes the maximum velocities reached 50–80 cm/sec.

In these flights, as in many others, the intensification or reduction of the vertical motions was sharp and sudden. At such times the velocities of the ascending and descending motions reached high values.

On the other hand, for long periods, as on May 31, for instance, the velocities were below the initial sensitivity of the measuring apparatus (about 1 cm/sec).

In many flights of free balloons the velocities of the vertical air motions reached 2.5–3.0 m/sec, but on the average were 30–50 cm/sec in the daytime and 2–15 cm/sec at night. The changes in sign (direction) of the velocity were usually irregular and only occasionally periodic, with a wide range of periods; in 80% of the cases the period did not exceed 120 sec. It was noted that high velocities retained their sign (direction) for a longer time.

Figure 5.2 shows the frequency of the velocities of vertical motions for developed turbulence at altitudes of 1–3 km, measured during the flight of two free balloons on July 25–26, 1949. The distribution curves for w_+ and w_- are generally symmetric. Similar distributions were obtained in other flights, although asymmetric distributions of w_+ and w_- were quite often obtained. The degree of asymmetry was determined, as investigations showed, by the magnitude and frequency of the maximum values of the ascending and descending air gusts. The magnitude and sign (direction) of the maximum gusts in the overwhelming majority of cases also determined the relation between the mean values of w_+ and w_-.

Table 5.3 gives the mean velocities of the ascending and descending vertical

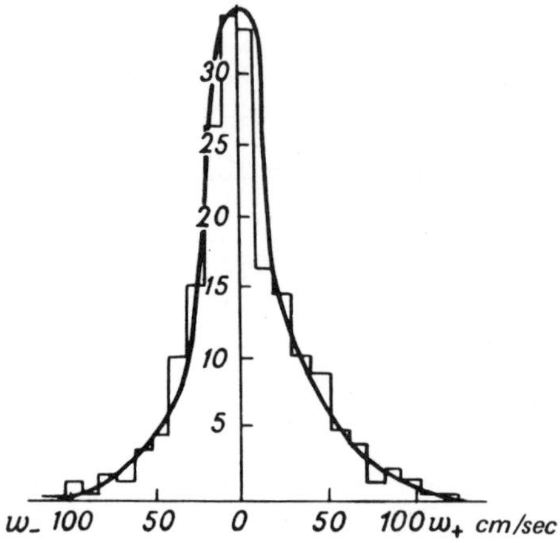

Figure 5.2. Frequency of occurrence of w_+ and w_- from balloon data.

TABLE 5.3. Vertical Distribution of Velocity of Vertical Air Motions In Cloudless Weather

Altitude (m)	$\bar{w}_+$ (cm/sec)	$\bar{w}_-$ (cm/sec)	$\frac{\bar{w}_+}{\bar{w}_-}$	σ_w (cm/sec)
200 – 600	49.3	22.2	2.22	45.0
600 – 1200	25.4	22.0	1.15	29.6
1200 – 1600	24.0	24.0	1.00	31.0
1600 – 2000	36.3	36.0	1.00	45.1
2000 – 2500	27.5	35.5	0.78	40.0
2500 – 3000	26,0	27.0	0.96	33.1
3000 – 4000	–	–	–	30.5
4000 – 5000	–	–	–	34.5
5000 – 6000	–	–	–	29.5
6000 – 7000	–	–	–	37.0
7000 – 8000	–	–	–	36.0
8000 – 9000	–	–	–	27.0

motions obtained from the results of 38 flights of free balloons and substratosphere balloons, and also the standard deviations σ_w of the velocity at different altitudes.

Table 5.3 shows that in the lower layer of the atmosphere $\overline{w}_+ > \overline{w}_-$ on the average. This difference is particularly pronounced in the layer from the earth's surface to an altitude of 500-600 m. In the 1200- to 2000-m layer $\overline{w}_+ \approx \overline{w}_-$.

An analysis of the experimental data showed that σ_w increases with increase in the vertical temperature gradients and mean wind velocity and is closely correlated with the Richardson number Ri. According to Verle's data, the coefficient of correlation between σ_w and Ri is 0.8.

2. Structure of Turbulent Layer of Atmosphere. Intermittency of Turbulence

Extensive zones of clear-air turbulence are encountered relatively rarely in the atmosphere. Only individual, spatially localized turbulent zones, moving along with the mean flow, are usually detected. Such local zones can appear spontaneously as "bursts" of turbulence. Bursts of turbulence can arise quite frequently, increase in size, and merge with one another as they move downstream. Turbulent zones in the form of spots are gradually formed along the flow, leading to an alternation of quasi-laminar and turbulent conditions. At a fixed point in space, over which the air flow moves, quasi-laminar conditions prevail at the start of the described process, and subsequently give way to turbulent conditions.

The spontaneous nature of the appearance of microturbulence in the free atmosphere was first discovered by Obukhov, Pinus, and Krechmer during investigations of atmospheric turbulence with free balloons.

In the laboratory, shadow photography can be used to visualize regions of turbulent motion, but in the atmosphere this is impossible. Some idea of the geometry of turbulent layers, particularly at low altitudes, can be obtained by the fraction of the flight path occupied by turbulent fluctuations in the case of an airplane flying on a horizontal course. The most accurate picture is obtained if simultaneous measurements of the wind-velocity fluctuations at different levels are made. In practice, however, flight experiments are usually performed by one airplane and, hence, in obtaining a vertical cross section of the atmosphere one has to combine the results of measurements made in several flights at different altitudes between fixed points on the earth's surface. In addition, in the assessment of the experimental evidence indicating the presence or absence of turbulence in the air flow, the frequency characteristics and initial sensitivity of the instruments must be taken into account.

The diurnal course of turbulence over level terrain is shown in Fig. 5.3, plotted from the data of a flight experiment conducted on September 29, 1946. A PO-2 airplane made horizontal flights at altitudes of 100, 300, 500, 1000, 1500, and 2000 m. The lines in Fig. 5.3 denote the regions in which the airplane experienced excess loads. The initial sensitivity of the accelerograph was about $0.05g$. During the first flight, which took place 15 min after sunrise, "continuous turbulence" was present at the altitude of 100 m, whereas at higher altitudes there were only local bursts of turbulence; individual bursts of turbulence were detected at altitudes of 1000 and even 1500 m. A similar picture was revealed by the second flight. Subsequently, as Fig. 5.3 shows, the turbulence rapidly became more intense and occupied the layer of atmosphere up to 500 m by 1100–1200 h, and up to 1000 m by 1200–1300 h. It reached its greatest intensity at 1400 h. The last flight took place at 1625 h, i.e., 1–2 h before sunset. The turbulence at all altitudes was slight, and the disturbed regions of flow consisted of intermittent spots with horizontal dimension from a few tens to several hundreds of meters.

Thus, at stages of development and decay, turbulence is of intermittent nature.

The experimental investigations showed that turbulent layers in the atmosphere usually have fairly distinct boundaries, outside which turbulence is absent. The boundary between turbulent and laminar regions of the air flow is irregular and changes in a random manner in time and space. This is due to the fact that the air can flow freely from the quasi-laminar region into the turbulent region, whereas the turbulent fluctuations become much weaker when they penetrate into the quasi-laminar flow region. According to Landau and Lifshits [63], the decay of the amplitude of velocity fluctuations penetrating into a quasi-laminar flow region can be represented in a first approximation by the formula

$$A_H = A_0 e^{-\Omega H}, \tag{5.2}$$

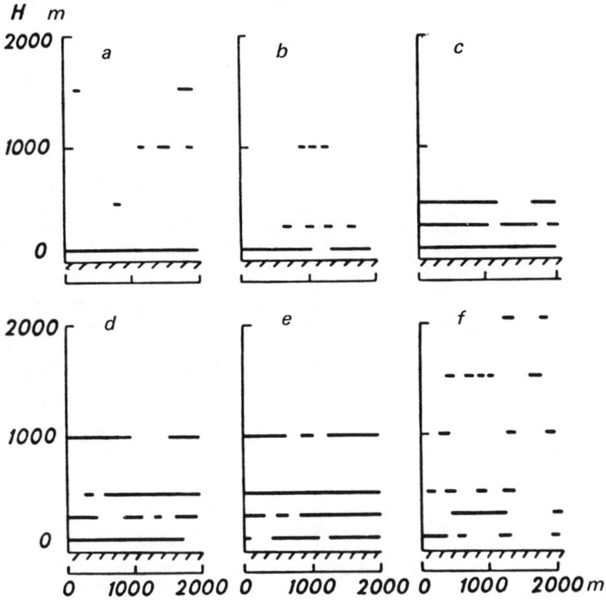

Figure 5.3. Diurnal course of development of turbulence. (A) 0640 h; (b) 0830 h; (c) 1029 h; (d) 1238 h; (e) 1433 h; (f) 1625 h.

where A_0 is the initial amplitude of the fluctuating velocity, Ω is the wave number, which is inversely proportional to the spatial scale of the turbulent fluctuations, and H is the coordinate along a normal to the turbulent layer. It follows from (5.2) that the smaller the scale of the turbulent fluctuations, the shorter the distance from the turbulent layer at which they decay. Hence, only relatively large-scale turbulent fluctuations can penetrate deeply into the quasi-laminar flow region. These fluctuations lead to irregular distortions of the boundary and to the alternation of turbulent and quasi-laminar flow conditions. As an example, Fig. 5.4 shows a vertical cross section of a turbulent boundary layer of the atmosphere. This cross section was constructed by combining the results of airplane load measurements in flights over level terrain.

The degree of intermittency is assessed from the value of the intermittency coefficient α, i.e., the ratio of the total horizontal extent of the turbulent flow regions to the total length of flow traversed by the airplane or passing through a given point. The vertical distribution of the intermittency coefficient α depends on the nature of the vertical distribution of air temperature and mean wind velocity, and, in the case of the boundary layer, also on the distance from the earth's surface. As an example, Fig. 5.5 shows three fairly typical vertical profiles of α [87]. The first type (curve 1) corresponds to morning conditions where the surface temperature inversion is being destroyed; the second type (curve 2)

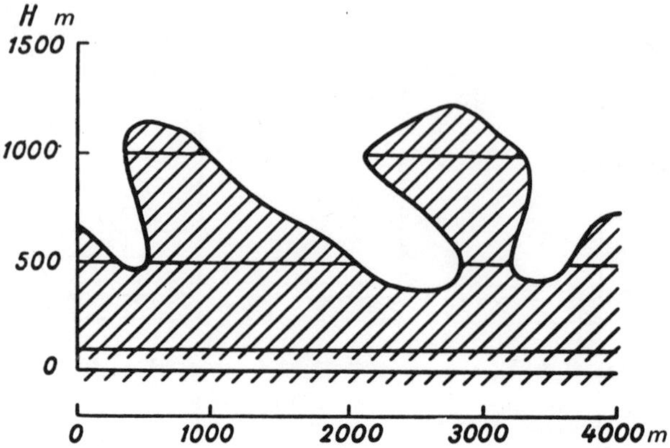

Figure 5.4. Vertical cross section of turbulent boundary layer of atmosphere.

corresponds to conditions where the vertical temperature gradients are small, but the vertical gradients of mean wind velocity are large; and the third type (curve 3) corresponds to developed convection. A comparison of the profiles shows that the vertical distributions of α in the upper part of turbulent layers are identical and can be represented by the normal distribution curve. From the physical viewpoint, this distribution is a consequence of the statistical nature of the changes in height of the swellings and the order in which they follow one another. Quantitative characteristics of these properties for turbulent boundary layers are given in Table 5.4, which shows that the thickness of the atmospheric layer in which there is intermittent turbulence due to the decay of small-scale fluctuations penetrating the quasi-laminar flow region can attain half the thickness of the parental layer, i.e., the main part of the turbulent boundary layer, and in certain conditions may be several times greater. In addition, curves 2 and 3 in Fig. 5.5 show that large vertical gradients of the mean wind velocity and, in particular, air temperature lead to an increase in the thickness of the turbulent boundary layer and give rise to bends on the vertical profile of α by producing and enhancing the intermittency of turbulence within the boundary layer.

Laboratory investigations in free turbulent flows have shown that the zone

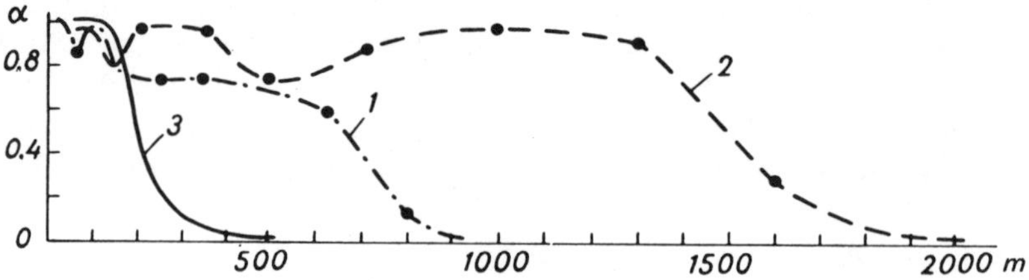

Figure 5.5. Vertical profiles of intermittency coefficient of turbulence.

TABLE 5.4. Parameters of Turbulent Boundary Layers of Atmosphere

Profile	Thickness of layers (m)		
	boundary	parent	intermittent part
1	600	100	500
2	1000	650	350
3	2000	1300	700

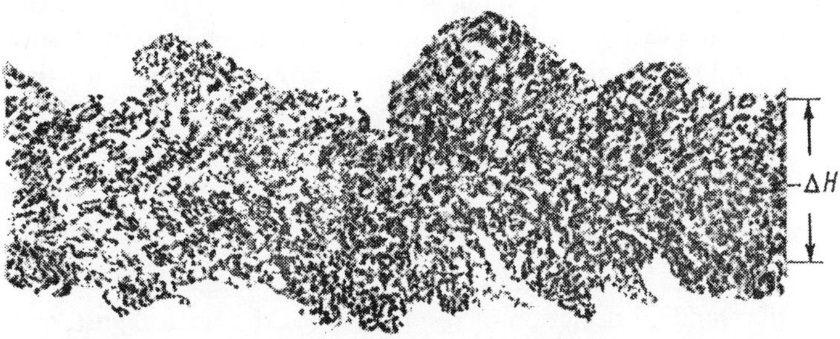

Figure 5.6. Shadowgraph of turbulent wake behind a bullet in flight.

of intermittent turbulence in them is much deeper than in boundary layers. In boundary layers (not thermally stratified) its depth can be 0.4–1.2 times the thickness of the boundary layer, whereas in free turbulent flows it can be as much as 3.2 times the transverse dimension of the turbulent flow region. Figure 5.6 shows a shadowgraph of the turbulent wake behind a bullet in flight in air [75]. It is apparent that the intermittent region is very sharp and distinct.

3. Experimental Values of Turbulent Viscosity Coefficient

The value of the turbulence coefficient contained in the equation of motion of a viscous fluid is often used as an index of turbulence intensity. For the free atmosphere this coefficient is calculated from formulas that take into account the methods of obtaining the empirical values of the fluctuations of the wind-velocity components. We refer, in particular, to the Lyapin–Dubov theoretical formula, improved by German [33], which can be used to calculate the turbulence coefficient from the load increments Δn of an airplane in flight in a turbulent atmosphere:

$$K = \frac{b\bar{\tau}\,|\,\Delta n\,|}{2\Delta\eta} \tag{5.3}$$

Here b is a coefficient that depends on flight-performance data and the speed of the airplane, τ is the mean lifetime of the sign of the load increment, and $\Delta = \rho_z/\rho_0$ is the ratio of the air density at flight altitude $z(\rho_z)$ to the density at the earth's surface (ρ_0).

The response of an airplane to turbulent disturbances, as indicated in Chapter 2, depends on the characteristics of the airplane and is characterized by the airplane transfer function. The correction factor η in expression (5.3) takes into account the difference in the transfer function for disturbances of different frequencies acting on the airplane. In the deduction of Eq. (5.3) it was assumed that the fluctuations of the vertical velocity component w' are equal to the fluctuations of the horizontal component u'. However, as observations show, this is not always the case; in a thermally unstable atmosphere (with convection), $w' > u'$, and in a stable atmosphere with a strong wind, $u' > w'$. In the first case the calculated values of K are underestimates, and in the second case they are overestimates. For this reason the error in determination of K may reach 20–25% [27].

$$K = \sigma_w^2 \bar{\tau}\,, \tag{5.4}$$

where σ_w is the standard deviation of the velocity of the vertical gust, and $\bar{\tau}$ is the mean lifetime of the sign (direction) of the fluctuating velocity.

The values of K obtained from airplane load data are usually higher than the values of K calculated from measurements of velocity fluctuations by free balloons. This is due to the fact that the initial sensitivity of the balloon–instrument system is greater than that of the airplane–instrument system, and the values of K obtained from airplane data are obtained for conditions where the airplane is subject to buffeting, i.e., for conditions of fairly intense turbulence.

We turn now to a consideration of empirical values of the turbulence coefficient. Zilitinkevich *et al.* [45] gave vertical profiles, obtained by several authors, for the bottom kilometer of the atmosphere. The scatter of the curves indicates great variation of K and the shape of the vertical profile in the atmospheric layer considered. In particular, even in the 200–500 m layer, where in most cases maximum values are observed, K lies in the range 10^{-2}–10^2 m²/sec, i.e., in a range of four orders of magnitude.

As experimental investigations [27, 87] showed, not only the vertical distribution of K, but also its spatial distribution in general, is extremely complex. The field of the turbulence coefficient is inhomogeneous and consists of local centers (nuclei) with increased turbulence adjoined by zones of light turbulence. The vertical dimensions of these centers are 50–100 to 300–500 m and the horizontal dimensions are 1–2 and 4–5 km.

From vertical cross sections of the field of the turbulence coefficient, we can deduce the mechanism of vertical propagation of turbulence and some special features of the multifocal structure of this field. Figure 5.7 shows a

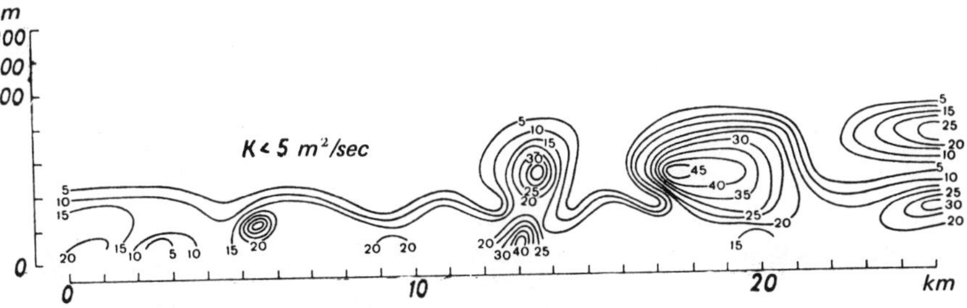

Figure 5.7. Vertical cross section of field of atmospheric turbulence coefficient. April 21, 1962.

vertical cross section of the turbulence field in the atmosphere according to measurements made on an LI-2 airplane flying over level terrain on April 21, 1962 in the period between 0800 h and 1100 h [87]. The weather was cloudless. In the surface layer there was temperature inversion, but by 1100 h the temperature gradients were almost adiabatic.

From Fig. 5.7 we can conclude that turbulence does not propagate vertically throughout space at once, but by upward "bursts" of separate "nuclei" with relatively high turbulence intensity in them. These "nuclei," rising to higher altitudes into the quasi-laminar part of the atmosphere, probably provide the "initial turbulence" that promotes the general development of turbulence at these higher levels.

As an illustration we show another cross section of the field of the turbulence coefficient (Fig. 5.8) for cloudless conditions and fairly large vertical temperature gradients in the layer up to an altitude of approximately 1200 m. This flight experiment showed interaction of the ordinary turbulent boundary layer with the turbulent layer of thermal origin. Up to an altitude of approximately 500 m the distribution of K is fairly irregular, and at altitudes of 600-700 m the horizontal distribution is fairly uniform. The main point, however, is that the vertical profiles of K over the various parts of the region in which the investigations were made differed significantly from one another. This is apparent from Fig. 5.9, which shows profiles of K, taken from Fig. 5.8, for distances of 5, 9, and 19 km from the start of the "test area" above which the flight experiment took place. The differences are particularly large up to an altitude of 400-500 m.

The local nature (in time and space) of the vertical profiles of K makes it difficult to choose an approximation of the K profile that most closely represents the turbulent conditions in the boundary layer in theoretical calculations for specific cases of practical importance. This is confirmed by the studies of Vorontsov [27] and Minervin [72], who proposed several types of K profiles, depending on the number of maxima on the profile and their height, and also

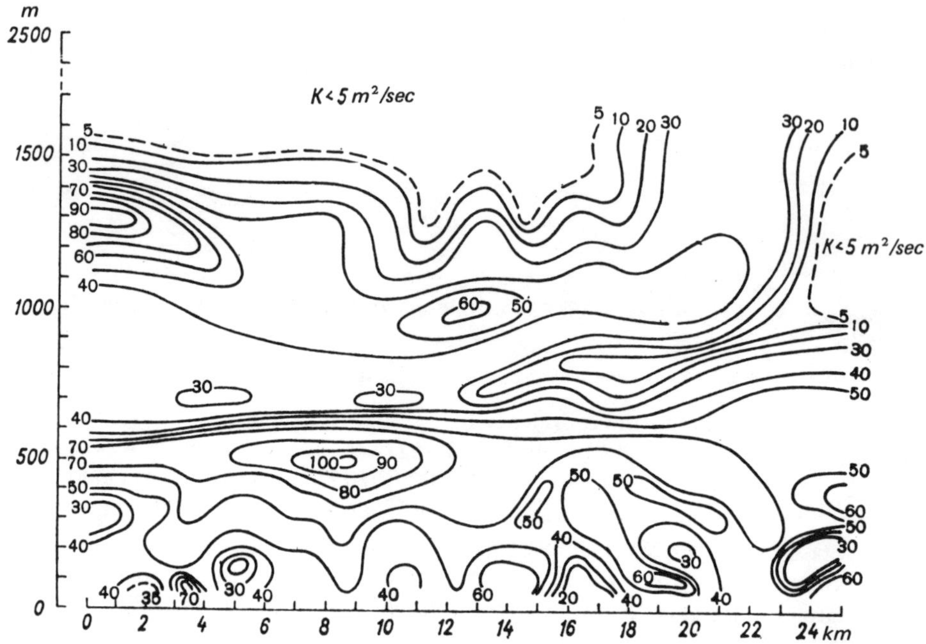

Figure 5.8. Vertical cross section of field of atmospheric turbulence coefficient. April 20, 1962.

on the vertical distribution of temperature (unstable, stable, and inversion) and wind velocity.

The turbulence coefficient in its diurnal course has maximum values around midday, especially in cloudless weather in summer [64]. The sharpest increase in K occurs between 1100 and 1300 h; the diurnal variations of K extend to an altitutde of 1200–1300 m.

Not only airplane, but also balloon, observations have shown that the turbulent transfer coefficient $A = \rho K$ (ρ is the air density) varies in a very wide range and depends significantly on the particular nature of the vertical profiles of temperature and mean wind velocity. Table 5.5 gives values of A obtained from the results of flights of free balloons over the European USSR (TsAO) and over Central Europe.

4. Spectral Characteristics of Turbulence

The energy characteristics of turbulence are of particular interest. We have already given values of ψ characterizing the integral turbulence intensity. It is obvious also that the energy of turbulent motion $\sigma^2_{u,w}$ is the sum of the energies

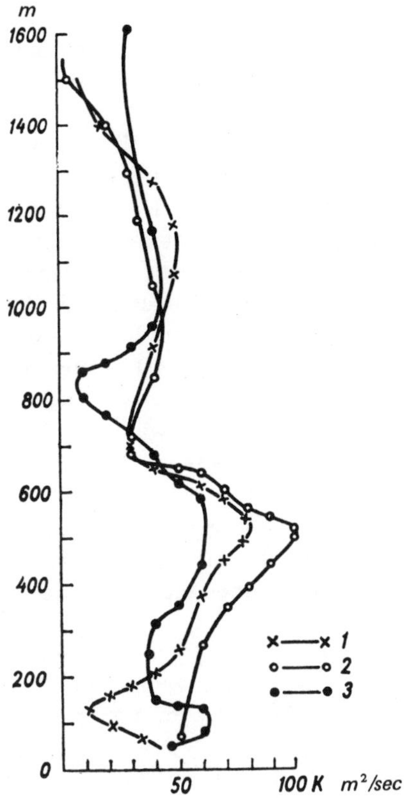

Figure 5.9. Vertical profiles of turbulence coefficient at different distances from origin of test area. April 20, 1962. (1) 5 km; (2) 9 km; (3) 19 km.

associated with the turbulent eddies that form the spectrum of atmospheric turbulence. We consider the special features of this distribution.

4.1. Spectral Characteristics of Fluctuations of Horizontal Wind-Velocity Component. At present the greatest amount of experimental data for the fluctuations of the horizontal wind-velocity component relates to the lower troposphere.

The experimental autocorrelation functions of the wind-velocity fluctua-

TABLE 5.5. Vertical Distribution of A ($g \cdot cm^{-1} \cdot sec^{-1}$)

Altitude (km)	European USSR			Central Europe		
	number of cases	A	$\bar{A}$	number of cases	A	$\bar{A}$
0–1	52	5–180	39	26	1–142	35
1–2	38	12–104	33	18	2–356	48
2–3	6	47–57	49	11	11–134	51
3–4	16	2–47	15	5	3–57	25

tions obtained by many investigators are usually of an exponential or exponential-oscillatory nature. This indicates that wind-velocity fluctuations at different altitudes often consist of fluctuations of turbulent nature and periodic oscillations.

Investigations have shown that in the lower half of the troposphere the spectra of the fluctuations of the longitudinal wind-velocity component depend on the thermal stratification of the atmosphere and the nature of the boundaries of the turbulent atmospheric layers [96].

The dependence of the energy spectra of the fluctuations of the horizontal wind-velocity component on the thermal and wind stratification of the atmosphere is illustrated by the typical examples shown in Figs. 5.10 and 5.11.

Curve 1 in Fig. 5.10 characterizes the relation $S(\Omega)$ for an altitude of 500 m over the earth's surface with neutral thermal stratification and a very small vertical gradient of mean wind velocity. It is apparent that for scales from a few hundred to a few tens of meters the exponent of Ω is $-\frac{5}{3}$, i.e., the theoretical value for the inertial subrange.

The second type of spectrum is observed at superadiabatic temperature gradients, where there is an influx of energy at certain scales due to the thermal instability of the atmosphere. Curve 2 in Fig. 5.10 characterizes $S(\Omega)$ for an

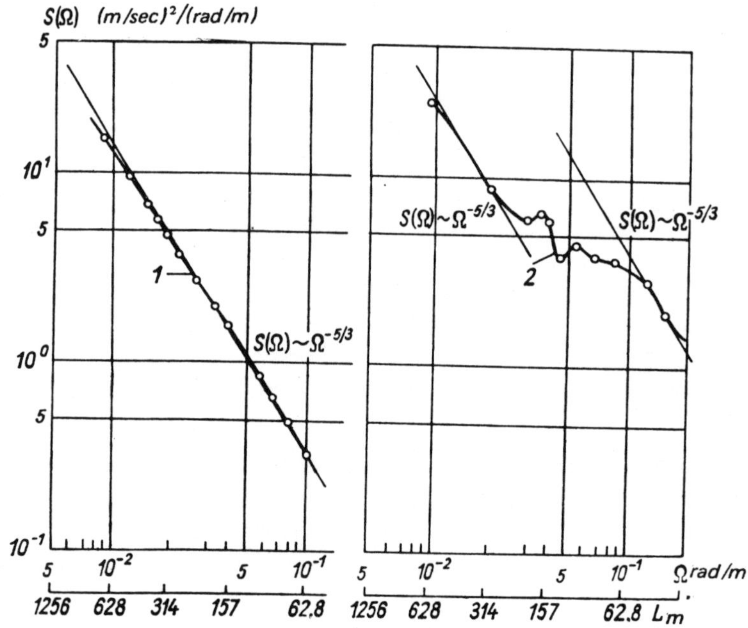

Figure 5.10. Typical spectral density curves for fluctuations of horizontal wind-velocity component.

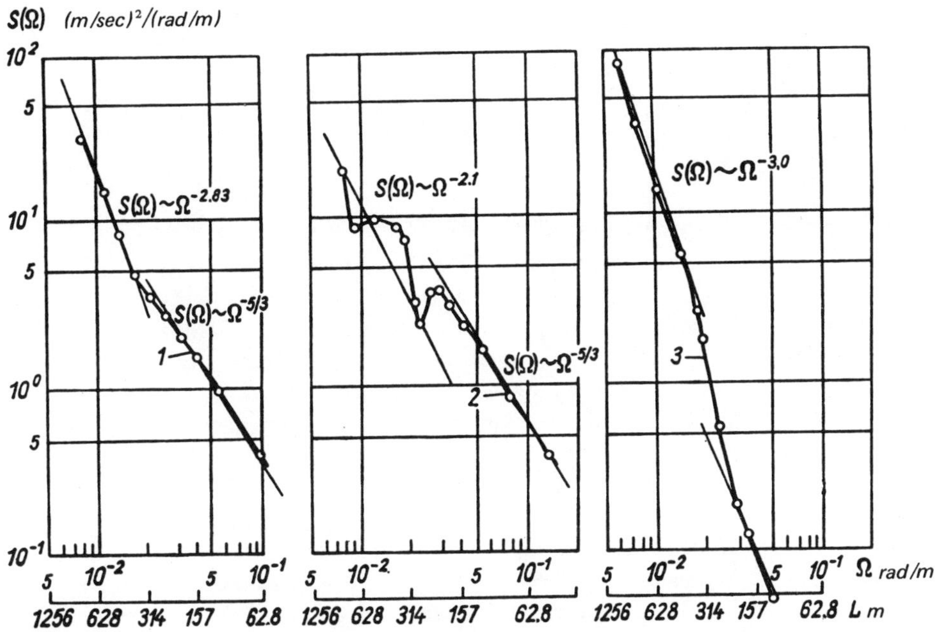

Figure 5.11. Typical spectral density curves for fluctuations of horizontal wind-velocity component.

altitude of 10 m, where the vertical temperature gradient in the bottom 100-m layer of the atmosphere was superadiabatic, and the vertical gradient of the mean wind velocity was zero. It is apparent that the left (low-frequency) and right (high-frequency) portions of the $S(\Omega)$ curve correspond to the "minus five-thirds" law. The right part of this curve, however, is in a region of higher spectral densities.

Curve 1 in Fig. 5.11, corresponding to the third type of $S(\Omega) = f(\Omega)$, characterizes the function $S(\Omega)$ for an altitude of 1000 m in conditions of stable thermal stratification and small vertical wind shear. This curve can be approximated, as can be seen, by two straight-line portions that conform to a power law. Up to scales of approximately 300 m the modulus of the exponent is 2.83, and for smaller scales it corresponds to the "minus five-thirds" law.

Curve 2 in Fig. 5.11 characterizes $S(\Omega)$ at an altitude of 80 m for very stable thermal stratification and large vertical wind shear. At scales in the range of a few hundred meters to approximately 150 m there are peaks on the $S(\Omega)$ curve, which are probably due to short-period oscillations of the wind velocity resulting from the large vertical gradient of mean wind velocity in a stably stratified atmosphere. A theoretical explanation of the physical nature of the described spectra has already been given (see Chapter 4).

TABLE 5.6. Exponent of Ω in Relation to Altitude

Altitude (m)	No. of spectra	γ (°C/100 m)	β (m/sec per 100 m)	Range of n	$\bar{n}$
400 – 700	9	0.65	1.84	2.0 – 2.9	2.43
700 – 1200	17	0.61	0.78	2.0 – 3.5	2.50
1200 – 1700	6	0.76	0.86	2.2 – 3.5	2.83
1700 – 2500	6	0.46	0.45	2.3 – 3.3	2.70
2500 – 3500	1	—	—	2.8	—
3500 – 4500	9	0.72	1.13	2.7 – 3.5	3.10

As investigations showed, the presence of large vertical gradients of mean wind velocity does not alter the nature of the $S(\Omega)$ curve in a stably stratified atmosphere, but shifts it into the region of higher spectral densities.

For a stable atmosphere the exponent can vary in a wide range. This is apparent from the data given in Table 5.6.

In addition, the modulus of the exponent n increases, on the average, with altitude.

The change in the shape of the spectrum of fluctuations of the horizontal wind-velocity component in relation to the diurnal course of thermal stratification is illustrated by the curves in Fig. 5.12 taken from [193]. Measurements of the wind-velocity fluctuations were made at an altitude of 330 m above the ground. At 0725 h the atmosphere was stably stratified, but by 0859 h the stratification had become neutral. This figure shows that the change in the gradient of the $S_u(\Omega)$ curve with change in thermal stratification, particularly the change from stable to neutral stratification, begins on the high-frequency side. In addition, with increase in vertical temperature gradient the functions $S(\Omega)$ gradually move into the region of higher energies.

We return to a consideration of Fig. 5.11. It follows from expression (5.2) that in intermittent turbulent parts of the flow small-scale turbulence can sometimes be present and sometimes be absent. Consequently, the spectral density function for these flow regions will be distinguished by very small values of spectral density for the region of small-scale turbulence. The experimental curve 3 in Fig. 5.11 is an example of the fourth type of spectral density function $S(\Omega)$. It is plotted for the case where at an altitude of 2000 m above the earth's surface the turbulence was intermittent, thermal stratification was stable (0.68°C/100 m), and the vertical gradient of the mean wind velocity was small (0.20 m/sec per 100 m). The gradient of the left part of the spectral density curve is characteristic of stable thermal stratification with a small vertical gradient of mean wind velocity. The right part of the $S(\Omega)$ curve has a very large gradient with very low spectral density for small-scale turbulence. Spectra possessing the same characteristic features as the spectra in Fig. 5.11 were obtained by Vinnichenko [19] at much greater altitudes (up to 10 km).

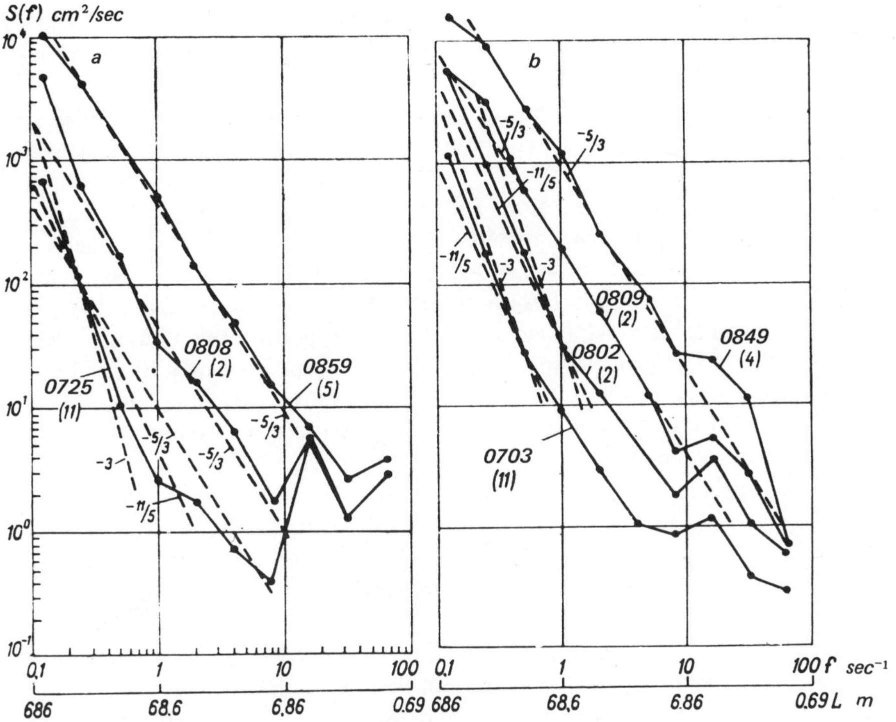

Figure 5.12. Time course of spectral density of fluctuations of horizontal (a) and vertical (b) wind-velocity components. The figures in the brackets give the number of spectra used in the averaging.

It should be noted that the fourth type of $S(\Omega)$ curve, shown in Fig. 5.11, can also characterize the special features of the spectrum of decaying turbulence in the final period of decay (see Chapter 1). The greater the wave number Ω, the more rapid the reduction of the spectral density $S(\Omega)$ during the decay of turbulence. Hence, the physical interpretation of the fourth type of spectrum of fluctuations of the flow velocity components for a free atmosphere is very difficult, since it is practically impossible to determine the nature of the boundaries of turbulent zones in a cloudless sky.

4.2. Spectral Characteristics of Fluctuations of Vertical Wind-Velocity Component. The distance from the ground has a significant effect on the fluctuations of the vertical wind-velocity component. In particular, the maximum vertical scale of the disturbances is limited by this distance. At a frequency corresponding to the maximum scale the turbulent energy distribution curve will have a maximum. This maximum of $S_w(\Omega)$, other conditions being equal, will move toward larger scales (lower frequencies) with increase in altitude. In addition, the buoyancy force will have a greater effect on fluctuations of the vertical wind-velocity component than on horizontal wind-velocity fluctuations.

Koprov [58] gives spectra of fluctuations of the vertical wind-velocity component for the scale region from a few meters to 2–3 km. The fluctuations were measured at altitudes of 50, 100, 200, 500, 1000, and 2000 m over level terrain on days with developed convection. The measurements were made by an airplane equipped with an acoustic anemometer. Figure 5.13 shows spectra obtained in experiments carried out between 1106 h and 1417 h on August 31, 1963. Up to an altitude of approximately 1000 m the spectra conformed satisfactorily to the "minus five-thirds" law. At higher altitudes, where the atmosphere was stably stratified, the gradient of the spectrum plotted in logarithmic coordinates for disturbance scales exceeding 100–200 m was more than $\frac{5}{3}$ (in modulus).

Burns [147] gives spectra of fluctuations of the vertical wind-velocity component measured on a Canberra airplane in low-altitude flights over different underlying surfaces in England and North Africa. The velocity fluctuations were measured with an anemoclinometer mounted on a boom in front of the nose of the airplane. The fluctuation spectra covered fluctuation scales in the range of 15–2500 m, approximately. Some characteristics of these spectra are given in Table 5.7.

Table 5.7 shows that the modulus of the exponent n has a fairly large spread, especially at altitudes of 60–150 m, but on the average $|n|$ is close to $\frac{5}{3}$.

At present there are very few spectral density data for turbulence scales exceeding 2.5–3.0 km. This is due to the fact that in order to obtain reliable spectral density values for fluctuations of the vertical wind-velocity component of such large scale measurements on a horizontal flight path of at least 90–100 km are necessary. The horizontal extent of turbulent zones, however, is usually less than 100 km. Moreover, one can hardly postulate that on a path of such length it would be possible to have conditions of even relative homogeneity, for which calculations of spectral density have sense.

The results of investigations of the effect of the nature of the underlying surface on the spectrum are of definite interest. According to Burns' experimental data [147], over hilly terrain the spectral density in the entire scale range is greater than over a level surface. This difference increases with increase in the scale of the disturbances, beginning at scales of approximately 100 m.

The spectra obtained by Burns in flights with and against the flow (wind) differed slightly from one another. This difference was very slight if the ground speed of the airplane, and not its airspeed, was used for conversion from time spectra to space spectra. This indicates that at low altitudes $S(\Omega)$ for the vertical wind-velocity component depends on the distance between orographic inhomogeneities. Spectra obtained in flights across the flow differed from spectra ob-

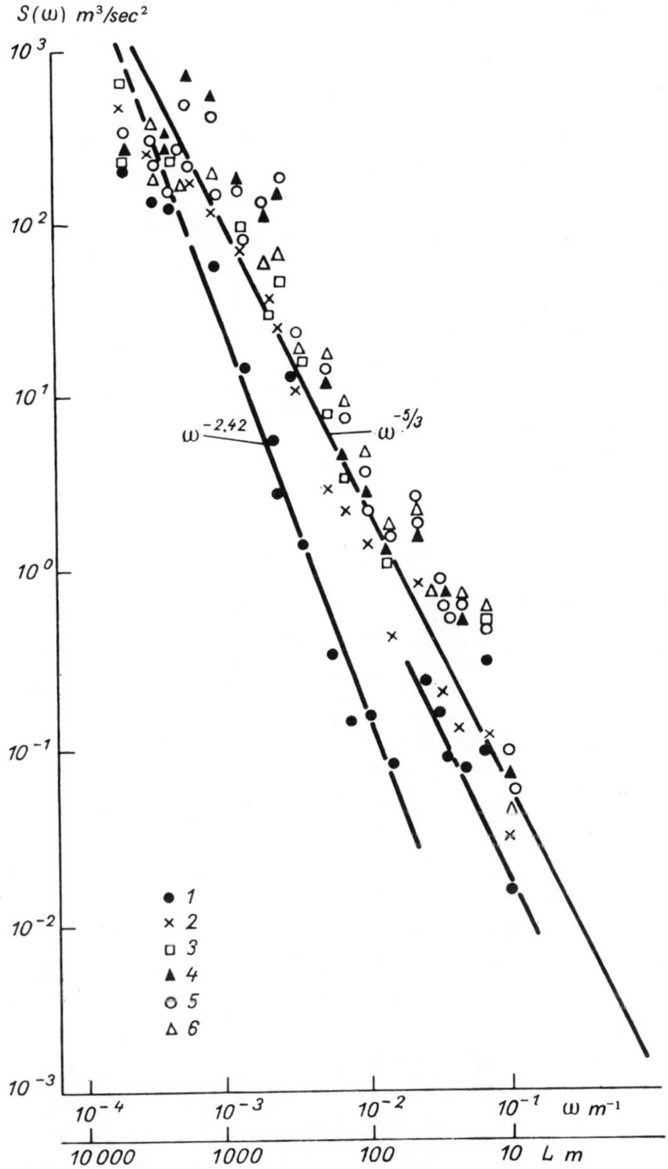

Figure 5.13. Spectral density curves for fluctuations of vertical wind-velocity component (at different altitudes). (1) 2000 m; (2) 1000 m; (3) 500 m; (4) 200 m; (5) 100 m; (6) 50 m.

TABLE 5.7. Characteristics of $S_w(\Omega)$ in Relation to Altitude

Altitude (m)	Number of spectra	Range of n	$\|\bar{n}\|$	σ_{n_i}	σ_w (w/sec)
60	14	1.20 – 2.03	1.58	±0.32	0.96
150	7	1.27 – 1.92	1.68	±0.23	1.19
300	4	1.53 – 1.80	1.65	±0.16	1.05

tained in flights with or against the flow. This difference was particularly pronounced in the region of small turbulence scales.

Not only the turbulence intensity but, to a certain extent, the nature of the spectrum, changes with change in altitude. In particular, for neutral or unstable stratification and low vertical gradients of the mean wind velocity the spectral density of relatively large disturbance scales increases with altitude, whereas that of small disturbance scales decreases. The $S(\Omega)$ curves for different altitudes intersected at scales of about 600 m.

4.3. Relations between Spectra of Fluctuations of Vertical and Horizontal Wind-Velocity Components. It was shown in Chapter 1 that the longitudinal wind-velocity fluctuations obtain energy directly from the mean flow, whereas transverse fluctuations obtain energy from the longitudinal fluctuations. Hence, it is to be expected that $\sigma_u > \sigma_w$. The failure of this relation can be due to various kinds of external factors affecting the turbulent energy balance. In particular, in a thermally unstable flow with vertical wind shear the vertical fluctuations can obtain energy not only from the longitudinal fluctuations, but also from the kinetic energy generated by buoyancy forces.

As experimental investigations show, the values of the vertical and longitudinal components of the fluctuating velocity depend on the altitude above the ground and on the thermal and wind stratification of the atmosphere. Hence, they are subject to diurnal and annual variation. Table 5.8 gives Vorontsov's [27] experimental data characterizing this relation for level terrain in temperate latitudes.

Table 5.8 shows that in summer and in daytime, irrespective of the time of year, the standard deviations of the fluctuations of the wind-velocity components σ_w and σ_u were greater than in winter and in the evening or night. In addition, in the bottom 100-m layer of the atmosphere, σ_u in the warm part of the year and σ_w throughout the year increase with increase in altitude, but at higher altitudes, on the other hand, they decrease with altitude. The diurnal amplitudes of σ_w and σ_u vary in a similar way with altitude. In the cold part of the year not only the values of σ_u and σ_w decrease with altitude throughout the layer, but their diurnal amplitudes also decrease (Table 5.9).

Table 5.8 also gives the ratio σ_w/σ_u, which characterizes the anisotropy of the fluctuations of the wind-velocity components. This table, which gives data

TABLE 5.8. Characteristics of Fluctuations of Horizontal (u') and Vertical (w') Wind-Velocity Components

Altitude (m)	Time of day (h)											
	9	13	17	21	1	5	9	13	17	21	1	5
	Warm part of year						Cold part of year					
	σ_u (m/sec)											
2	0.45	0.84	0.51	0.34	0.16	0.21	0.86	1.04	0.76	0.51	0.37	0.34
100	0,68	1.35	0.11	0.58	0.48	0.25	0.71	1.00	0.65	0.37	0.39	0.32
200	0.65	1.17	0.70	0.41	0.31	0.22	0.46	0.72	0.50	0.29	0.31	0.22
300	0.71	0.81	0.61	0.45	0.25	0.35	0.36	0.45	0.27	0.26	0.20	0.19
	σ_w (m/sec)											
2	0.26	0.61	0.34	0.10	0.05	0.06	0.20	0.34	0.21	0.08	0.10	0.10
100	0.41	0.98	0.40	0.18	0.18	0.09	0.30	0.60	0.29	0.12	0.12	0.16
200	0.39	0.81	0.34	0.14	0.10	0.10	0.26	0.41	0.20	0.09	0.14	0.09
300	0.30	0.60	0.29	0.12	0.07	0.12	0.21	0.26	0.09	0.07	0.09	0.07
	σ_w/σ_u											
2	0.58	0.72	0.65	0.36	0.30	0.29	0.23	0.31	0.28	0.29	0.30	0.30
100	0.62	0.76	0.56	0.32	0.36	0.35	0.42	0.49	0.44	0.32	0.31	0.42
200	0.60	0.75	0.48	0.31	0.32	0.34	0.46	0.52	0.38	0.30	0.36	0.38
300	0.42	0.74	0.47	0.27	0.30	0.35	0.42	0.52	0.39	0.32	0.37	0.40
	γ (°C/100 m)											
2	–	–	–	–	–	–	–	–	–	–	–	–
100	0.9	1.5	0.9	−1.3	−3.4	−2.4	−0.0	0.6	0.4	−1.1	−2.3	−2.0
200	0.9	1.2	0.9	0.1	−0,2	−0.5	−0.4	0.2	0,3	−0.3	−0.8	−1,1
300	0.8	1.1	0.8	0.4	0.2	−0.2	−0.3	−0.2	−0.0	−0.1	−0.5	−0.7
	β (m/sec per 100 m)											
2	–	–	–	–	–	–	–	–	–	–	–	–
100	1,2	1.2	1.2	3.3	3.6	3.1	2.6	1.9	2.5	4.9	4.8	4.3
200	0.4	0.2	0.2	1.1	1.4	0.5	0.9	0.5	1.9	1.3	1.5	1.2
300	0.3	0.3	0.6	0.0	0.0	−0.1	1.0	0.4	0.0	0.9	0.6	0.0

TABLE 5.9. Diurnal Amplitudes of σ_u and σ_w (m/sec)

Time of year	Altitude (m)							
	2	100	200	300	2	100	200	300
	σ_u				σ_w			
Warm	0.68	1.10	0.95	0.56	0.56	0.89	0.71	0.53
Cold	0.70	0.68	0.50	0.26	0.26	0.48	0.32	0.19

for the wind-velocity fluctuations without correction for the vertical gradient of the mean wind velocity, shows that even for a neutral or thermally unstable state of the atmosphere the variance of the vertical wind-velocity fluctuations on the average is less than that of the longitudinal fluctuations. The anisotropy of the wind-velocity fluctuations is particularly pronounced in winter, and also in the evening and night in summer, when temperature inversions accompanied by relatively large vertical gradients of mean wind velocity are very often observed in the atmospheric layer considered. At midday in summer the anisotropy coefficient σ_w/σ_u is 0.72–0.76 on the average, and at midnight it is 0.30–0.36, whereas in winter at these times it is 0.31–0.52 and 0.29–0.32, respectively.

Correlation of the results of measurements of the vectoral wind shear in the 10- to 40-m layer above the ground in different countries showed that the median value in the empirical distribution curve was a vertical wind shear of $5.1 \cdot 10^{-2}$ $\sec^{-1}$, the 98% cumulative probability value was $1.41 \cdot 10^{-1}$ $\sec^{-1}$, and the 99.6% cumulative probability value was $1.72 \cdot 10^{-1}$ $\sec^{-1}$.

Vectoral wind shears exceeding $1.41 \cdot 10^{-1}$ $\sec^{-1}$ were observed at night in the case of thermal stratification under inversion layers associated with jet streams at low altitude in the atmospheric boundary layer, or in daytime in the case of unstable thermal stratification and clear air.

It should be borne in mind that in the case of weak winds and small vertical gradients of mean wind velocity, where free convection conditions may exist, the anisotropy coefficient σ_w/σ_u can differ significantly from the mean values given in Table 5.8.

We consider in more detail the relations between σ_w and σ_u, taking into account the special features of the spectral density functions for fluctuations of the vertical and longitudinal wind-velocity components. As already mentioned, the spectra of the wind-velocity fluctuations depend on the nature of the thermal stratification of the atmosphere, but the spectra of the vertical wind-velocity fluctuations at low altitudes also depend on the distance above the ground. With increase in the distance from the ground the observed bend in the $S_w(\Omega)$ curve moves, as Fig. 5.14 shows, toward larger scales (smaller Ω). In addition, as Burns' investigations showed, in the region of small scales the spectral density in general decreases with altitude, whereas in the region of large scales, after the start of the bend in the $S_w(\Omega)$ curve, it increases. Thus, the increase in the total energy of

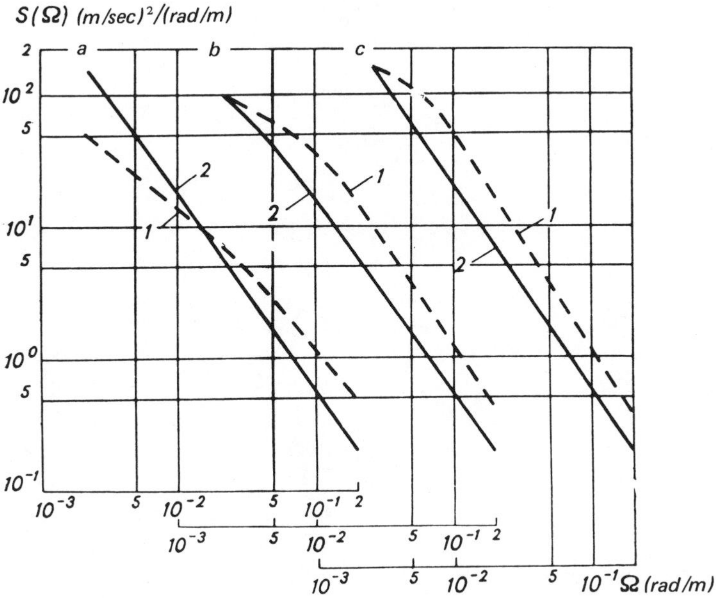

Figure 5.14. Spectral density curves for vertical (1) and horizontal (2) wind-velocity components at altitudes of 60 m (a), 150 m (b), and 300 m (c).

the vertical wind-velocity fluctuations with altitude is due to the increase in the long-wave part of the spectrum. At the same time, the spectrum for the horizontal wind-velocity fluctuations has the same gradient for a wind range of scales, and in this case the spectral density decreases with altitude. For these conditions the relation between σ_w and σ_u at low altitudes has a characteristic vertical profile: Up to altitudes of 50–70 m $\sigma_w < \sigma_u$, and at higher altitudes $\sigma_w > \sigma_u$.

It should be noted that radar investigations of the structural characteristics of fluctuations of the longitudinal and transverse wind-velocity components by Mel'nichuk [68] showed that the structure functions of these components were almost the same for horizontal eddy scales less than the altitude of observation. Beginning at scales close in magnitude to the altitude of observation, the structure function of the transverse fluctuations leveled out, whereas the structure function of the longitudinal fluctuations continued to increase. This indicates that the anisotropy of the horizontal fluctuations moves toward larger disturbance scales with increase in altitude.

The investigations carried out by Rider *et al.* [207] on a "Mirage" airplane in conditions of very severe buffeting ($\Delta n = \pm 2.5g$) in a cold front region showed that in the long-wave part of the energy spectrum $S_u(\Omega) \gg S_w(\Omega)$. This is indicated by the data in Table 5.10.

At present there are no reliable data on the nature of the anisotropy for high

TABLE 5.10. Relation between σ_u and σ_w for Region of Cold Atmospheric Front

Altitude (m)	σ_u	σ_w	σ_u/σ_w
	m/sec		
910	9.8	6.9	1.42
960	12.8	9.2	1.39
1160	12.2	10.0	1.32

altitudes. We can postulate, however, that at high altitudes, especially where stratification becomes stable, $\sigma_w < \sigma_u$.

4.4. Conversion of Kinetic Energy of Turbulence to Heat. In the atmosphere the kinetic energy of the mean motion is converted to turbulent energy and then to heat. The amount of turbulent energy converted to heat in unit mass of air is equal to the dissipation rate ϵ. Experimental investigations have shown that the values of ϵ are greatest in the immediate vicinity of the earth's surface, and ϵ decreases with altitude. In the bottom kilometer of the atmosphere the average reduction of ϵ with altitude is inversely proportional to the altitude:

$$\varepsilon_z = \varepsilon_0 z^{-1}; \tag{5.5}$$

$\epsilon_0 = 10^3$ cm^2 · sec^{-3} at altitude $z = 1$ m. Heating of the air (in °C/h) due to the dissipation of turbulent energy can be calculated from the formula

$$\frac{\Delta T}{\Delta t} = 3.6 \cdot 10^{-4}\,\varepsilon, \tag{5.6}$$

where ϵ is in cm^2 · sec^{-3}. It is easy to calculate that the turbulent energy expended in the turbulent surface layer of air at an altitude of 1 m can heat the air by approximately 0.36 °C/h.

In each specific case the change of ϵ with altitude depends on the particular nature of the vertical temperature and wind-velocity profiles. In particular, according to [202], in the subinversion layers, which are distinguished by large vertical wind-velocity gradients, ϵ increases with altitude and reaches a maximum at the lower boundary of the inversion. The vertical gradient of ϵ in the subinversion layer of the atmosphere can exceed 10^{-3} cm · sec^{-3}. In the inversion layer itself ϵ decreases sharply to the upper boundary, and immediately above the inversion layer $\epsilon \ll 1$ cm^2 · sec^{-3}. As is known, temperature inversion is a fairly common phenomenon in the lower troposphere. The diurnal variation of ϵ in the layer up to an altitude of 1–2 km, which is particularly pronounced in the warm part of the year, with a maximum at about noon, should also be noted.

6

Atmospheric Convection and Thermal Turbulence

One of the commonest forms of turbulent motion is the so-called thermal turbulence, which consists of fluctuations of the vertical air velocity due to the buoyancy forces acting on the air particles. Thus, thermal turbulence results from the development of atmospheric convection, i.e., vertical displacements of individual batches of air due to the difference between their density and that of the surrounding air.

Convection is produced either by unstable stratification of the atmosphere or by nonuniform heating of the underlying surface. In the first case thermal turbulence is due to spontaneous convection, and in the second case it is due to contact convection. It is obvious that, in principle, spontaneous convection can occur at any height, whereas contact convection can only occur at the earth's surface.

These two mechanisms usually act in the boundary layer, and thermal turbulence can be associated with unstable, neutral, or even weakly stable stratification. In the last case turbulence is caused only by contact convection and, hence, its intensity increases with increase in the patchiness of the temperature distribution over the underlying surface.

We will point out, finally, that the development of thermal turbulence is usually triggered by mechanical turbulence, which is of particular importance in the development of spontaneous convection.

We can distinguish ordered and disordered spatial forms of convection. The first includes various forms of cellular circulation, where the flow is broken up into "regular" circulation cells, i.e., cells which are approximately the same in size and structure. These cells are sometimes called closed cells (e.g., Bénard cells), and are sometimes in the form of parallel bands of ascending and descending streams oriented along the wind shear vector or at an angle to it. In ordered convection the flow is laminar, and hence it cannot be directly equated to thermal turbulence. Destruction of the elements of ordered convection due to loss of thermodynamic stability by the flow gives rise to a turbulent flow.

In disordered convection the spatial disposition of the individual convective streams (elements) is of a more or less random nature. Outside the clouds the convection elements (elements of thermal turbulence) have linear dimensions ranging from a few centimeters to hundreds of meters, and the most intense convective streams within cumulus and, in particular, cumulonimbus clouds sometimes have transverse dimensions of several kilometers.

The main energy source for convection in clouds is the latent heat associated with phase changes. Hence, in this case the parameters of the vertical motions depend greatly on phase changes. Condensation leads to an increase in the velocity of the ascending stream and the height to which it ascends. Evaporation of droplets, on the other hand, slows down the ascending streams, but accelerates the descending streams.

1. Disordered Atmospheric Convection

1.1. Structure of Convection Elements. The elements of disordered convection ("thermals") can apparently take two forms. The first form consists of individual air masses, or bubbles, separated from one another. The second form consists of approximately vertical air jets. The form of thermal actually observed probably depends on the thermodynamic conditions in the atmosphere and the nature of the underlying surface.

The suggestion that the primary convection elements are isolated masses of air was first put forward by Molchanov and Selezneva, who identified thermals with large eddies formed during the day in the upper part of the surface layer. A similar hypothesis was later proposed and developed by Ludlam, Scorer, and other investigators (see [125]).

The front part of the bubble is hemispherical, and the hind part consists of a train of relatively cold air (Fig. 6.1a). The representation of convection elements as bubbles is an extension of the "particle method" in atmospheric thermodynamics but, as distinct from this method, it takes into account the interaction of the ascending air mass with the surrounding air, which makes the process nonadiabatic. This interaction is manifested in the form of turbulent transfer of momentum and other characteristics between the bubble and the surrounding air. This process is most intense in the front part of the bubble. The turbulence "erodes" the bubble, and during its ascent the boundary layer is washed away into the wake region behind the thermal.

An important property of bubbles, according to laboratory and theoretical investigations (see [187], for instance), is the quasi-stationary circulation of the "Hill annular eddy" type† in their front part (Fig. 6.1b). This circulation plays a stabilizing role, preventing the mixing of the bubble with the surrounding air.

†For a fuller account of the Hill eddy, see H. Lamb, *Hydrodynamics*, Dover, New York (1932).

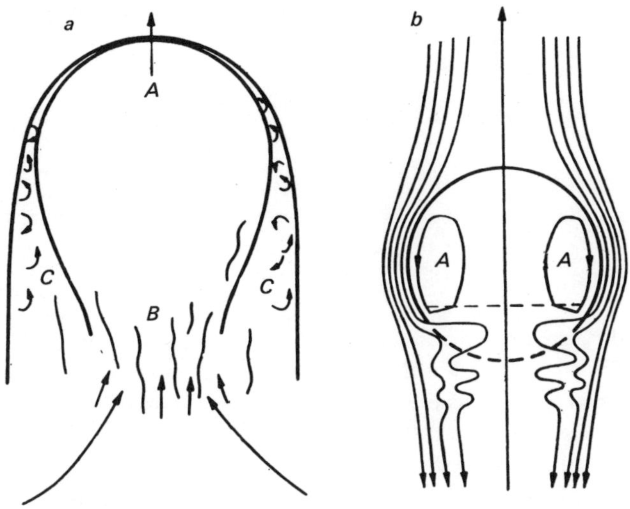

Figure 6.1. Structure of "bubble." (a) Diagram of bubble. A is heated air mass, B is wake region, C is erosion layer. (b) Streamlines in ascending bubble (according to Levine); A is annular eddy.

The bubbles usually have dimensions from a few meters to tens of meters. The largest are formed by the coalescence of several small bubbles. This process is assisted by the local reduction of pressure in the wake of the bubble, which leads to entrainment of thermals into the wake of a previously ascending ("maternal") bubble from an area many times greater than its cross section.

The acceleration of the ascent of the bubble can be represented approximately by the equation

$$\frac{dw}{dt} = g\frac{\theta_v - \theta_{0v}}{\theta_{0v}} - \frac{3}{8}\left(\frac{3}{2}K^* + c_d\right)\frac{w^2}{R}, \tag{6.1}$$

where θ_v and θ_{0v} are the virtual potential temperatures† of the thermal and the surrounding air, K^* is the fraction of mass of the thermal involved in exchange with the surrounding atmosphere during a half-cycle ($K^* \leqslant 1$), c_d is the drag coefficient, and R is the bubble radius.

Priestley [34] showed that if $dT_e/dz = \text{const}$ (the subscript e denotes values

†The virtual temperature is the hypothetical temperature which dry air must have to ensure that its density at the same pressure is equal to the density of moist air. The potential temperature $\theta = T(1000/p)^{0.288}$ (T and p are the temperature and pressure at the initial level) is the temperature which the air particle would acquire if it is displaced in dry adiabatic conditions from the initial level to a level where $p = 1000$ mb (see [67]).

of the meteorological elements outside the thermal) and $w_e = 0$, then, neglecting the square term, which is very small, we can obtain for w the equation

$$\frac{d^2w}{dt^2} + (K_1 + K_2)\frac{dw}{dt} + \left[\frac{g}{T_e}\left(\frac{\partial T_e}{\partial z} + \gamma_a\right) + K_1K_2\right]w = 0. \tag{6.2}$$

Here

$$K_1 = \frac{c_1 \nu}{r_0^2}, \qquad K_2 = \frac{c_2 a^2}{r_0^2},$$

where c_1 and c_2 are numerical factors that depend on the shape of the particle (for a sphere $c_1 \approx c_2 \approx 8$), ν is the coefficient of turbulent viscosity, a^2 is the thermal diffusivity, and r_0 is the characteristic dimension of the ascending particle.

An analysis of Eq. (6.2) shows that as long as the thermal is situated within an unstably stratified layer, its height will increase exponentially. If the thermal is ascending in a stable layer its vertical velocity gradually decreases, and near the equilibrium level decaying harmonic oscillations of the Brent–Weisel frequency will take place. The period of such oscillations is

$$\tau = 2\pi\left|\frac{g}{T_e}\left(\frac{\partial T_e}{\partial z} + \gamma_a\right) - \frac{(K_1 - K_2)^2}{4}\right|^{-1/2} \tag{6.3}$$

In the troposphere τ is usually 10–20 min.

Equation (6.2) can be used to find the greatest vertical distance L_b which a "superheated" air particle can traverse for a given $\partial T_e/\partial z, K_1, K_2, T_0'$, and w_0:

$$L_b = \frac{\left|g\dfrac{T_0'}{T_e} + K_2 w_0\right|}{\dfrac{g}{T_e}\left(\dfrac{\partial T_e}{\partial z} + \gamma_a\right) + K_1K_2}, \tag{6.4}$$

where T_0' and w_0 are the initial superheating and initial ascent velocity of the thermal.

Since K_1 and K_2 are proportional to $1/r_0^2$, it follows from Eq. (6.4) that, other conditions being equal, the greater the size of the convection element, the greater the height to which it can ascend within a stably stratified layer.

Convective bubbles can be classed as free-floating thermals, since their motion in the atmosphere is not directly related to the position at which they were formed. In many cases the level at which a bubble is created is not ground level, but somewhere within the unstably stratified part of the atmospheric surface layer, and below this level stratification may even be stable.

Another form of convective element in the atmosphere, in addition to that described, is convective jets, whose vertical extent can be more than 10 times their horizontal extent. Although such jets are typical mainly of the interior of cumuliform clouds (see Chapter 8) they can also be formed below the condensation level. This can be deduced, for instant, from the results of synchronous measurements of vertical air velocities (w) at different heights on high meteorological towers. The fluctuations of w in a wide range of heights are found to correlate with one another, which demonstrates the jetlike nature of the convective elements. Above well-heated regions of the underlying surface, if the wind velocity on the ground is low, the base of the jet is attached, as it were, to the source of heat, above which a "convective tube" is formed. Such tubes, which can sometimes be discerned by eye (the air inside the jet is more turbid), are often observed above heated rocks, forest fires, volcanoes, etc. The vertical extent of "attached" jets can be hundreds, or even thousands, of meters. If there is great instability in the convection layer, the convective jets often rotate in a clockwise or anticlockwise direction around their vertical axis. An example of rotating convective jets is provided by the dust or sand swirls that are frequently observed in summer in desert and steppe regions.

If the wind near the ground becomes stronger, the convective jets are drawn in its direction and subsequently break away from the earth to become free-floating thermals. When the thermal breaks away, the heated air that has risen is replaced by colder air from above and from the sides. After a certain time this "new" air is again heated and the evolution of the thermal is repeated.

Thus, if there is any wind, convective elements in the surface layer sooner or later become bubbles, even if they were initially jets.

Figure 6.2 depicts schematically the formation and evolution of thermals.

1.2. Experimental Data on Disordered Atmospheric Convection. Thermal turbulence (disordered convection) has a distinct diurnal course. Over land it is characterized by a maximum around noon and a minimum at night. Over large bodies of water the diurnal course is the reverse, since the stratification in the bottom part of the boundary layer is least stable at night. In the following account we will not be concerned with the characteristics of thermal turbulence over expanses of water.

Convection usually develops in summer when there is little cloud. Early in the morning there is usually inversion, sometimes very pronounced, in the bottom 200–300 m.

The inversion is destroyed from the bottom in the morning. By 0800–0900 h the stratification near the ground becomes unstable and convection develops, whereas at a certain height inversion still prevails. So long as this containing layer exists, the air under it, owing to radiative heating from the earth, becomes warmer and warmer, while the air above the inversion continues to cool due to radiation. These processes lead to a sharp increase in instability and, sooner or later, to a very rapid ("explosive") increase in the thickness of the convective

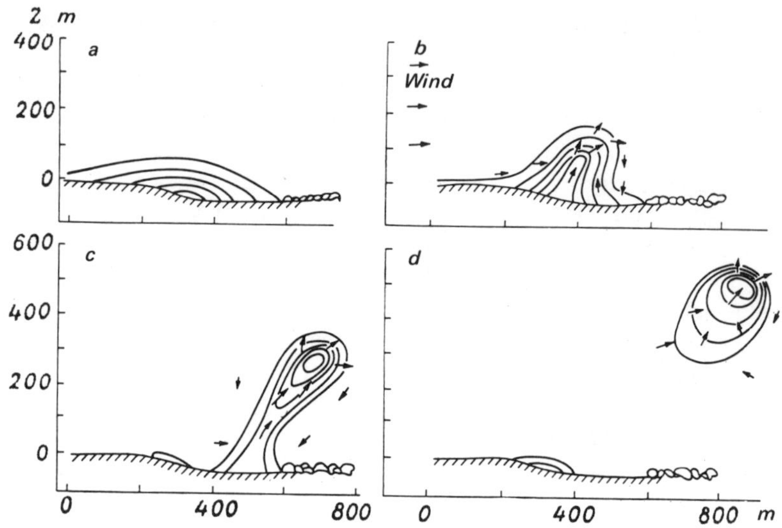

Figure 6.2. Evolution of a "bubble."

layer. By 1000 h the entire bottom half of the boundary layer has become unstable, while near-neutral stratification prevails in the top half. Around midday individual thermals (elements of "penetrating" convection) can be found up to a height of several kilometers. After 1600–1700 h convection usually diminishes, at different rates at different heights.

This accounts for the fact that, according to pilots' observations, several thin buffeting layers are usually encountered in flights in the boundary layer in the evening.

Skvortsov proposed the "layer" scheme of development of convection. Initially (usually in the morning) small-scale turbulence carries water vapor and heat from the earth to a height of 50–100 m. At this height eddies (thermals) of larger size are then formed and carry heat and moisture to greater heights.

Experimental data indicate that, depending on where the unstable layer (i.e., the layer in which $\gamma > \gamma_a$) begins, the lower boundary of the convective layer can be located either at the earth's surface, or at some height above it.

Table 6.1 gives empirical data on temperature gradients during vigorous development of convection, as indicated by the presence of Cu in the daytime.

As Table 6.1 shows, the thermal stratification in the bottom part of the boundary layer during the convection period is sometimes stable ($\gamma < \gamma_a$). An analysis of the individual soundings shows that on days with developed convection, even an isothermal or inversion temperature distribution can occasionally be observed in the bottom 100 m. Such a situation is usually observed in the morning, but is also occasionally encountered around midday.

TABLE 6.1. Mean Vertical Temperature Gradients γ (°C/100 m) over European USSR on Days with Cumulus Clouds

Time (h)	Layer (mb)						
	ground–950	950–900	900–850	850–800	800–700	700–600	600–500
0400	−0.34	0.57	0.61	0.74	0.59	0.52	0.64
1000	0.91	0.75	0.71	0.68	0.56	0.53	0.62
1300	0.99	0.94	0.90	0.73	0.55	0.53	0.64
1600	1.01	1.00	0.95	0.81	0.59	0.53	0.63
For all times	0.64	0.83	0.79	0.74	0.57	0.53	0.64

The data presented indicate that the conditions most favorable for the development of convection can sometimes arise not at the earth's surface, but a little higher. Convection in this case develops in an elevated (not at the earth's surface) thermally unstable layer and is mainly of a spontaneous nature. It owes its origin to dynamic turbulence, which causes initial vertical displacements of the air particles, leading to an increase in kinetic energy of the vertical motions at the expense of the potential energy of density stratification.

In addition to the elements of spontaneous convection, thermals can also penetrate into the elevated convective layer; these thermals are formed at the earth's surface in areas that are relatively warm. This occurs in cases where the "superheat" of the air particles (Δt) and their size are sufficiently great to ensure that the characteristic buoyancy scale L_b [see (6.4)] is greater than the thickness of the stable surface layer.

According to [124], most of the large thermals ($r_0 > 100$ m) within convective layers are of spontaneous origin. On the other hand, as Vul'fson's investigations [29] showed, small-scale elements of thermal convection obey the laws of spontaneous convection.

A possible reason for the development of convection in the free atmosphere is the presence of local regions of increased air humidity, i.e., regions of the atmosphere with relatively low density.† The role of humidity in the onset of convection is greatest over bodies of water.

We note, finally, that the conditions for development of convection are greatly affected by the vertical wind shear (gradient) $U'_z \equiv \partial U/\partial z$ in the convective layer. This factor plays a dual role. On the one hand, an increase in U'_z increases the intensity of turbulence and, hence, the probability of formation of large "turbulent bodies" which can initiate convection. On the other hand, an increase in U'_z suppresses the development of existing convection elements

†The density of moist air is $\rho_w = p/RT(1 + 0.378e/p)$, where p is the pressure, R the gas constant, T the temperature, and e the water vapor pressure. It is obvious that the larger e, the smaller ρ_w.

[125]. Thus, in each specific case the actual development of convection depends on the relation between these effects of vertial wind shear.

According to the data of Yates, who investigated thermals by means of gliders, one thermal (usually in the form of a bubble) is formed every 5–15 min over areas of heated soil of about 2 km^2. The frequency of occurrence of thermals depends on the terrain, the wind velocity, the thermophysical characteristics of the soil, and so on. Thermals are formed particularly frequently over areas where the characteristics of the underlying surface change sharply.

The physical characteristics of elements of atmospheric convection include their dimensions, distribution of temperature and vertical wind velocity within them, and so on.

Investigations of the structure of the fields of meteorological elements in the presence of convection [49] showed that in the 25- and 50-m-thick surface layer the fluctuations were of a random nature, and the dimensions of the convective elements did not exceed 100–200 m. With further rise above the earth the maximum diameters of the thermals increased, and above 50 m they usually exceeded 1 km. The largest values of ΔT and w were associated with these large thermals. In addition, they showed a close correlation, in sign and amplitude, of the fluctuations of the meteorological elements at adjacent levels. It appeared as if these thermals were in the form of extended jets.

Large-scale fluctuations are more pronounced, the smaller the wind velocity.

Figure 6.3 shows probability density distributions for horizontal dimensions of thermals according to Vul'fson's data [29] obtained from aircraft measurements of ΔT. Since it was impossible to determine the form of thermals in flights on a single airplane, two types of calculations were made—one for bubbles (ellipsoids of revolution) and one for jets. The measurements showed that the horizontal dimensions of thermals vary in a wide range. The "diameters" of thermals usually lie in the range 30–40 to 70–80 m, but in certain cases they may be 10–15 times greater than these values.

The relative area or relative volume of the ascending streams is approximately 0.21 and, hence, if we assume that the descending compensating streams are situated outside the zones of ascending streams uniformly throughout space,

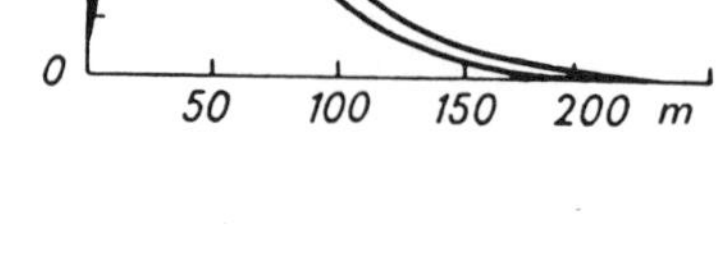

Figure 6.3. Size distribution of convective steams. (1) "Bubbles"; (2) jets.

TABLE 6.2. Mean Excess Temperatures $\Delta\overline{T}$ (°C) at Centers of Thermals at Different Altitudes

Form of thermal	Altitude (m)								
	10	30	50	100	300	500	1000	2000	3000
Jets	0.64	0.23	0.23	0.23	0.15	0.12	0.12	0.10	0.19
Bubbles	0.54	0.18	0.19	0.18	0.14	0.11	0.10	0.09	0.17

their velocity will be on the average 3.5–4 times less than the velocity of the ascending flows.

In the 10- to 50-m layer the mean dimensions of the thermals increase fairly rapidly with height, but become practically constant higher up. The concentration of thermals decreases fairly rapidly with altitude. In the layer from the earth's surface up to 1 km the concentration of thermals can be reduced by a factor of 10 or more. The relative area occupied by the thermals is accordingly altered.

Table 6.2 gives data on the excess temperature at the center of the thermals (i.e., the average maximum values of $\Delta\overline{T}$) in comparison with the outside temperature. The calculation was made for two forms of thermals—jets and bubbles.

The data given in Table 6.2 were calculated from the results of daytime, morning, and evening measurements. Hence, the values of $\Delta\overline{T}$ given in the table are slightly below the maximum values, since ΔT is maximum about midday.

It is of interest that the ascending streams are not only warmer, but moister, than the surrounding air. This is probably why the sizes of regions of nonuniformity of the refractive index of the atmosphere and excess temperature vary in approximately the same way with altitude.

The results of the measurements enabled Vul'fson to determine the dependence of ΔT (°C) and w (m/sec) in a jet of radius R (m) on altitude z (m) above the earth's surface and the distance r (m) from the thermal axis:

$$\Delta T \approx 1.1 z^{-1/3}\left(1-\frac{r^2}{R^2}\right)^{1/2}, \tag{6.5}$$

$$w \approx 0.2 z^{1/3}\left(1-\frac{r^2}{R^2}\right)^{1/2}. \tag{6.6}$$

The forms of Eqs. (6.5) and (6.6) are in good agreement with the laws $w = f(z, r)$ and $\Delta T = \varphi(z, r)$ obtained from theoretical models of spontaneous turbulent jets [76].

Numerous investigations of the largest convective streams have been made on gliders. Unfortunately, these investigations have usually been confined to qualitative estimates of the width and (more rarely) the vertical velocity of the thermals, within which the glider rose. The few instrumental investigations include Chernov's study [115].

More than 200 traversals of thermals were made by a glider equipped with instruments for measuring the velocity and temperature of the vertical streams. The traversals were made, with the controls locked, in one of the planes of symmetry of the thermal (axial symmetry of the ascending streams was adopted as a working hypothesis). The research glider was directed toward the center of the thermal by another glider of the same type. To determine the center of the flow, thermometers were mounted on the tips of the wings and recorded their temperature difference ΔT^*. It is obvious that when the glider passed through the center of the stream, $\Delta T^* = 0$.

An investigation of the relation between the maximum vertical velocity w_{max} and the diameter D of the thermal showed that an increase in D was usually accompanied by an increase in w.

Maximum velocities were observed in the center of the streams; alongside the ascending stream there was usually a broad train of descending streams. In 79% of the cases there was a region of approximately constant vertical velocity instead of a sharp maximum of w, on the thermal axis. In narrow streams it occupied 10% of the cross section of the thermal, and in broad thermals it occupied 32%.

Figure 6.4 shows the cumulative frequencies of the maximum velocities of ascending ($w^{\uparrow\ max}$) and descending ($w^{\downarrow\ max}$) streams per hour of flight and per 100 km of flight path. The high frequency of ascending streams, distinctly revealed by the 1962 data, was due, according to Chernov, to the fact that the gliders in this year were flown by more experienced pilots, who were better at selecting regions with $w > 0$.

The results of aircraft measurements of the characteristics of 130 thermals with diameters exceeding 100 m are given in [6, 124]. The flights took place over steppe, at altitudes of less than 1 km in 72% of the cases. In the 0–200-m layer the stratification on the average was dry-unstable, in the 201–400-m layer it was neutral, and higher up $\gamma = 0.7$–$0.9°C/100$ m. The wind velocity at the earth's surface usually did not exceed 5 m/sec, and there were no distinct fronts in the flight region. The clouds were cumuliform with their lower boundary at an altitude of more than 1.5 km.

The main characteristics of the thermals are given in Table 6.3. Columns 1, 2, 4, and 6 give the absolute maxima of w, ΔT, $\partial \Delta T/\partial x$, and $\partial v/\partial x$ (v is the wind velocity). Columns 3, 5, and 7 give the mean maximum values of the same characteristics (the mean values of the absolute maxima for the individual thermals).

Sampling analysis of the $\partial w/\partial x$ data showed that they were of the same order as $\partial v/\partial x$. The velocities of the ascending streams reached 7 m/sec, and those of descending streams reached 4 m/sec.

A combination of large temperature and wind-velocity gradients (which

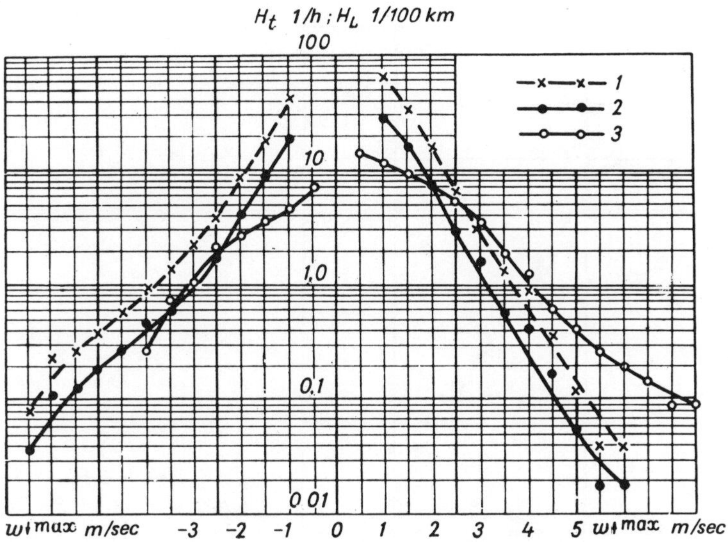

Figure 6.4. Cumulative frequencies of ascending and descending air streams per hour of flight and per 100 km of flight path from data of 1963 experiment in comparison with results of 1962 investigations. (1) Frequency of streams H_L per 100 km of path (data obtained in 1963 experiments for a calculated flight path of 2546 km); (2) frequency of streams H_t per hour of flight (results of 1963 experiment, total flight time 54.5 h); (3) the same (results of 1962 experiment, the total flight time 45.0 h).

is particularly characteristic of the region at the boundaries of thermals) must lead to intense turbulent mixing with the surrounding air.

In addition, the results of wind-fluctuation measurements showed that disordered convection was present in the region of the thermal in more than half of the cases. In the layer where $\partial w/\partial z > 0$ the outer air flowed into the thermal, and where $\partial w/\partial z < 0$ (i.e., at the top of the thermal) the air flowed out of it.

The air inflow (outflow) velocities did not exceed 0.2–0.5 m/sec, i.e., they were approximately $\frac{1}{10}$ of the maximum vertical air velocities within the thermal.

1.3. Statistical Characteristics of Thermal Turbulence. Figures 6.5 and 6.6

TABLE 6.3. Characteristics of Large Thermals

w_{max} (m/sec)	ΔT_{max} (°C)	$\Delta \overline{T}_{max}$ (°C)	$\left(\frac{\partial \Delta T}{\partial x}\right)_{max}$ °C/m	$\left(\frac{\partial \Delta \overline{T}}{\partial x}\right)_{max}$ °C/m	$\left(\frac{\partial v}{\partial x}\right)_{max}$ sec^{-1}	$\left(\frac{\partial \overline{v}}{\partial x}\right)_{max}$ sec^{-1}
1	2	3	4	5	6	7
From +7.2 to −4.1	1.49	0.49	$5.3 \cdot 10^{-2}$	$1.8 \cdot 10^{-2}$	$2.1 \cdot 10^{-1}$	$6 \cdot 10^{-2}$

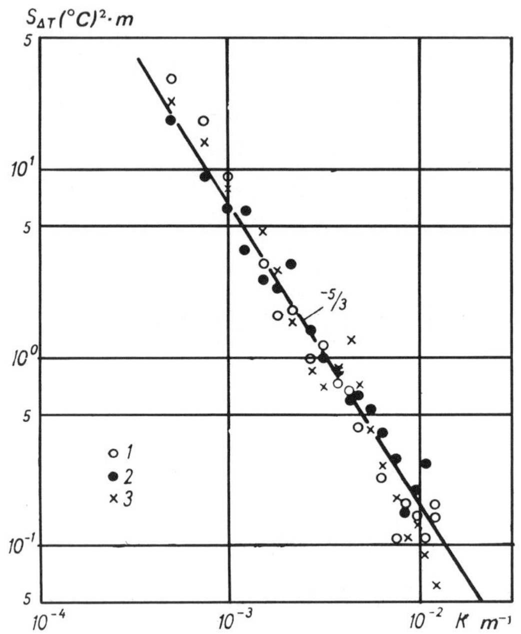

Fig. 6.5. Spectral density of temperature fluctuations $S_{\Delta T}(K)$ in convective layer. Krivoi Rog, June 10, 1969. $H = 350$ m. Time of measurements: (1) 1538–1542 h; (2) 1543–1548 h; (3) 1548–1553 h.

show typical spectral density curves for temperature and wind fluctuations in the scale range from 70 m to 1.5 km. It is apparent that both the horizontal (Fig. 6.6a) and vertical (Fig. 6.6b) wind fluctuations, and also the temperature fluctuations (Fig. 6.5), conform satisfactorily with the "minus five-thirds" law. This law holds for the whole convective layer. Outside the surface layer the values of $S_u(K)$ and $S_w(K)$ up to the top of the convective layer do not vary greatly with altitude [6, 27, 147].

The results of investigations (see [124]) have also shown that within the convective layer, i.e., up to an altitude of 400–500 m above level terrain, the conditions of quasi-stationarity and homogeneity are fairly rigidly fulfilled for periods of 20–30 min and distances of 40–50 km. This is also shown by Figs. 6.5 and 6.6, which give the spectral density curves for horizontal gusts, calculated from the results of successive series of measurements separated by distances of up to 40 km and by times of up to 24 min.

Table 6.4 gives the standard deviations of the wind fluctuations in the convective layer, obtained for complete realizations, i.e., for all scales, and also separately for characteristic inhomogeneity scales L of 300, 1000, and 2000 m.

As Table 6.4 shows, the main contribution to σ is made by large-scale fluctuations. The values of σ in the individual realizations varied in the range 0.6–2.4 m/sec for all the wind components, and the amplitudes of the individual fluctuations, according to the results of the measurements, reached 7–8 m/sec.

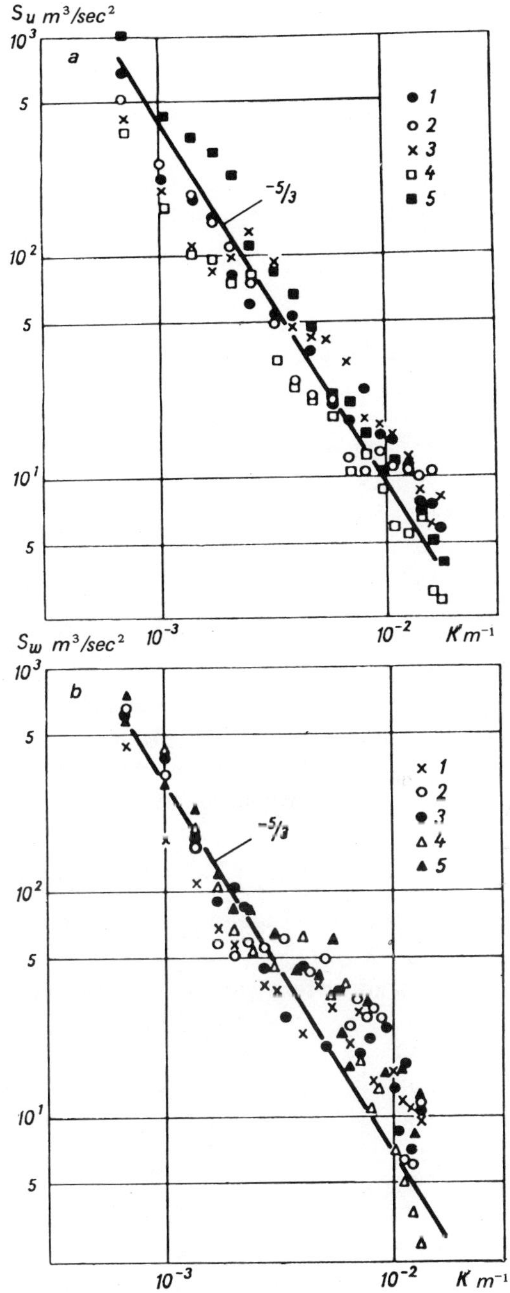

Fig. 6.6. Spectral density of wind fluctuations in convective layer. (a) Horizontal component (u). Krivoi Rog, July 27, 1968; H = 160–190 m. Times of measurement: (1) 1257–1300 h; (2) 1251–1255 h; (3) 1246–1249 h; (4) 1239–1244 h; (5) 1235–1237 h. (b) vertical component (w). Krivoi Rog, July 21, 1968, H = 800 m. Times of measurement: (1) 1149–1157 h; (2) 1202–1206 h; (3) 1207–1212 h; (4) 1214–1220 h; (5) 1221–1226 h.

TABLE 6.4. Standard Deviations of Wind Fluctuations $\bar{\sigma}$ (m/sec) due to Thermal Turbulence

σ	All scales	L (m) 300	1000	2000
$\bar{\sigma}_u$	1.25	0.36	0.56	0.73
$\bar{\sigma}_v$	1.13			
$\bar{\sigma}_w$	1.13	0.39		0.82

To estimate the linear scales of the fluctuations, we can use the integral turbulence scale L_n, defined as the distance at which the initial velocity correlation decreases by a factor e. As Taylor *et al.* [215] showed, if the energy spectrum of the fluctuations is plotted in coordinates $KF(K)$ and $\log K$, then in a first approximation $L_n K_{\max} = 0.216$, where $K_{\max}$ is the wave number, in cycles per meter, corresponding to the maximum kinetic energy of turbulence. The quantity L_n is usually used as a characteristic parameter in numerical models of turbulence spectra.

The values of $\overline{L}_n$ for longitudinal ($\overline{L}_{n,u}$) and transverse ($\overline{L}_{n,v}$) components of the fluctuations, according to [124], are close to one another and in 67% of the cases do not exceed 400 m. On the average $\overline{L}_{n,u}/\overline{L}_{n,v} = 1.2$, i.e., the dimensions of individual wind disturbances are maximum in the direction of the main flow. At present there are no reliable experimental values of $L_{n,w}$.

Many of the physical characteristics of fluctuations of different scales are different. For instance, large disturbances are characterized by a long lifetime and a high degree of correlation of the fluctuations of temperature and wind (particularly its vertical component). This correlation is a feature not only of the average, but also of individual disturbances.

Owing to the increase, with increase in L, of the coefficients of correlation for the fluctuations of the wind components (u', v', w) with one another and with temperature fluctuations the vertical fluxes of momentum ($\rho\overline{u'w}$, $\rho\overline{v'w}$) and heat ($\overline{\rho w \Delta T}$) ($\overline{u'w}$, $\overline{v'w}$, etc. are autocorrelation functions) also increase with increase in L.

The different degrees of correlation of fluctuations of the meteorological elements in disturbances of different scales is due to the fact that the statistical regime of small-scale fluctuations is almost isotropic and, hence, the values of $\overline{u'w}$ and $\overline{v'w}$ in this case do not differ greatly from zero, whereas large disturbances are anisotropic, which ensures a relatively high degree of correlation of u', v', w, and ΔT with one another.

In conclusion we note that owing to the dependence of the degree of correlation of w with T' on the fluctuation scale the average coefficient of correlation of w with T' for convective elements with dimensions $10 < L < 1000$ m, i.e., for scales of disturbances responsible for aircraft buffeting, usually does not exceed 0.3–0.4. Hence (at least in the troposphere) attempts to devise a method of warning the crew of the intensity of buffeting to be expected on a flight path from the values of T' in front of the aircraft, measured by, say, an aircraft lidar or radiometer, will not be particularly successful.

2. Cellular Convection

Photographs of cloud fields taken from aircraft and, in particular, from artificial earth satellites show that the spatial arrangement of clouds in several cases is of a "regular" (ordered) nature. The clouds (usually cumulus) either form cells, separated from one another by almost equal distances in all directions, or are arranged in parallel banks oriented in a particular way relative to the wind. The process responsible for the ordered spatial arrangement of clouds is cellular convection. It should be noted that this process, like gravity-shear waves, is one of the forms of quasi-ordered mesoscale atmospheric motions whose loss of stability leads to intense generation of turbulence. It should also be noted that aircraft passing through a region of cellular convection can experience alternating cyclic load increments, whose effect on the handling and stability of airplanes and helicopters can in some cases exceed the effect of "ordinary" turbulence.

Although cellular convection is fairly common above the atmospheric boundary layer,† we still have insufficient information about it. This is mainly due to the difficulty of finding convective zones if there are no clouds in them.

In addition, even in cases where cloud is present, it is practically impossible from external features to distinguish between gravity-shear waves and the so-called two-dimensional cells, which are characterized by bands of ascending and descending streams extending in the direction of the wind shear or at an angle to it.

The main data on the physical relations characterizing cellular convection have been obtained from laboratory experiments and from observations of cellular cloud forms. It has been found that (see [110]):

(1) For the existence of cellular convection the vertical temperature gradient in the convective layer must exceed a certain critical value which depends mainly on the layer thickness and the turbulence intensity in it.

(2) The shape and orientation of the cells formed, their dimensions, direction of vertical motions in different parts, and other circulation characteristics depend on the thickness of the convective layer and the physical

†Ordered convection usually cannot develop at the earth's surface owing to the disturbing effect of the underlying surface and the strong development of dynamic turbulence [29].

properties of the liquid (gas); the vertical distribution of temperature and wind in the convective layer is also a significant factor.

(3) The time from onset of cellular convection to attainment of its greatest development, and also the horizontal dimensions of the cells, depend on the rate of heating of the lower boundary, or of cooling of the upper boundary, of the convective layer (β in °C/sec). According to Foster [169], the time required for conversion of a laminar regime to developed convection is proportional to $\beta^{2/5}$, and the cell length is proportional to $\beta^{1/5}$.

The usual approach in the theory of cellular convection is to consider a thin layer of incompressible viscous fluid bounded above and below by free or rigid boundary surfaces. By free boundary surfaces we mean surfaces on which there is no viscous tangential friction. By rigid boundary surfaces we mean surfaces on which there is no slip. Another common assumption is the absence of vertical velocities and temperature disturbances (ϑ) on the boundary surfaces.

The boundary conditions for free surfaces have the form

$$\frac{\partial u}{\partial z} = \frac{\partial v}{\partial z} = w = 0, \quad \vartheta = 0, \tag{6.7}$$

and for rigid surfaces are

$$u = v = w = 0, \quad \vartheta = 0, \tag{6.8}$$

The system of equations of atmospheric hydrothermodynamics is used and includes the simplifications of the theory of free convection, which consist essentially in the assumption that the density fluctuations are due to changes in temperature, and not atmospheric pressure.

In addition, it is assumed that:

(1) The velocities of convective motions are small in comparison with the main-flow velocity and hence, their higher powers can be ignored.
(2) The vertical dimensions of the convective layer are smaller than the total height scale Δz, which in this case is the height of the homogeneous atmosphere, equal to approximately 8 km.

When these assumptions are made, the equation for the vertical velocity can be written in the form

$$\left(\frac{1}{\mathrm{Pr}}\frac{\partial}{\partial t} - \nabla^2\right)\left(\frac{\partial}{\partial t} - \nabla^2\right)\nabla^2 w - \mathrm{Ra}\,\nabla_1^2 w = 0, \tag{6.9}$$

where $\mathrm{Pr} = \nu/\kappa$ is the Prandtl number (ν is the coefficient of turbulent viscosity,

κ is the thermal diffusivity),

$$\nabla^2 = \vec{i}\,\frac{\partial^2}{\partial x^2} + \vec{j}\,\frac{\partial^2}{\partial y^2} + \vec{k}\,\frac{\partial^2}{\partial z^2}\,, \qquad \nabla_1^2 = \nabla^2 - \vec{k}\,\frac{\partial^2}{\partial z^2}\,;$$

and

$$\mathrm{Ra} = g\,\frac{(\gamma - \gamma_a)^4}{\nu \varkappa T}$$

is the Rayleigh number.

Since Eq. (6.9) is of sixth order, its solution requires the assignment of six boundary conditions in relation to height.

If the thickness of the convective layer is h, and the boundary surfaces are free, these conditions have the form

$$w\,|_{z=0} = w\,|_{z=h} = \left.\frac{\partial^2 w}{\partial z^2}\right|_{z=0} = \left.\frac{\partial^2 w}{\partial z^2}\right|_{z=h} = \left.\frac{\partial^4 w}{\partial z^4}\right|_{z=0} = \left.\frac{\partial^4 w}{\partial z^4}\right|_{z=h} = 0\,, \qquad (6.10)$$

Equation (6.9) without boundary conditions (6.10) is used to find $w = f(x, y, z)$ in cells of different shape. For instance, for rectangular (in plan) cells

$$w = w_0 \sin\left(\frac{\pi z}{h}\right)\cos(\Omega_x x)\cos(\Omega_y y)\exp(nt)\,, \qquad (6.11)$$

where $\Omega_x = 2\pi/L_x$, $\Omega_y = 2\pi/L_y$ are the wave numbers along the axes $0x$ and $0y$, (L_x and L_y are the distances between the cell centers), and w_0 is the velocity at the cell center. Substituting (6.11) in Eq. (6.9) we obtain a quadratic equation for the eigenvalues n:

$$n^2 + (\mathrm{Pr} + 1)(a^2 + \pi^2)\,n + \mathrm{Pr}\left[(a^2 + \pi^2)^2 - \frac{a^2\,\mathrm{Ra}}{a^2 + \pi^2}\right] = 0\,, \qquad (6.12)$$

where

$$a^2 = \Omega_x^2 + \Omega_y^2\,.$$

It is known from the theory of hydrodynamic stability that the solution of (6.11) will be stable with respect to small variations if all the eigenvalues have negative real parts. According to Eq. (6.12), the so-called boundary motion corresponding to $n = 0$ and separating the regions of stable and unstable disturbances is possible if

$$(a^2 + \pi^2)^3 = a^2\,\mathrm{Ra}\,. \qquad (6.13)$$

For fixed h, κ, ν, and $\overline{T}$, Eq. (6.13) can be used to determine uniquely the distance between the cell centers and the minimum temperature gradient γ_{cr}, on the attainment of which the flow becomes divided into cells. To determine γ_{cr} we use the fact that γ_{cr} corresponds to the minimum of Ra and, hence, according to Eq. (6.13), to the condition $d\,\text{Ra}/d(a^2) = 0$, from which in our case $\text{Ra}_{cr} = \frac{27}{4}\pi^4 \approx 657.5$. Hence

$$\gamma_{cr} = \gamma_a + \frac{27}{4}\pi^4 \frac{\nu\varkappa\overline{T}}{gh^4} \approx \gamma_a + 657{,}5\frac{\nu\varkappa\overline{T}}{gh^4} \tag{6.14}$$

Thus, in the presence of two free boundaries, cellular circulation is possible when

$$\Delta\gamma = \gamma - \gamma_a > 657.5\frac{\nu\varkappa\overline{T}}{gh^4} \tag{6.15}$$

If, for instance $\overline{T} = 250°\text{K}$, $h = 100$ m, $\nu = \kappa = 10^4$ cm^2/sec, then $\gamma_{cr} - \gamma_a = 0.02°\text{C}/100$ m. When $h \to 0$ criterion (6.5) ceases to have any physical meaning.

In stationary conditions (and, hence, $\text{Ra}_{cr} = \text{const}$) each cell is an isolated region with closed circulation. For instance, in hexagonal cells (Bénard cells) the horizontal distribution of w is given by the equation

$$w = \frac{w_0}{3}\sin\left(\frac{\pi z}{h}\right)\left[2\cos\left(\frac{2\pi}{\lambda\sqrt{3}}x\right)\cos\left(\frac{2\pi}{3\lambda}y\right) + \cos\left(\frac{4\pi}{3\lambda}y\right)\right], \tag{6.16}$$

where $\lambda/2$ is the length of the side of the regular hexagon.†

In round (cylindrical) cells

$$w = w_0 \sin\left(\frac{\pi z}{h}\right) J_0\left(\frac{a\lambda}{2h}\right), \tag{6.17}$$

where λ is the diameter of the circle, and J_0 is a cylindrical function of zero order.

Owing to the linearization of the initial system of equations and the indeterminacy of the relation between Ω_x and Ω_y, the classical theory of cellular convection does not allow simultaneous unambiguous determination of the shape of the developed cells and the magnitude and sign of w_0. According to laboratory experiments, hexagonal and round cells are usually formed, and the direction of the vertical motions at the center of the cell depends on the me-

†Formula (6.16) was obtained for $n = 1$, which corresponds to the fundamental oscillation excited at minimum Ra_{cr}. In this case there is one convective layer, at whose boundaries $w = 0$.

dium. In liquids, air rises at the center of the cells ("closed" cells), while in gases it sinks ("open" cells).

The values of Ra_{cr} are determined exclusively by the nature of the boundary surfaces and do not depend on the cell geometry. With two free boundary surfaces, $Ra_{cr} = 657.5$; with one free and one rigid boundary surface $Ra_{cr} = 1100.7$; finally, with two rigid boundary surfaces, $Ra_{cr} = 1707.8$. It follows from Eq. (6.14) that when Ra_{cr} changes, $\Delta\gamma$, which is proportional to it, also changes.

With two free boundary surfaces the wavelength of hexagonal cells $\lambda = 3.77h$; for round cells $\lambda = 2.16h$ (in the first case λ is twice the length of the side of the hexagon, and in the second case it is the diameter of the circle).

According to results of observations of clouds formed in zones of cellular convection, the horizontal dimensions of atmospheric cells are usually several kilometers,† and the ratio $\lambda/h \approx 3$. The cells in plan are usually round; air rises in their center and descends at the edges. Special cellular clouds, Cu mammatus, can be observed on the lower surface of cumulonimbus clouds. They are mammiform and have a descending stream at the center of the cell.

We discussed above the characteristics of convection for the case where there is no wind in the convection cell or its velocity and direction do not change with altitude. This is a rather rare situation in atmospheric layers with $h \geqslant 500$–1000 m. Gandin points out‡ that if the wind direction in the convective layer remains constant (or changes slightly), and its velocity depends on the altitude, then two-dimensional cells, consisting of longitudinal bands of ascending and descending motions, extending in the direction of the wind, are formed. If the velocity and direction of the wind in the convective layer vary sharply with altitude, cellular circulation probably cannot arise.

According to the calculations of Vorob'eva and Trubnikov [26], two-dimensional cells can lie at an angle to the wind, extending in the direction of the component whose profile curvature (U_z'') is not zero.

According to empirical data [125], transverse two-dimensional cells in the atmosphere are usually associated with an average shear, while longitudinal cells are associated with greater shears. Considerable rotation of the wind with increase in altitude also promotes the formation of cells that make large angles (up to a limit of 90°) with the wind direction.

The laboratory experiments of Kimura *et al.* [181] enabled them to distinguish three distinct types of cellular convection: (1) "purely cellular" convection, developing at high Ra and small vertical wind shears; (2) longitudinal

†We are not dealing here with large-scale cellular convection, whose elements, as photographs of cloud cover from artificial earth satellites show, have horizontal dimensions of several tens of kilometers.

‡Gandin did not give rigorous proof of his deductions and confined himself to qualitative considerations.

bands of ascending motions observed at low Ra and large shears; (3) a form of convection intermediate between these two. The critical wind shear, which separates the longitudinal bands from the isolated cells, increases with increase in Ra. The formation of transverse convective bands was not investigated in this experiment.

We point out in conclusion that when cells become unstable the energy of cellular circulation can provide a source of turbulent energy. Since cellular circulation can be regarded as a result of the interference of several standing gravity waves, then the stability criteria obtained for these waves (see Chapter 7) are probably valid in a first approximation for this process. This means, in particular, that the shorter the waves (i.e., the size of the cells), the less stable they are.

It follows from Eq. (6.13) that for boundary motion the cell size (a^2) is uniquely determined by $\mathrm{Ra}_{\mathrm{cr}}$. The situation will be different if $\mathrm{Ra} > \mathrm{Ra}_{\mathrm{cr}}$, i.e., $\gamma > \gamma_{\mathrm{cr}}$. This gives rise to supercritical conditions, where each value of Ra corresponds to two cell systems, one of which has a smaller wavelength λ than in the boundary regime. In this case the transition from cellular circulation to small-scale turbulent fluctuations of w will be facilitated. This hypotheses requires experimental verification.

7

Waves and Atmospheric Turbulence

Periodic motions, which include, in particular, waves and cellular convection, are one of the commonest forms of mesoscale atmospheric flows. They are of great interest not only in themselves, but also because their loss of stability leads to the formation of turbulent eddies. Relatively small eddies, which cause buffeting of aircraft, are mainly due to the breakup of short waves, whose length does not exceed a few kilometers and whose dynamics is practically unaffected by the deflecting rotational force of the Earth.

If we leave aside sound waves, whose formation is due to the compressibility of the air, atmospheric short waves can be subdivided into gravity and gravity-shear waves. Waves of the first type are due to the effect of the gravitational force and draw upon the potential energy of position of air particles, if the latter are displaced from the equilibrium position by some factor or another. Gravity-shear waves (or "Helmholtz waves") are formed if there is a tangential discontinuity of wind velocity near an interface at which there is a sharp change (a discontinuity at the limit) of air density.†

The commonest type of gravity waves are the waves formed when air flows over mountains—mountain waves or obstacle waves. In the free atmosphere a density discontinuity is usually observed in zones where the wind velocity also changes sharply, which is typical, for instance, of atmospheric fronts, the tropopause layer, inversions above clouds, etc. Hence, the waves formed are not gravity waves, but gravity-shear waves, which we will now describe.

1. Gravity-Shear Waves

1.1. Main Characteristics of Stationary Waves. We are interested in the characteristics of stationary waves, and also in their stability, i.e., the direct relation

†Waves can also arise when there is only a discontinuity of wind velocity ("shear waves"). As will be shown later, however, they are unstable, i.e., they cannot be observed in the stationary state.

between wave and turbulent motions. We will consider first the relation between the main characteristics of waves—their length (λ), period (T), phase velocity (c), and amplitude (A)—and discontinuities of temperature or density in the atmosphere, or a wind-velocity discontinuity near an interface. We will restrict ourselves to a consideration of internal waves, since waves on a free surface are not typical of the atmosphere.

In theoretical investigations of gravity-shear waves it is usually assumed that the waves are due to the superposition of small wavelike perturbations on a main horizontal flow of ideal fluid. In the treatment of small waves the deflecting force of the Earth's rotation is ignored in the derivation of the equations of motion.† In addition, the process is assumed to be adiabatic.

If we assume that the main flow is stationary and, in addition, its characteristics are independent of the horizontal coordinates, and vary only with altitude z, then we can write the equations of the small oscillations in the form

$$\frac{\partial u'}{\partial t}+\bar{u}\frac{\partial u'}{\partial x}+\bar{v}\frac{\partial u'}{\partial y}+\bar{w}\frac{\partial u'}{\partial z}=-\frac{1}{\bar{\rho}}\frac{\partial p'}{\partial x},$$

$$\frac{\partial v'}{\partial t}+\bar{u}\frac{\partial v'}{\partial x}+\bar{v}\frac{\partial v'}{\partial y}+\bar{w}\frac{\partial v'}{\partial z}=-\frac{1}{\bar{\rho}}\frac{\partial p'}{\partial y},$$

$$\frac{\partial w'}{\partial t}+\bar{u}\frac{\partial w'}{\partial x}+\bar{v}\frac{\partial w'}{\partial y}+\bar{w}\frac{\partial w'}{\partial z}=-\frac{1}{\bar{\rho}}\frac{\partial p'}{\partial z}-\frac{g}{\bar{\rho}}\rho', \tag{7.1}$$

$$\frac{\partial \rho'}{\partial t}+\bar{u}\frac{\partial \rho'}{\partial x}+\bar{v}\frac{\partial \rho'}{\partial y}+w'\frac{d\bar{\rho}}{dz}+\bar{\rho}\left(\frac{\partial u'}{\partial x}+\frac{\partial v'}{\partial y}+\frac{\partial w'}{\partial z}\right)=0,$$

$$\frac{\partial p'}{\partial t}+\bar{u}\frac{\partial p'}{\partial x}+\bar{v}\frac{\partial p'}{\partial y}+w'\frac{d\bar{p}}{dz}-\varkappa\frac{\bar{p}}{\bar{\rho}}\left(\frac{\partial \rho'}{\partial t}+\bar{u}\frac{\partial \rho'}{\partial x}+\bar{v}\frac{\partial \rho'}{\partial y}+\bar{w}\frac{d\rho}{dz}\right)=0.$$

The first three equations of system (7.1) are obtained from the equations of motion, and the fourth and fifth from the equations of continuity and heat flux, respectively. Here u, v, and w are the components of the motion, t is the time, ρ is the air density, p is the pressure, $\varkappa = c_p/c_v$ is the ratio of specific heats, and g is the acceleration of gravity. The overbar denotes characteristics of the main flow, and the dashes denote characteristics of the small perturbations. We note that system (7.1) consists of linear equations for the unknown u', v', w', p', etc.‡

The following boundary conditions are usually used in the solution of (7.1).

†This assumption means, in particular, that the interface must be horizontal in the unperturbed state.

‡In addition to the linear theory of atmospheric waves, nonlinear problems have also been considered in recent years (see the review [117]). The assumption of smallness of the wavelike perturbations is dropped.

First, it is assumed that the velocity components normal to the interface and on different sides of the latter are equal, i.e., there is no mixing. Second, the pressure is assumed to be continuous on the interface. In addition to these conditions, all the functions must be bounded at infinity.

Laikhtman determined the characteristics of plane waves for the case where the main flow on both sides of the interface is directed along the axis $0x$, and the potential temperature θ varies linearly with altitude. At the interface ($z = 0$) there are discontinuities of wind velocity, temperature, and potential temperature gradient.

The wave characteristics were sought in the form

$$\begin{aligned}
u' &= u_1(z)\sin 2\pi\left(\frac{t}{\tau} - \frac{x}{\lambda}\right),\\
w' &= w_1(z)\cos 2\pi\left(\frac{t}{\tau} - \frac{x}{\lambda}\right),\\
p' &= p_1(z)\sin 2\pi\left(\frac{t}{\tau} - \frac{x}{\lambda}\right),\\
\rho' &= \rho_1(z)\sin 2\pi\left(\frac{t}{\tau} - \frac{x}{\lambda}\right),\\
f' &= A\sin 2\pi\left(\frac{t}{\tau} - \frac{x}{\lambda}\right),
\end{aligned} \tag{7.2}$$

where τ is the period, λ is the wavelength, and A is the amplitude of the oscillations.

Substituting (7.2) in system (7.1) and neglecting the infinitely small higher-order terms, we can obtain an ordinary differential equation for the momentum $\tilde{w} = \bar{\rho}w$:

$$\frac{d^2\tilde{w}}{dz^2} + \beta\frac{d\tilde{w}}{dz} + \tilde{w}\left[-k^2 + \frac{g\beta}{(c-\bar{u})^2}\right] = 0, \tag{7.3}$$

where $\beta = d\ln\theta/dz$, $k = 2\pi/\lambda$ is the wave number, and $c = \lambda/t$ is the phase velocity.

Using the boundary conditions, we find from (7.3) that

$$\begin{aligned}
\tilde{w}_B(z) &= A\exp\left\{\left[-\frac{\beta_B}{2} + \sqrt{\frac{\beta_B^2}{4} + k^2 - \frac{g\beta_B}{(c-u_B)^2}}\right]z\right\}k\bar{\rho}_B(0)(c-u_B),\\
\tilde{w}_A(z) &= A\exp\left\{\left[-\frac{\beta_A}{2} - \sqrt{\frac{\beta_A^2}{4} + k^2 - \frac{g\beta_A}{(c-u_A)^2}}\right]z\right\}k\bar{\rho}_A(0)(c-u_A).
\end{aligned} \tag{7.4}$$

The subscripts B and A denote layers situated below and above the interface. The connection between c and k is given by the dispersion equation

$$\bar{\rho}_B(0)(c-u_B)^2\left[\frac{\beta_B}{2}+\sqrt{\frac{\beta_B^2}{4}+k^2-\frac{g\beta_B}{(c-u_B)^2}}\right]-$$
$$-\bar{\rho}_A(0)(c-u_A)^2\left[\frac{\beta_A}{2}-\sqrt{\frac{\beta_A^2}{4}+k^2-\frac{g\beta_A}{(c-u_A)^2}}\right]=g\left[\bar{\rho}_B(0)-\rho_A(0)\right]. \quad (7.5)$$

Equations (7.4) and (7.5) show that the vertical velocities at the interface, the phase velocity of the waves, and the wavelength, other conditions being equal, depend on the flow velocities and air density on the two sides of the interface and on the stability of stratification of the atmosphere.

In the special case† where $\beta_B = \beta_A = 0$, i.e., the atmospheric stratification on both sides of the interface is neutral, it follows from (7.5) that

$$\begin{aligned}\tilde{w}_B(z) &= Ak\bar{\rho}_B(0)(c-u_B)e^{kz},\\ \tilde{w}_A(z) &= Ak\bar{\rho}_A(0)(c-u_A)e^{-kz},\end{aligned} \quad (7.6)$$

and

$$c=\frac{\bar{\rho}_B(0)u_B+\bar{\rho}_A(0)u_A}{\bar{\rho}_B(0)+\bar{\rho}_A(0)}\pm$$
$$\pm\sqrt{\frac{g}{k}\,\frac{\bar{\rho}_B(0)-\bar{\rho}_A(0)}{\bar{\rho}_B(0)+\bar{\rho}_A(0)}-\frac{\bar{\rho}_B(0)\bar{\rho}_A(0)(u_B-u_A)^2}{[\bar{\rho}_B(0)+\bar{\rho}_A(0)]^2}} \quad (7.7)$$

According to Eq. (7.6), the vertical velocities associated with waves decrease exponentially with increasing distance from the interface.

In Eq. (7.7), which gives the value of the phase velocity, the first term is equal to the weighted mean velocity as a whole, while the second describes how the value of c is affected by the discontinuities of air density and wind velocity on passage through the interface. It is apparent that the phase velocity increases with reduction of k, i.e., with increase in λ. It can be shown that at values of λ not exceeding a few kilometers the phase velocity does not exceed a few meters per second.

It is obvious that for some ratio k and density and velocity discontinuities ($\Delta\rho$ and Δu) at the interface the radicand becomes negative and, hence, c is a complex quantity. In this case the motion becomes unstable and the waves will break up. We will discuss the stability of waves more thoroughly later.

If $c = 0$, i.e., if we consider standing waves, then, as Eq. (7.2) shows, there is

†In the more general case where $\beta_B = \beta_A = \beta \neq 0$, Eq. (7.5) can be solved for c by using the fact that $\beta/k \ll 1$.

an alternation of zones of ascending and descending flows near the interface; the position of these zones does not change with time. If $\beta_B = \beta_A = \beta$, the length of such waves is

$$\lambda_{stat} = \frac{2\pi}{g}(\Delta u)^2 \frac{T_0}{\sqrt{4(\Delta T)^2 + \frac{(\Delta u)^2 \beta T_0^2}{g^2}\left[4g - \frac{1}{4}\beta(\Delta u)^2\right]}}. \tag{7.8}$$

Here $\Delta u = u_A - u_B$, $\Delta T = T_A - T_B$, T_0 is the temperature at the interface, and

$$\beta = \frac{d \ln \theta}{dz} = \frac{2\gamma_a - \gamma}{T_A + T_B}.$$

Equation (7.8) shows that the wavelength increases with increase in the wind discontinuity Δu and with reduction of the temperature discontinuity ΔT. In addition, since $\beta(\Delta u)^2 \ll g$ ($\beta \approx 1.3 \cdot 10^{-5}$ m^{-1}, $\Delta u \simeq 1$ m/sec), an increase in thermal stability leads to a reduction of λ_{stat}. In real conditions λ_{stat} can vary from hundreds of meters to several kilometers.

Naito calculated the phase velocity and amplitudes of plane gravity–shear waves for two- and three-layer models of the atmosphere on the assumption that within each of these layers the air density and wind velocity are constant. The assumption that the air density in a particular layer is independent of altitude, i.e., that the air is quasi-incompressible, is very rough, but if the layers are not very thick, the quantitative results of calculations based on a quasi-incompressible model do not contain significant errors.

The calculations showed that the amplitudes of the perturbations are greatest at the interface, and the smaller λ, the more rapidly they decrease with increase in distance from the interface. This accounts for the fact, known from observations, that short gravity–shear waves are very rare near the ground.

At the interface an ordered circulation develops, consisting of closed circulation cells carried in a horizontal direction along with the flow present in the region of the discontinuity surface.

According to calculations, the shorter the wave, the greater the value of A/λ; this ratio also increases with increase in u'. With reduction of λ the steepness of the wave front increases. For instance, for $\lambda = 100$ m and $u' = 1$ m/sec the transition from crest to trough of the wave occurs in a distance of 50 m ($\lambda/2$) in approximately 30 sec ($\tau/2$) and is accompanied by a drop in height of 60 m (twice the amplitude). When u' is larger the steepness of the wave front is, of course, even greater. Because of this the wave crests even in cases where c is real and, thus, the waves are stable, show a tendency to break periodically (like sea waves) and to form isolated rolls with a horizontal axis, which are carried along by the flow—the so-called rotors. The general flow, so long as c

is real, is laminar. The described process will be observed more often, of course, in atmospheric layers where the "background" turbulence is reduced.

In the models discussed above, the authors considered plane waves and neglected perturbations along the axis $0y$. Hence, the wave crests were perpendicular to the direction of the general motion (transverse waves). Observations have shown, however, that bands of clouds due to ascending motions in the wave crests are usually oriented at an acute angle to the wind, or may even be parallel to it. To account for these empirical data it is essential to solve the three-dimensional, and not the plane, problem. In this case the perturbations in the wind field are assigned in the form

$$u_i' = \bar{u}_i \exp \imath (kx + ly - mt), \tag{7.9}$$

where u_i' is the ith component of the flow velocity perturbation, $k = 2\pi/L_x$, $l = 2\pi/L_y$ (L_x and L_y are the length scales in the direction of the axes $0x$ and $0y$), and m is the frequency.

Investigations of three-dimensional waves have been made in [7, 161] and other studies. According to Sekera, if a thin transitional layer in which the wind velocity changes suddenly is considered instead of an interface, then from system (7.1), using (7.9), it is possible to determine the necessary conditions for the existence of waves, provided that the vertical wind shear in each of the layers does not vary with height.

This requires that

$$2\bar{T}(U_z')^2 \left[g\left(1 + \frac{l^2}{k^2}\right)\right]^{-1} \leqslant \gamma - \gamma_a, \tag{7.10}$$

where $U_z' = \partial U/\partial z$ is the vertical wind gradient (shear). Thus, near the level where the wind velocity is equal to the wave velocity,[†] the stratification of the atmosphere must be neutral at least. This conclusion is consistent with empirical data, according to which cloud bands parallel to the wind are usually observed when $\gamma \geqslant \gamma_a$ immediately under an inversion layer.

Since $L_x = 2\pi/k$, $L_y = 2\pi/l$ (L_x and L_y are the wavelengths), it follows from (7.10) that the spatial position of the wave for a prescribed γ is determined by U_z'. It is obvious that the formation of transverse waves requires the presence of small wind shears, while the formation of longitudinal waves requires large wind shears.

The upper limit of $\gamma_a - \gamma$ in (7.10) must give the conditions in which the circulation becomes disordered and the spontaneous development of turbulence begins. It should be noted that the range of $\gamma_a - \gamma$, lying between the lower and upper limits, is small and, hence, even a slight change in the external parameters,

[†]Sekera derived Eq. (7.10) on this assumption

e.g., U_z' or T, can lead to the destruction of ordered circulation. This, in particular, probably accounts for the relative scarcity of billow clouds and the shortness of their life.

1.2. Empirical Data on Gravity–Shear Waves in the Atmosphere. Since investigations of waves in clear skies, i.e., when there are no visible signs of their existence, are extremely difficult, we have little information about such waves at present.

Gossard and Munk measured the parameters of gravity–shear waves in clear skies over the Californian coast. They used synchronous records of highly sensitive ground microbarographs, barovariographs, and anemorhumbographs mounted along the coast, and also the results of vertical temperature and wind soundings in the atmosphere.

An example of a record of the fluctuations of wind and atmospheric pressure for one case is shown in Fig. 7.1.

In the period in which the waves were observed, there was an inversion layer above the region of observations, and the wind velocity was relatively low. The parameters of the waves and the characteristics of the inversion layer are given in Table 7.1. The azimuth of the point from which the waves moved was taken as the direction of motion of the waves.

An analysis of the experimental data showed that the wave phase velocities were close to 10 m/sec, i.e., greatly exceeded the wind velocity. The magnitude of the wind fluctuations caused by the waves was usually also very large, especially on July 27, October 2 and 16, and May 12. The wavelengths varied from 4 to 10 km, and the wave crests were oriented approximately perpendicular to the wind shear vector.

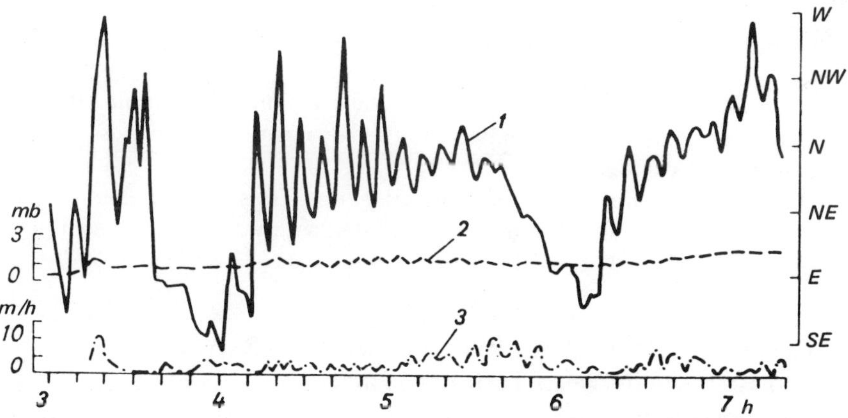

Figure 7.1. Changes in atmospheric pressure, wind velocity, and wind direction in gravity waves on October 16, 1952 (from Gossard and Munk). (1) Wind direction; (2) pressure; (3) wind velocity.

TABLE 7.1. **Parameters of Waves Observed above Californian Coast**[a]

Date	Weather	Wind velocity (m/sec)	Wind direction	Amplitude of wind fluctuations		Oscillation period (min)	Frequency (sec^{-1})	Altitude of inversion (m)		Change in potential temperature in inversion (°C)
				m/sec	deg			lower boundary	upper boundary	
Apr. 4, 1952	Fog	1.4	NNW	1.0	200	14.0	0.0075	170	600	15.0
July 27	Clear	0.9	NE	1.3	78	7.5	0.014	170	670	14.0
Aug. 4	Fog	2.9	W	1.2	175	6.5	0.016	300	1190	20.0
Sept. 4	Clear	2.2	N	0.5	318	5.5	0.019	260	740	10.5
Oct. 2	Fog	1.2	WSW	3.2	240	8.5	0.012	210	1000	11.5
Oct. 16	»	0.8	NNE	1.4	268	7.5	0.014	180	670	17.5
	»	0.9	NW	0.9	282	7.5	0.014	180	670	17.5
May 12, 1953	Clear	0.7	ENE	1.6	280	11.0	0.0095	100	400	6.0

[a]On October 16, 1952 the waves were recorded twice.

The values of c_i and λ_i obtained in Gossard and Munk's investigations were in good agreement with the results of Naito's calculations.

Georgii, who used gliders to investigate vertical motions in gravity–shear waves formed near inversion layers in the lower half of the troposphere, found that $\lambda = 500$ m and $w_{max} = \pm(1\text{–}2)$ m/sec. From data on the horizontal distribution of air temperature above an inversion layer at an altitude of 500 m, Vasil'chenko and Ledokhovich [15] found periodic temperature fluctuations with $\lambda = 1.0\text{–}1.2$ km and with temperature oscillations up to 1.2°C.

Waves in clear air have also been investigated by following the motion and deformation of jets of smoke discharged into the atmosphere by rocket flares or other methods. As an example, we can cite the data of Cooke, in whose experiment different parts of the investigated smoke trail moved faster or slower than the air, although the trail as a whole moved at the same speed as the wind.

The largest amount of empirical data has been obtained for cases where clouds were observed in the wave crests. A particularly large amount of data has been obtained by observations from airplanes and by photographs from meteorological satellites [7, 189, etc.]. Owing to the inadequate resolution, satellite photographs mainly give information about comparatively long waves with λ equal to several kilometers.

An analysis of photographs of cloudy sky from Tiros satellites showed that geometrically regular cloud formation associated with waves were present in 13% of the cases.

2. Criteria of Stability of Internal Gravity–Shear Waves and Relation between Turbulence and Disintegrating Waves

As Eqs. (7.7) and (7.8) show, for particular values of ΔT, $\Delta U = u_A - u_B$, and β the radicand becomes negative, i.e., c is a complex quantity. The size of the perturbations then increases indefinitely with time, i.e., the motion becomes unstable. In the special case where $\beta_B = \beta_A = \beta \neq 0$, all waves with length $\lambda < \lambda_{cr}$ are unstable, where

$$\lambda_{cr} = \frac{2\pi (\Delta U)^2 \bar{T}}{g\,\Delta T + \sqrt{(g\,\Delta T)^2 + \dfrac{g\beta \bar{T}^2 (\Delta U)^2}{8}}}. \tag{7.11}$$

Equation (7.11) shows that the instability region is broader (i.e., λ_{cr} is greater) the greater ΔU and the smaller ΔT.

To assess the stability of waves, Gandin proposed a dimensionless parameter Γ, given by the equation

$$\Gamma = \frac{\bar{T}^2}{g}\bar{\beta}\left(\frac{\Delta U}{\Delta T}\right)^2 \tag{7.12}$$

This parameter describes the effect of static stability of the atmosphere on the dynamic stability of flows at an interface. The critical wavelength λ_{cr} is expressed in terms of Γ in the following way:

$$\lambda_{cr} = \frac{\lambda_0}{\sqrt{\Gamma + 1}}, \tag{7.13}$$

where

$$\lambda_0 = \frac{\pi \bar{T}}{g}\,\frac{(\Delta U)^2}{\Delta T}. \tag{7.14}$$

As Eq. (7.11) shows, λ_0 is the critical wavelength for the case of neutral stratification ($\beta = 0$).

According to (7.13), with increase in Γ the critical wavelength decreases, i.e., the instability zone contracts. Since $\Gamma > 0$ corresponds to $\bar{\beta} > 0$, this means that with increase in static stability the dynamic stability also increases. When $\Gamma \leqslant -1$ there is absolute dynamic instability, i.e., instability of perturbations of all wavelengths. The same situation occurs if $\Delta T \leqslant 0$.

Gandin determined the relation between the stability of waves and the direction of their propagation within the framework of a two-layer model in which the air density ρ_B and ρ_A and the horizontal velocity components u_B, v_B and u_A, v_A in each of the layers are constant.

We put $u_B = w_B \cos \varphi_B$, $u_A = w_A \cos \varphi_A$, $v_B = \sin \varphi_B$, $v_A = w \sin \varphi_A$, where φ_B and φ_A are the angles between the wind and the $0x$ axis in each layer.

In the considered case the phase velocity

$$c = \frac{\rho_B w_B \cos(\varphi_B - \alpha) + \rho_A w_A \cos(\varphi_A - \alpha)}{\rho_B + \rho_A} \pm$$

$$\pm \sqrt{\frac{g\lambda}{2\pi}\frac{\rho_B - \rho_A}{\rho_B + \rho_A} - \frac{\rho_B \rho_A}{(\rho_B + \rho_A)^2}[w_B \cos(\varphi_B - \alpha) - w_A \cos(\varphi_A - \alpha)]^2}\,, \tag{7.15}$$

where α is the angle between the $0x$ axis and the direction of propagation of the wave.

Equation (7.15) shows that

$$\lambda_{cr} = \frac{2\pi\rho_A\rho_B}{\rho_B^2 - \rho_B^2}[w_B \cos(\varphi_B - \alpha) - w_A \cos(\varphi_A - \alpha)]^2, \tag{7.16}$$

i.e., the stability of a wave of prescribed length depends on the direction of its propagation. If ξ is the angle of deviation of the direction of the flows from the mean direction, and φ^* is the angle between this mean direction and the direction of propagation of the wave, then

$$\lambda_{cr} = \frac{2\pi\rho_B\rho_A}{\rho_B^2 - \rho_A^2}[w_B \cos(\varphi^* + \xi) - w_A \cos(\varphi^* - \xi)]^2 \tag{7.17}$$

Equation (7.17) shows that if the main flows are collinear, i.e., $\xi = 0$, waves propagating along these flows are least stable, while waves propagating in a perpendicular direction are always stable ($\lambda_{cr} = 0$). If the velocities of the main flow are the same, and $\xi \neq 0$, waves propagating in the direction of the bisectrix of the angle between the directions of the main flows are always stable, and waves in a direction perpendicular to it are the most unstable.

Although Eqs. (7.13), (7.17), etc., given above were obtained for strictly horizontal interfaces, they can also be used for an approximate estimate of the stability of waves on fronts and the tropopause. This is due to the fact that the slope of these interfaces is small and has no significant effect on the dynamics of short waves. In the treatment of the stability of the latter on fronts, Γ and λ_0 can be expressed in terms of the angle of inclination of the front α_f and the Coriolis parameter l:

$$\Gamma = \frac{\bar{\beta} g}{l^2}\tan^2\alpha_F, \quad \lambda_0 = \frac{\pi g}{l^2}\frac{\Delta T}{\bar{T}}\tan^2\alpha_F. \tag{7.18}$$

The wind-velocity component perpendicular to the front experiences a relatively small discontinuity and, hence, the direction of the vector of the wind-velocity difference on the two sides of the front is close to the direction of the front. Hence, the manifestation of instability is greatest for waves propagating along the front and, thus, wind gusts in this direction will have the greatest amplitude.

We mentioned above that the loss of stability by atmospheric waves is accompanied by the generation of small-scale turbulence.

We can distinguish "primary" and "secondary" instability zones. The "primary" instability zones include regions where the Richardson number $Ri < Ri_{cr}$. The amplitudes of perturbations occurring in such a layer increase rapidly, and the turbulence arising when the waves break up will occupy the whole instability region. "Secondary" instability zones are formed in the crests of waves with sufficiently large (for breaking) amplitude within the zones where stratification on the average is stable. According to Kogan's theoretical investigations, the formation of turbulent regions in zones of secondary instability occurs most easily in layers with sharp changes in dT/dz. This may account for the increased probability of turbulence under the tropopause, where a jetlike wind profile is usually accompanied by rapid changes in vertical temperature gradients with height.

As theoretical [117] and experimental (mainly radar) [136, 175, 206] investigations have shown, the disintegration of waves occurs in the following way. Their amplitude increases rapidly and the waves "break." This converts the waves to a system of large eddies (rotors) with horizontal axes. The rotors, in turn, gradually break up into smaller eddies with randomly oriented axes. Waves are converted to a system of rotors fairly rapidly (in less than 30 min, according to Kogan and Shakina's estimate [54]), provided that $Ri < \frac{1}{4}$ throughout the flow or at least in the regions close to the crests.

An interesting example of the conversion of wave motions to turbulent motions is given in a paper by Vinnichenko [19], who used a highly sensitive hot-

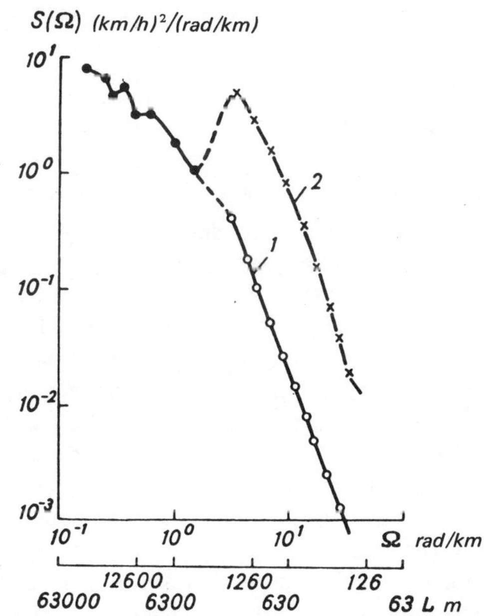

Figure 7.2. Spectra of fluctuations of horizontal wind-velocity component. $H = 5400$ m, $\overline{U} = 99$ km/h. (1) 16 h 46 min 00 sec to 17 h 01 min 30 sec, calm flight; (2) 17 h 01 min 30 sec to 17 h 11 min 00 sec, slight buffeting, wave motion present.

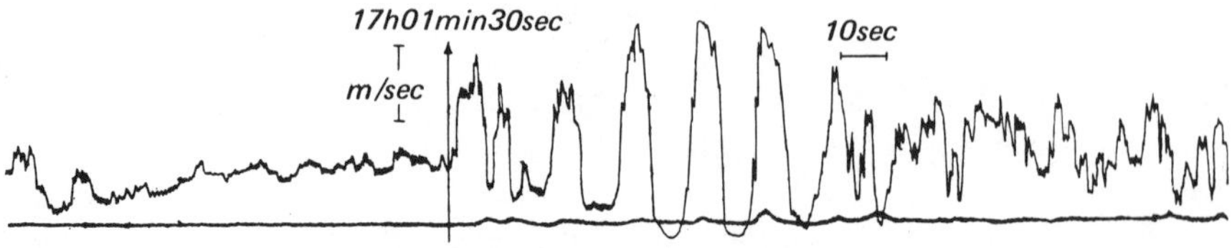

Figure 7.3. Part of trace of aircraft hot-wire anemometer showing wavelike disturbances.

wire anemometer (see Chapter 2), mounted on a research airplane, for the investigations.

The flight took place at an altitude of 5400 m in the region of an almost stationary secondary front with waves under the axis of a jet stream oriented southwest to northeast. The mean wind velocity at the flight altitude was about 28 m/sec. Near the flight altitude $\partial U/\partial z = 0.61$ m/sec per 100 m, and the change in wind direction with altitude was $\partial \varphi/\partial z = 3.4°/100$ m, and $\gamma = 0.63°C/100$ m. In places there was weak buffeting with $|\Delta n| \leqslant 0.2g$.

Figure 7.2 shows the spectra of fluctuations of the horizontal wind-velocity component for a calm and disturbed region. Figure 7.3 shows a part of the hot-wire anemometer trace corresponding to the end of the calm part and the start of the turbulent part of the flight. The figure shows how at a certain instant the waves disappeared and the wind-velocity fluctuations became turbulent. Figure 7.3 shows that at scales close to 2 km the spectral density curve has a sharp local maximum, which probably owes its origin to the disintegration of atmospheric waves.

3. Mountain Waves

When air flows over mountains the structure of the flow is sharply altered. This is caused primarily by the distortion of the streamlines due to the ascent of the air on the windward side of the obstacle and its descent on the lee side. Superimposed on this "ordered" picture there may be a specific mountain-valley circulation, which is one of the forms of convective motion, and also "mountain waves." The latter are formed in particular conditions, usually on the lee side of the ridge. Sometimes (though much more rarely) waves are observed above the ridge and even on the windward side. In addition to these mesoscale motions, whose characteristic scales are a few kilometers to tens of kilometers, in the region of mountains there is usually strongly developed turbulence due to loss of stability of the mesoscale motions and to dynamic deformation of the flow due to its interaction with the underlying surface.

The existence and intensity of all types of disturbances in the wind field in mountain regions depend on the wind velocity, the thermal stratification of the atmosphere, the height and shape of the ridge, its orientation relative to the flow, and so on. A change in just one of those factors can appreciably alter the general nature of the flow over the obstacle.

3.1. Characteristics of Mountain Waves and Their Relation to Atmospheric Conditions. Mountain waves are gravity waves of a special kind formed (Fig. 7.4) when wind blows across, or at a large angle to, an obstacle. The wavelength λ can vary from 2-3 to 40-45 km, but is usually close to 10 km.

Vertical oscillations of the air take place around the equilibrium level of the undisturbed flow. The source of disturbances is the mountain, and the static stability of the atmosphere acts as a restoring force. The presence of wind means that an air particle displaced upward by the mountain not only oscillates in a vertical plane (when stratification is stable), but also moves horizontally and at successive times is located over different points on the ground. This gives rise to an air wave, whose length is determined by the wind velocity (the horizontal distance traversed by the air particle in unit time) and the thermal stability, which determines the period (frequency) of the oscillations.

Because of the presence of lenticular (Sc and Ac lenticularis) and billow (Sc and Ac undulatus) clouds, which are formed in the crests of mountain waves owing to ascent of air in them, mountain waves can sometimes be observed visually.

The amplitudes A of mountain waves are close to the height of the obstacle over which the air flows, and this greatly exceeds the amplitudes typical of internal atmospheric gravity-shear waves. Even above relatively low mountains, such as the Carpathians and Tatra, A is sometimes more than 1 km. Above high ranges (Alps, Rockies, etc.) A can reach 2-3 km [76, 80].

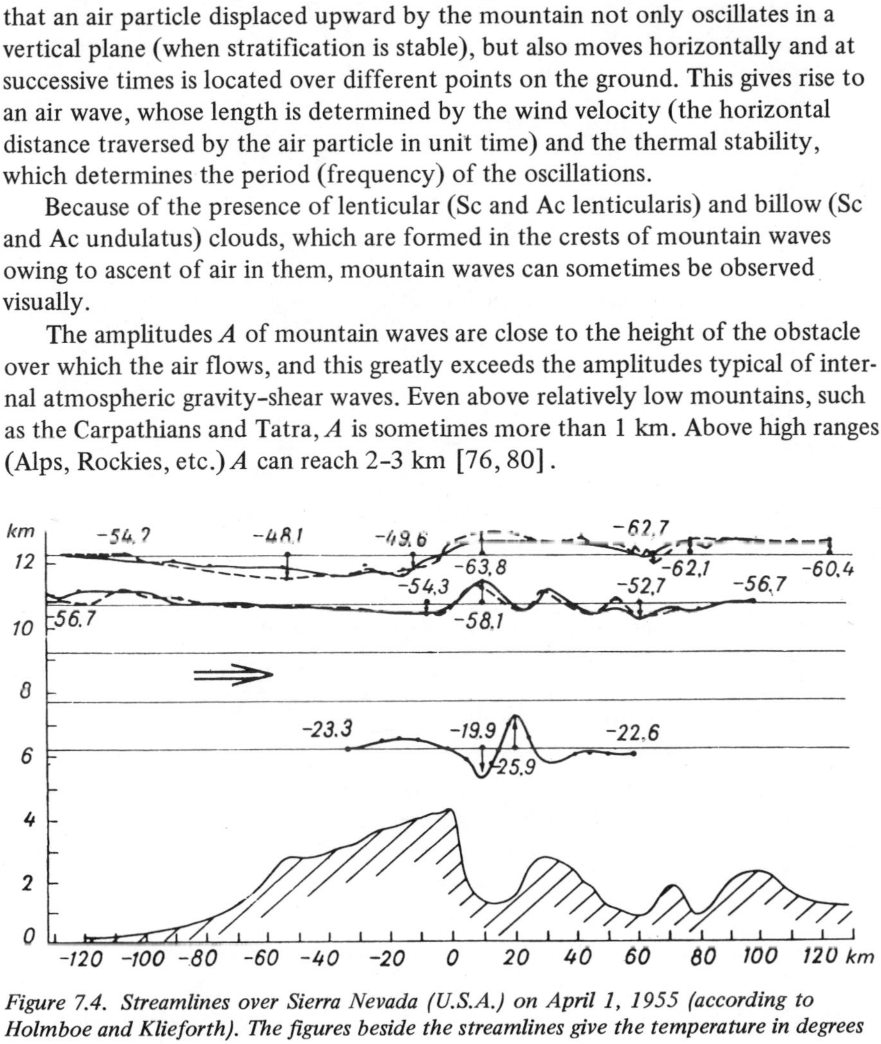

Figure 7.4. Streamlines over Sierra Nevada (U.S.A.) on April 1, 1955 (according to Holmboe and Klieforth). The figures beside the streamlines give the temperature in degrees Celsius.

The vertical velocities w in the crests and troughs of mountain waves reach 10–25 m/sec [76, 80, 185]. Both A and w are usually greatest in the middle troposphere.

The presence of mountain waves leads not only to vertical motions of the air, but also to great horizontal wind-velocity fluctuations with amplitudes sometimes 3–4 times greater than w.

The theory of mountain waves is fairly well developed. Dorodnitsyn, Lira, and others investigated steady flow of an incompressible (baroclinic) ideal fluid over obstacles of different size and shape. The initial system of equations included the equations of motion, continuity, and heat flux. This system of equations was linearized on the assumption of smallness of the perturbations introduced by irregularity of the relief in relation to the main flow. The system was solved by the short-wave method†; the Coriolis force was ignored. According to Dorodnitsyn, the linear theory is suitable if the height of the mountain is less than the characteristic length $\xi = (U/g)\,[T/(\gamma_a - \gamma)]^{1/2}$, which corresponds to the "natural frequency" of the flow. In real conditions $\xi \leqslant 2$ km usually. Experimental data, however, confirm the main deductions from linear theory even for higher obstacles.

The effect of baroclinicity of the atmosphere on the characteristics of mountain waves is that λ and A depend on the Dorodnitsyn parameter

$$D^2 = \frac{g}{U^2 T}(\gamma_a - \gamma), \tag{7.19}$$

given that

$$\lambda_i = \frac{2\pi}{\sqrt{D^2 - \dfrac{n_i \pi^2}{H^2}}}, \tag{7.20}$$

where H is the thickness of the flow, and n_i is the number of waves formed behind the ridge. The quantity n_i depends on the wind velocity and the stability of stratification. Usually $n_i = 1, 2, 3$, i.e., a system of three waves of different length is formed behind the ridge.

The wave amplitudes increase with reduction of γ, i.e., with increase in D^2. In addition, A decreases rapidly with increase in the wave number i. Hence, the qualitative flow picture is almost completely determined by the first ($n_i = 1$) wave system, and the remaining waves affect only the quantitative characteristics.

It follows from Eq. (7.20) that waves can exist only when $D^2 > 0$, i.e., $\gamma < \gamma_a$ and, hence, the stratification of the atmosphere is stable. We will show below that this conclusion is correct only in the case where the wind velocity or gradient does not vary with altitude.

†Kibel' investigated the properties of air flows over mountains by the long-wave method, which reduces the problem to an ordinary (though nonlinear) differential equation.

The investigations of Dorodnitsyn, Lira, Scorer, and others revealed two extremely important properties, subsequently confirmed by experimental investigations, of flows over mountains–inversions of streamlines and the presence of a distinctive resonance.

The first of these properties consists in the fact that the streamlines are not parallel to one another at all heights.† Above and below the "nodal surfaces" (inversion levels) the streamlines are mirror images of one another, with the result that above regions of ascending currents there are regions of descending motions, and vice versa (Fig. 7.5). A consequence of streamline inversion is that the locus of the wave troughs and crests is not vertical, but inclined in, or against, the flow direction, which can be seen if multilayered lenticular clouds are observed.

According to [76], the vertical air velocity above the mountain can be represented by the equation

$$w(z) = \bar{U}\left(1 - \frac{\gamma}{T_0} z\right)^{\frac{\varkappa R(\gamma_a - \gamma) + g}{2\varkappa R\gamma}} \frac{\sin D(H - z)}{\sin DH} \frac{d\zeta}{dx}, \tag{7.21}$$

where $\bar{U}$ is the mean wind velocity, $z = \zeta(x)$ is the equation of the surface of the obstacle in the flow, $\kappa = c_p/c_v = 1.41$ (c_p and c_v are the specific heats at constant pressure and volume), and H is the altitude of the tropopause, at which w is assumed to be zero.

It is obvious that where $\sin D(H-z) > 0$, the signs of the vertical velocity $w(z)$ and $d\zeta/dx$ are the same, i.e., the streamlines reproduce the shape of the mountain profile. Above the level where $\sin D(H-z) = 0$, i.e., above the nodal surface, $\sin D(H-z) < 0$, and, hence, there is inversion of the streamlines, which acquire a configuration opposite to that which they had under the surface $w = 0$. Within the troposphere, $\sin d(H-z)$ can have several zeros, i.e., here there can be several nodal surfaces, given by the equation

$$z_{0n} = H - \frac{n\pi}{D}, \tag{7.22}$$

where $n_i = 0, 1, 2, \ldots$. It is apparent that $z_{0n} = f(D)$.

The thickness of the layers between adjacent nodal surfaces, i.e., zones of ascending and descending streams, is given by the relation

$$\delta = \pi U \sqrt{\frac{\bar{T}}{g(\gamma_a - \gamma)}}. \tag{7.23}$$

For typical conditions in the troposphere $\delta = 2$–4 km.

†Inversion of streamlines occurs both in the wave region and in the mean flow, corresponding to the "main" picture of flow over the obstacle.

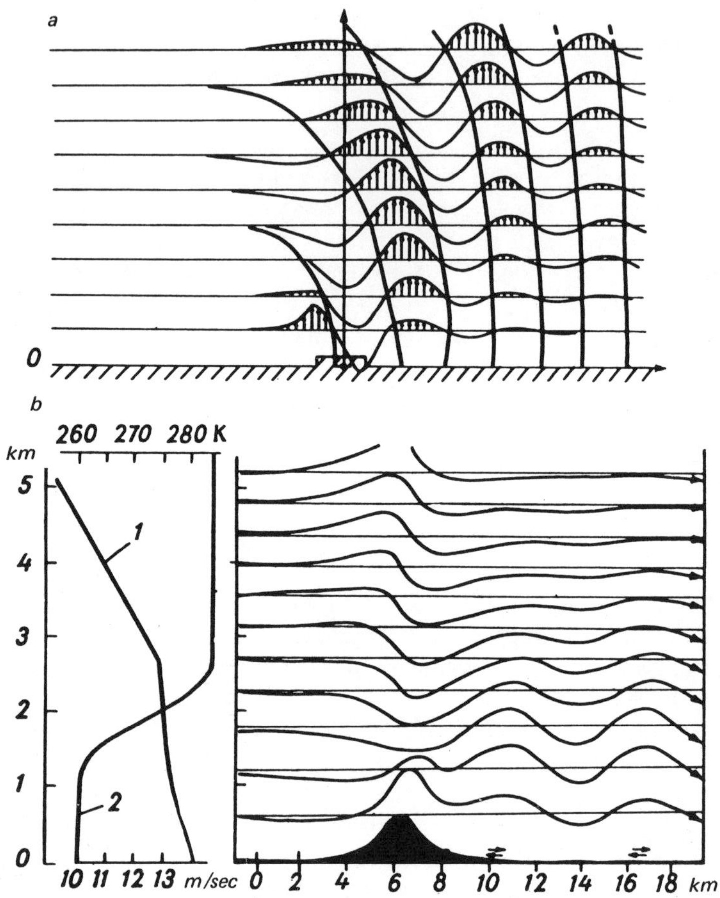

Figure 7.5. Diagram of streamlines in mountain waves. (a) According to Lira; (b) according to Scorer. (1) Variation of temperature with altitude; (2) variation of wind velocity with altitude.

If the stratification of the atmosphere is not neutral ($\gamma \neq \gamma_a$), the vertical streams generated by the mountains cause temperature changes ΔT. In the case of adiabatic ascent and descent of the air

$$\Delta T \equiv T - T_\infty = -\gamma_a \zeta(x, z), \tag{7.24}$$

where ζ is the vertical displacment of the air (streamlines).

Since thermal stratification is usually stable ($\gamma < \gamma_a$) in the region of mountain waves, the sign of the temperature disturbances is the opposite of that of w, i.e., the ascent of air ($w > 0$) is accompanied by reduction of temperature ($\Delta T < 0$) and vice versa. If instead of the ordinary temperature we use the values of the potential temperature θ, then, knowing the spatial distribution $\theta(x, z)$, we can find the distribution $w(x, z)$ approximately if we use the condition of conservation of the potential temperature when the air is displaced,

$$u\frac{\partial\theta}{\partial x} + w\frac{\partial\theta}{\partial z} = 0, \tag{7.25}$$

according to which

$$w = -u\frac{\dfrac{\partial\theta}{\partial x}}{\dfrac{\partial\theta}{\partial z}}. \tag{7.26}$$

Measurements of T (and, hence, determination of θ) can be made much more accurately and simply than those of w, with the result that in several papers (particularly American papers) potential temperature isolines (isentropes) (Fig. 7.6) are given instead of the streamlines in the wave region.

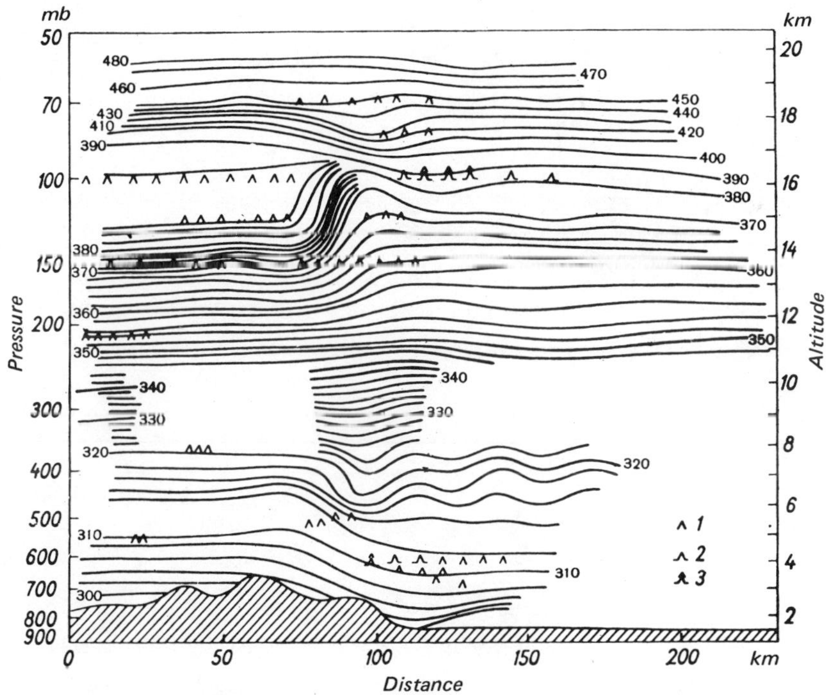

Figure 7.6. Isolines of potential temperature (°K) above Rockies (U.S.A.) on February 17, 1970. (1) Light buffeting; (2) moderate buffeting; (3) severe buffeting. The vertical profile of the terrain is hatched.

The resonance effect consists in the fact that if the wavelength is close to the width of the mountain, the amplitude of the vertical motions, other conditions being equal, is a maximum and can even exceed the height of the obstacle. Hence, large-amplitude waves can be observed even above mountains that are not very high.

One of the main simplifications introduced by Dorodnitsyn in dealing with problems of mountain waves was the assumption of constancy of the wind velocity at different altitudes. The role of vertical wind shear was clarified by Scorer [210].

Although the problem of air flow over mountains was later solved in a much more complete form, we will discuss only Scorer's results, since they show particularly clearly the relation between the wave characteristics and the thermodynamic conditions in the atmosphere.

Scorer considered laminar isentropic flow over an obstacle given by the formula

$$\zeta = \frac{hb^2}{b^2 + x^2}, \tag{7.27}$$

where h is the height of the ridge, $2b$ is its width, and x is the distance from the ridge to the given point. The effect of the Earth's rotation was ignored.

After a number of simplifications the following equation was obtained for the stream function ψ:

$$\frac{\partial^2\psi}{\partial z^2} - \left(\frac{g}{c_s^2} + \beta\right)\frac{\partial\psi}{\partial z} + \left(\frac{g\beta}{U^2} - k^2 - \frac{1}{U}\frac{\partial^2 U}{\partial z^2}\right)\psi = 0, \tag{7.28}$$

where c_s is the velocity of sound, $\beta = (1/T)(\gamma_a - \gamma)$, and $k = 2\pi/\lambda$ is the wave number in the expansion of function (7.27) in a Fourier series. The quantity

$$l^2 \equiv \frac{g\beta}{U^2} - \frac{1}{U}\frac{\partial^2 U}{\partial z^2} \tag{7.29}$$

is called the Scorer parameter. Since†

$$\frac{1}{U}\frac{\partial^2 U}{\partial z^2} \ll \frac{g\beta}{U^2},$$

$l^2 \approx D^2$ usually.

Neglecting in (7.28) the terms with $\partial\psi/\partial z$, which reflects the effect of compressibility of the air on the amplitude and becomes significant only at high

†According to [11], in the boundary layer the terms in (7.29) are sometimes close in magnitude. Similar data for jet streams are given in [168].

altitudes, we obtain

$$\frac{\partial\psi}{\partial z} + (l^2 - k^2)\psi = 0. \tag{7.30}$$

To discover the effect of changes of U with z on the wave characteristics, Scorer considered the case where the flow over the ridge consisted of two layers in each of which l_i^2 = const, but $l_1^2 \neq l_2^2$ (the subscripts 1 and 2 refer to the lower and upper layers, respectively).

The solution showed that for the existence of waves it is essential that the parameter l^2 is positive and decreases with altitude. The latter is usually caused by increase in U with z, since the range of variation of β in the atmosphere is much narrower than the range of variation of U.

It follows from Eq. (7.29) that in a flow where $\partial^2 U/\partial z^2 \neq 0$, waves are possible in principle and, if stratification is unstable, provided that

$$\frac{\partial^2 U}{\partial z^2} < 0,$$

and

$$\frac{1}{U}\left|\frac{\partial^2 U}{\partial z^2}\right| > g\left|\frac{\gamma_a - \gamma}{TU^2}\right|.$$

Empirical data confirmed the theoretical conclusions regarding the role of thermal stratification in wave formation. Thus, it was found that waves are usually formed when stratification is stable, especially if there are layers with an isothermal or inversion temperature distribution above the mountain. Sometimes, however, waves were observed in the lower part of the troposphere when stratification was stable. Their amplitudes were usually very small, but their length was fairly large.

According to Scorer, the wavelength λ lies between the maximum and minimum values of $2\pi/l$. As Eq. (7.29) shows, this means that λ is directly proportional to the wind velocity and inversely proportional to the stability of stratification. From Scorer's theory it is easy to obtain an equation for estimation of the length (in kilometers) of a mountain wave:

$$\lambda = 6.3U\sqrt{\frac{T}{10(\gamma_a - \gamma)}} \cdot 10^2. \tag{7.31}$$

Corby and Wallington showed that the deflection ζ_z of the streamlines at altitude z from their equilibrium position in the undisturbed part of the flow is given by the equation

$$\zeta_z = -2\pi hbe^{-kb}\left(\frac{U_1}{U_z}\right)\psi_{z,k}\left(\frac{\partial\psi_{1,k}}{\partial k}\right)\sin kx, \tag{7.32}$$

where U_1 and U_z are the wind velocities at the ground and at height z; $\psi_{1,k}$ and $\psi_{z,k}$ are the stream functions for the corresponding level satisfying Eq. (7.30). The maximum vertical velocity w_{max} in the wave corresponds to the point of inflection of ζ_z:

$$w_{\max} = \left[U_z \frac{\partial \zeta_z}{\partial x}\right]_{\max} = -2\pi hbe^{-kb}U_1 a_n k, \tag{7.33}$$

where

$$a_n = \psi_{z,k}\left(\frac{\partial \psi_{1,k}}{\partial k}\right)^{-1}.$$

Since ψ and k are functions of l^2, the wave amplitudes depend on the vertical wind distribution in the atmosphere, i.e., on the wind velocity at different altitudes. For the same profile of l^2 the greatest amplitudes will occur in the cases where U_1 is greatest, i.e., strong winds on the ground promote the formation of intense waves. Empirical data show that waves are usually formed only in cases where the wind velocity at the ridge level exceeds some critical value $U_{1,cr}$. Above not very high mountains (mountains in England, Crimean mountains, French Alps, etc.) $U_{1,cr} \approx 8$–10 m/sec. Above higher ridges, such as the Cordilleras, for instance, $U_{1,cr} \geqslant 12$ m/sec.

The factor hbe^{-kb} represents the effect of height and width of the ridge on the amplitude of the displacements of the streamlines. The higher the mountain, i.e., the greater h, the greater the amplitude. In addition, be^{-kb} is a maximum when $b = k^{-1}$.

The described characteristics of mountain waves were obtained for the case where the motion can be regarded as two-dimensional. This assumption, strictly speaking, is valid only for waves developing behind long ridges of fairly uniform height. If the width of the ridge is not very different from its height, three-dimensional waves have to be considered. The theory of these waves was developed in [76, 81, 82] and elsewhere. It was found that near short ridges and, in particular, near isolated mountains, waves will be formed not only in a vertical, but also in a horizontal, plane extending sideways from the mountain and downstream. The horizontal-wave field also contains nodal surfaces on which the amplitude of the wind disturbances is zero. In the case of three-dimensional flow, as distinct from two-dimensional flow, waves are formed not only along the flow, but also across it (particularly in the immediate vicinity of the mountain).

The specific nature of three-dimensional flow over an obstacle is a consequence of the fact that in this case the impinging stream flows partially over the obstacle, and partially round it. As the numerical calculations of Pekelis [81, 82] showed, when the flow is split the disturbances behind the mountain interfere, intensifying and reducing each other in different places. The partial flow around the sides of the mountain, i.e., the reduction of the air mass piling up over it, leads to a reduction of the amplitudes of the waves and their rapid decay downstream. This process is intensified with increase in l^2. The greatest amplitudes will be observed over the lee slope. They increase with increase in height of the obstacle, and more rapidly than the latter.

Empirical data have shown that near not very high mountains the frequency of occurrence and the amplitudes of mountain waves are subject to diurnal and annual variation, mainly due to the corresponding changes in the parameter l^2 and its vertical gradient. Maximum amplitudes are usually observed in the morning and evening. The daily minimum is due mainly to the fact that at this time the stability of stratification near the surface of a mountain decreases so rapidly that, despite the wind intensification that occurs here at this time, l^2 decreases sharply. At the same time $\partial l^2/\partial z$ decreases, so that l^2 sometimes begins even to increase with altitude and, hence, waves cannot exist at all. The night minimum of occurrence of waves is due to sharp reduction of the wind at the ground at this time, with the result that its velocity sometimes drops below the critical level.

Above high mountains waves do not usually show a distinct diurnal variation. The variation of their intensity is mainly due to processes of synoptic scale. The same processes are mainly responsible for the annual variation of the occurrence of mountain waves over mountains of any height.

According to Trubnikov's theoretical estimates [108], the diurnal variation of the difference between the temperatures of the mountain surface and the surrounding air must also have some effect on the diurnal course of mountain waves. If the mountain is warmer than the surrounding air, which is typical of rocky summits in daytime, the wave amplitude is reduced. Above rocky summits at night, and above snow-covered summits by day, on the other hand, the waves are amplified because the mountain at this time is colder than the surrounding air. These features are confirmed by the diurnal behavior of lenticular clouds located in the wave crests.

The spatial extent of the wave zone can be characterized by the following data. Downstream behind the ridge the wave amplitudes rapidly decrease: Theoretical estimates and experimental data both show that the amplitudes of the third or fourth wave are negligibly small. The height of the propagation of waves even over low mountains is very great. For instance, in the Northern Alps, gliders rose in mountain waves to altitudes of 9–10 km, and over the Sierra Nevada (U.S.A.) they rose to 13–14 km.

Mountain waves are observed not only in the troposphere, but also in the stratosphere. In Colorado in 1970 a series of flights by six research airplanes and a glider were made in wave regions. In the region of the flights the mean height of the top of the ridge above its base was about 2 km. During the research flights the wind at the 500-mb level ($H \approx 5.5$ km) blew perpendicular to the ridge. In six cases waves were observed up to the level of the tropopause (altitude about 12 km) and in one case reached an altitude of 15.5 km. The maximum height of occurrence of mountain waves was less than 8 km in only four cases. Investigations carried out in the same region in the winter of 1968 [185] gave similar results.

The cited heights are not the greatest. For instance, over Scandinavia and Alaska, where the height of the mountains does not exceed 2.5 km, mother-of-pearl clouds reminiscent of mountain clouds in shape have been observed at altitudes of more than 22 km.

A theoretical estimate of the altitude at which waves die out is extremely difficult, since its results depend strongly on the adopted boundary conditions on the $0z$ axis, the method of correction for the presence of the stratosphere, and so on.

Palm [196], Trubnikov [109], and others have shown that propagation of gravity waves in the region of atmospheric interfaces conforms approximately to the laws of geometric optics. In particular, for certain values of l^2 on the two sides of the interface (the tropopause, for instance) waves can be partially reflected from the latter.

The "transmissivity" of the tropopause is

$$\mu = \frac{2\cos\vartheta}{\cos\vartheta + \sqrt{\left(\frac{l_2}{l_1}\right)^2 - \sin^2\vartheta}}, \tag{7.34}$$

where ϑ is the angle of incidence of the wave on the tropopause, and the subscripts 1 and 2 refer to the troposphere and stratosphere, respectively. The quantity $(l_2/l_1)^2$ is the "refractive index" of the tropopause. In accordance with (7.34) the vertical wind velocity above the tropopause can be estimated by means of the relation

$$w_2 = A\mu e^{ixl_1 \sin\vartheta + iz\sqrt{l_2^2 - l_1^2 \sin^2\vartheta}}, \tag{7.35}$$

where z is measured from the tropopause. Calculations show that in typical conditions 75% of the energy of gravity waves incident perpendicularly on the tropopause passes through it. Thus, although mountain waves are weakened by passage into the stratosphere, the tropopause is not absolutely opaque for them.

4. Rotor Motions and Turbulence in Mountain-Wave Region

4.1. Rotors and Their Formation. Mountain wave can be accompanied by turbulence, which is sometimes even very intense. It is characteristic that strong

Figure 7.7. Diagram illustrating onset of eddies due to "breaking" of unstable mountain waves.

turbulence is sometimes observed above a wave region (Fig. 7.6). In the case illustrated in Fig. 7.6 the lee region of the lower stratosphere at altitudes of 14–17 km was turbulent; near the 16-km level buffeting was very strong. Very large horizontal temperature and wind gradients were characteristic of the buffeting zone.

In the lower part of the wave region in many cases there are "rotors," which are usually accompanied by extremely strong horizontal and vertical air gusts. Rotors are quasi-ordered cylindrical eddies with an approximately horizontal axis. They are observed even when the main wave flow is practically laminar. The largest rotors (with diameter up to 0.5–1.0 km) are formed on the lee slope of the ridge.† Rotors periodically break away and, being carried along with the flow, produce turbulence in the air far from the mountains. They are replaced by new rotors. Near the "main" rotors there are usually large (though smaller than the rotors themselves) eddies, whose axis of rotation is not necessarily horizontal.

It should be noted that lee rotors are not always formed at the ground. For instance, above the Rockies (U.S.A.) there were cases where, in the lower layer 2–2.5 km thick, the isentropes and, hence, the flow followed the profile of the terrain, i.e., there were neither wave motions nor rotors. Above the described layer there were well-developed waves with $\lambda \approx 10$ km.

The reasons for the formation of lee rotors have not been conclusively established. According to one hypothesis they are formed from waves by the action of a hydraulic discontinuity close to the ground, similar to the shock wave in supersonic flow.

Another mechanism of rotor formation is that postulated by Long, who proposed that rotors are formed if

$$\frac{\partial \zeta_z}{\partial z} > 1 \quad \text{or} \quad \frac{\partial \zeta_z}{\partial z} < -1, \tag{7.36}$$

i.e., the slope of the streamlines becomes very large and is reversed [210] (Fig. 7.7). This leads to loss of stability and the waves break up and form large eddies. Conditions favoring the fulfillment of criterion (7.36) are usually found near wave crests and troughs, and usually near the underlying surface. In particular cases, as Fig. 7.8 shows, rotors are formed in the free atmosphere.

†According to Förchtgott, the main stream flows over such rotors like a solid obstacle. The "effective" width of the mountain in this case is naturally increased.

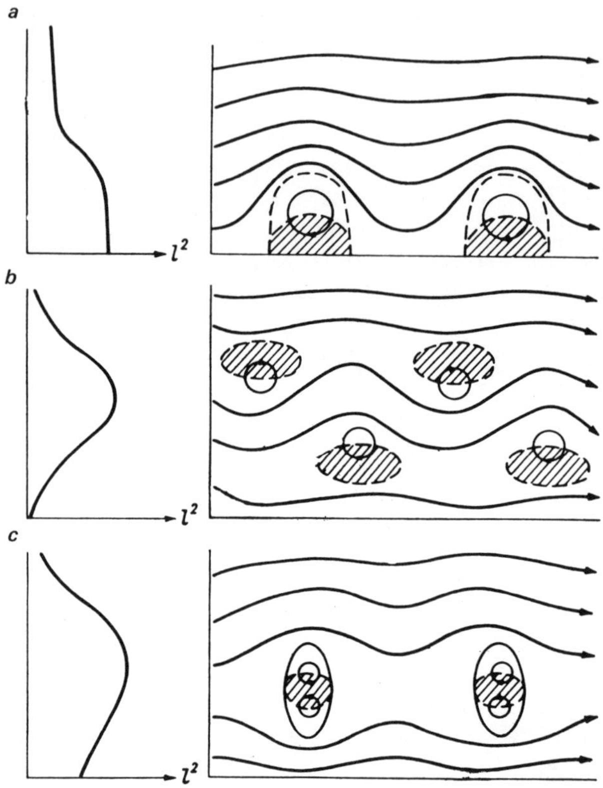

Figure 7.8. Relation between rotor zones and vertical profile of l^2.

The hatched areas in Fig. 7.8 correspond to the zone of statically unstable stratification of the air. It is apparent that if the rotors are formed not at the earth's surface or a nodal surface, the region of static instability is situated outside the closed rotor circulation. Unstably stratified air can be carried downstream and cause turbulence of the flow.

Above a rotor zone near the ground the turbulence is usually reduced, although here, as Fig. 7.6, for instance, shows, it may be considerable. For instance, in 20 out of 66 flights in a wave region, conducted in England by Pilsbury, turbulence was recorded. It was probably due to wind shears caused by wave motions in a horizontal plane, superposed on the undisturbed wind profile.

From the results of observations above the French Alps Berenger and Gervier established that if the wave region at some altitude contains a thin layer in which the wind is much weaker, or the direction of flow is reversed, then the waves within it break up into large eddies, and this layer becomes the upper boundary of wave propagation. Turbulence also arises in layers where the wind-velocity component normal to the ridge increases sharply. Figure 7.9 shows diagrams of lee waves for various vertical wind profiles.

A comparison of Figs. 7.9 and 7.8 shows that the experimental data agree quite well with the predictions of theory.

Wave motion is often converted to turbulence very rapidly. This change is due to loss of stability by the waves, as is indirectly indicated by the fact that it is usually observed in those cases where the waves are very short (usually $\lambda < 2$ km), i.e., least stable.

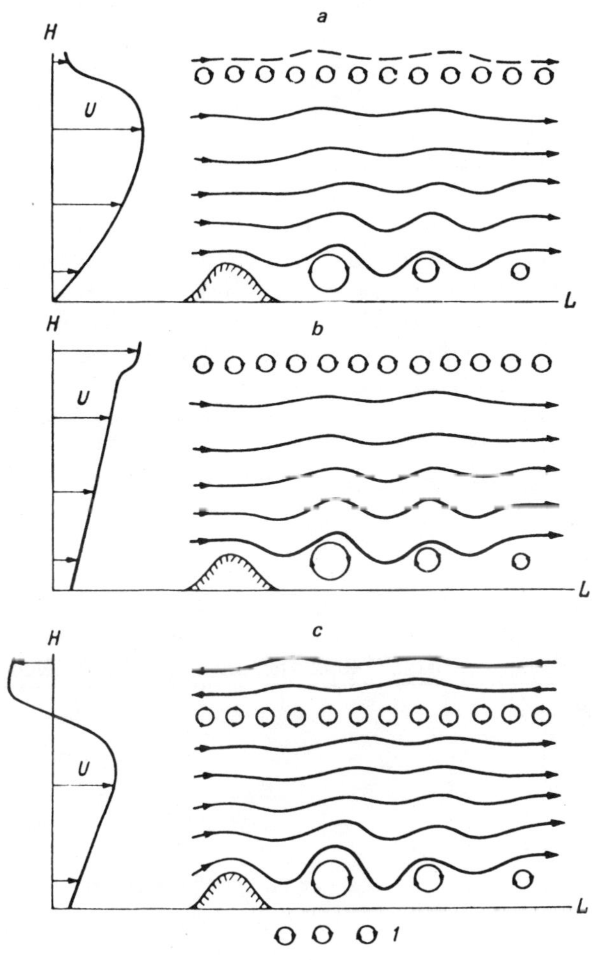

Figure 7.9. Lee waves and turbulence. (1) Disturbed layer. The vertical wind profiles are shown on the left.

The generally known local intensification of turbulence near the tropopause in mountain regions can be attributed (at least partially) to reflection of the waves from the tropopause. The interference of the direct wave with the reflected wave can lead to a sharp increase in the amplitude of the oscillations, and in some cases to loss of stability of the "resultant" wave.

4.2. Turbulence on the Lee Side of Mountains. We have discussed turbulence due to loss of stability by mountain waves. Turbulence above mountains, however, is much more often due to other factors. The main one is the formation of eddies due to dynamic interaction of the flow with the relief (and, in summer, to thermal turbulence). The intensity of dynamic turbulence increases with increase in wind and increase in the steepness of the slopes. This is because these two factors promote convergence of wind (increase in U) above a ridge, reducing Ri and thus increasing the probability of the formation of turbulence in the flow incident on the ridge. Even over such a relatively flat ridge, as the Suram Range, the vertical wind shears arising in the convergence layer can exceed 5 m/sec per 100 m, according to Khrgian's data.

If the flow is incident not on a single mountain, but on a barrier consisting of a large number of mountains, the latter behave like a turbulence-inducing grid. The generation of turbulent energy in this case occurs mainly in the layer between some middle level and the summit, and turbulence higher up is due to upward diffusion of the eddies. When stratification is stable, the rising turbulent energy is reduced by work against buoyancy forces and dissipation as heat.

On the basis of dimensional considerations, Laikhtman and Byutner [60] found that at an altitude z above not very high mountains the variance of the gustiness of the flow is

$$\bar{U}'^2 = U^2 \Phi\left(\frac{z}{L}, \frac{L}{\lambda^*}\right), \tag{7.37}$$

where $\Phi(z/L, L/\lambda^*)$ is a function that can be determined experimentally, L is the characteristic scale of the eddies, which depends on the size of the mountain, and λ^* is a parameter that has the dimension of length and is given by

$$\lambda^* = \frac{U}{R\dfrac{g}{\rho c_p T}}. \tag{7.38}$$

Here R is the radiation balance.

Equation (7.37) shows that at a given altitude the turbulence intensity depends on the wind velocity, the scale of the eddies, and the radiation balance. The dependence on the latter is due to the relation between the radiation balance and atmospheric stratification and, hence, the intensity of thermal turbulence, whose contribution is particularly large when winds are weak. The lower the thermal stability, the greater the altitude to which eddies can rise by upward diffusion.

The thickness of zones of orographic turbulence can exceed the height of the obstacle by a factor of 3–4, and the horizontal dimensions of the turbulent regions even behind mountains 1–2 km high reach tens of kilometers. This is because wind transfer of eddies formed close to the ridge can be accompanied by loss of stability of the flow at a considerable distance from the mountain mass. If the flow is situated on a surface of unstable equilibrium, which occurs particularly often on hot summer days, such eddies begin to act as the "initial turbulence" leading to turbulent motions that can no longer be regarded as entirely orographic.

Some idea of the distance at which the disturbing effect of high mountains affects air flows can be obtained from data on aircraft buffeting over Erevan. They show that buffeting is usually encountered at altitudes up to 4–5 km when strong southwest winds blow. This is presumably due to the deformation of the flow above the mountains Great Ararat (5156 m) and Little Ararat (3914 m), which are about 50 km SSW of the town.

Figure 7.10 illustrates the typical structure of a turbulent zone above mountains. A region of very intense turbulence with $w \geqslant 5$ m/sec is clearly seen above the lee slope. Above it and to the sides of it the values of w gradually diminish. The "calm" interlayers observed in turbulent zones over level ground are often absent over mountains.

Regions of orographic turbulence extend in the direction of the wind and lie mainly over the lee side of the mountain, where a distinctive "turbulent banner," similar to the widely known "cloud banners," is formed.

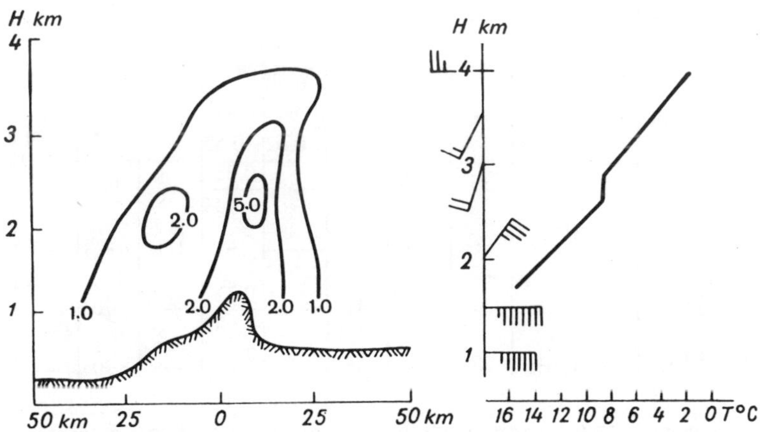

Figure 7.10. Diagram of turbulent zone above Suram Pass, September 3, 1955. Isolines of w in m/sec. The temperature and wind profiles are shown on the right.

Turbulent zones of the structure described are characteristic of isolated mountains or parts of a ridge where the height and shape do not vary greatly. If the structure of the mountain system is more complex, the disturbances produced by individual peaks are superposed on one another. In this case the thickness of the turbulent zone and the turbulence intensity are usually increased.

Interesting data on the structure of turbulence above the Black Sea coast on the lee side of the Caucasus during the "Novorossiskii Bora" (wind) between October 31 and November 2, 1963 were given by Reschikova [102]. When the wind was most intense the entire layer from sea level to an altitude of 1500 m was occupied by turbulence. A comparison of the characteristics of the turbulent zones with the profile of the Caucasian Range in the region of the flights (the Gelendzhik-Bett portion) showed that only a small part of the turbulent zone was situated above the ridge, and that the upper boundary of the zone was located at the level of the summit or lower. Table 7.2 gives turbulence characteristics from the results of measurements on October 31 and November 1. An interesting features is the small variation of the turbulence coefficient K not only with altitude, but also in the horizontal direction. For instance, the maximum values of K for the whole flight region hardly differ from the mean value $\overline{K}$. Only once (October 31, 1963), at $H = 400$ m, was $K_{max} \approx 2\overline{K}$. A possible

TABLE 7.2. Data on Turbulence Associated with Novorossiiskii Bora in Gelendzhik-Bett Region [a, b]

Height above sea level (m)	Relative extent of calm regions (%)		Mean turbulence coefficient (m²/sec)		Maximum turbulence coefficient (m²/sec)		Maximum load in g		Density of load with intensity			
									$>0.2g$		$>0.5g$	
	10/31	11/1	10/31	11/1	10/31	11/1	10/31	11/1	10/31	11/1	10/31	11/1
400	31	—	61	—	128	—	0,48 / −0.60	—	—	—	1	—
750–800	10	7	51	51	87	67	0.64 / −0.62	0.48 / −0.52	400	582	5	1
900	—	25	—	36	—	56	—	0.38 / −0.44	—	380	—	—
1000	35	—	49	—	69	—	0,38 / −0.35	—	211	—	—	—
1230	—	38	—	30	—	41	—	0.31 / −0.38	—	70	—	—
1500	100	52	—	28	—	36	—	0.24 / −0.25	—	30	—	—

[a] A load increment $\Delta n = 0.1g$ corresponds to $w = 1$ m/sec. The load density is the number of peak loads per hour of flight.

[b] 10/31 = Oct. 31; 11/1 = Nov. 1.

reason for this is that reduction of Δn by a factor of several units increases the linear dimensions of the gusts by a similar factor.

We now discuss the intensity of orographic turbulence in relation to general thermodynamic conditions in the atmosphere. Such investigations have mainly been conducted near mountain ridges whose height did not exceed 2-3 km [11, 27, 28, etc.].

Although there were some specific features in the development of turbulence above different mountain systems, some features common to all the geographic regions were discovered. It was found that movements over a ridge, and particularly over its lee slope, depend on the characteristics of the wind and temperature fields in the undisturbed (weather-side) part of the flow. These characteristics include the wind-velocity component perpendicular to the ridge at the level of the ridge top (U_H), and the vertical gradients of temperature $(\partial T/\partial z)_H$ and wind $(\partial U_H/\partial z)_H$ in the layer from the top to a level several hundred meters higher.

When $\partial T/\partial z < \gamma_a$ and the wind is weak (for most of the investigated regions, when $U_H \leqslant 5$ m/sec), there is little turbulence and the streamlines over the mountain are approximately parallel to the profile of the obstacle; the amplitude of their deviations from the horizontal position decreases rapidly with altitude.

When the wind is moderate or strong (usually when $U_H > 10$ m/sec) the nature of the flow depends greatly on $(\partial T/\partial z)_H$ and $(\partial U_H/\partial z)_H$.

If $(\partial T/\partial z)_H < \gamma_a$ and $(\partial U_H/\partial z)_H \geqslant 0$,† lee waves are formed above and behind the ridge, and above them there is a system of rotors, in whose zone turbulence is very strong.

When $(\partial U_H/\partial z)_H < 0$ the flow becomes unstable and turbulence develops, becoming stronger approximately parallel with the increase in U_H. The thickness of the turbulent layer can greatly exceed the ridge height, especially if $\gamma > \gamma_a$ above it. In the case of strong winds, however, orographic turbulence is also associated with stable thermal stratification and even with the presence of isothermal and inversion layers above the mountains.

Turbulence of orographic origin can be intensified by thermal turbulence. The latter may sometimes even predominate above rocky and tree-covered mountains in summer, when wind velocities are low. Thermal turbulence arises in the morning and increases till midday, gradually occupying a thicker and thicker layer of the atmosphere. Its development begins above mountain rocks facing the sun, and takes place somewhat later over flat ground.

The upper boundary of thermal turbulence is elevated above summits and is lower above valleys. Under the action of the wind the zones of thermal turbulence formed above mountains move a little in the direction of the flow. The evening decay of turbulence begins at the ground. This is accompanied by the

†When $\partial U/\partial z > 0$ the wind velocity increases with altitude.

appearance of layers of sharply reduced turbulence at certain altitudes. They usually occur where temperature gradients are low.

In conclusion we note that flow over mountains is the main factor causing turbulence in the lower stratosphere. For instance, measurements on the high-altitude U-2 probing airplanes, carried out in 1964–1971 over Europe, North America, and Australia, showed that in the lower half of the stratosphere the total probability of buffeting is 3.6 times greater above mountains than above oceans and plains. The great majority (90%) of cases of severe buffeting at altitudes $H > 11$ km were recorded above mountain ridges whose tops were more than 2 km above the surrounding terrain.

8

Turbulence and Vertical Flows in Clouds

Turbulence is usually more intense in all types of clouds than in the surrounding air. This is mainly due to the fact that in clouds, particularly cumuliform clouds, thermal stratification is often neutral or even unstable. Turbulence is also appreciably intensified by the pronounced horizontal inhomogeneity of the air temperature and density fields in clouds.

Turbulence is intensified at the upper boundary of clouds by the sharp local increase in vertical temperature gradients resulting from cooling of the air due to evaporation of cloud elements, and also by radiation at night. In addition, in this layer there is sometimes strong rotation of the wind, which increases the dynamic instability of the flow.

Finally, it should be borne in mind that turbulence itself is a cloud-forming factor and that clouds are usually formed in regions where the general "turbulent background" is increased.

1. Turbulence in Stratiform Clouds

The intensity of turbulence in stratiform clouds has so far been determined mainly from data on the intensity of aircraft buffeting. Table 8.1 sums up the data on the occurrence of buffeting of high-speed airplanes in different clouds. Buffeting in Ci and Cs is approximately 10 times more frequent than in cloudless sky at the same altitudes. In flights near the upper boundary of Cs, buffeting (usually light) is encountered 65–70% of the time.

The maximum velocities of vertical air gusts in stratiform clouds do not exceed 5 m/sec in 90% of the cases. It is only in clouds over mountains and in intense jet streams that individual gusts with velocity up to 15–20 m/sec may be encountered. Turbulence is most intense near the boundaries of the cloud layer, particularly if these boundaries are irregular. The intensity of turbulence within the cloud layer depends on the cloud structure. In clouds of uniform density,

TABLE 8.1. Frequency (%) of Aircraft Buffeting in Clouds of Various Types

Type of cloud				Unspecified type
St fr., St, Sc	Ns – As	Ac	Ci, Cs, Cc	
34	30	29	34	40

turbulence is usually light. If the clouds are inhomogeneous, i.e., if dense regions alternate with less dense regions, the intensity of turbulence can be considerable. In very rarefied clouds the turbulence is usually the same as in the surrounding clear air.

No systematic data on the frequency of gusts of different size in stratiform clouds in the lower and middle levels have yet been published. Table 8.2 gives such data for clouds of the upper level, obtained by Shmeter in aircraft investigations in 1959–1964.

The length of the regions within which $w > 4$/sec does not exceed 10 km. Regions with $w \leqslant 2$ m/sec can be up to several tens of kilometers long. The size of the disturbed zones is sometimes close to that of the whole cloud field, but this usually occurs only near the upper boundary of strongly undulating Cs.

Table 8.3 gives data on the relative extent of disturbed zones in clouds.

1.1. Coefficient of Turbulent Exchange in Stratiform Clouds. In the investigation of turbulence in stratiform clouds of the lower layer, great attention has been given to determination of the turbulence coefficient K in them, since its value is required for theoretical calculation of cloud formation.

TABLE 8.2. Velocity w of Vertical Gusts in Turbulent Zones within Ci and Cs

w (m/sec)	< 1	1–2	2.1–4	4.1–6	> 6
Frequency of occurrence, %	42	38	16	3	1

TABLE 8.3. Relative Extent of Buffeting Regions within Clouds

Type of cloud	Total flight length in clouds (km)	Flight length with buffeting (km/%)	Greatest loads in g
Ac, As, continuous	900	$\frac{7.37}{9.3}$	±0.25
Ac, As, stratified	700	$\frac{21.0}{30.0}$	±0.56
Ci, Cc	180	$\frac{28.8}{16.0}$	±0.22
Cu, Sc	860	$\frac{395.6}{46}$	±0.82

TABLE 8.4. Mean Values of Turbulence Coefficient K in Friction Layer

Type of cloud	Season	K (m^2/sec)	U_g	Number of cases
Sc	Summer	28.2	10.0	25
St	Winter	21.2	13.0	28
Sc	»	20.9	13.0	111
Ns	»	18.3	13.5	43
Cloudless	»	17.8	13.5	56

Churinova calculated K from pilot-balloon data by using D. L. Laikhtman's formula

$$K = \frac{2\,(\lg e)^2 l}{\frac{d}{dz}\lg\left[(U_g - u)^2 + v^2\right]^2}, \tag{8.1}$$

where l is the Coriolis parameter, e is the base of natural logarithms, U_g is the velocity of the geostrophic wind, and u and v are the velocity components of the actual wind.

The results of the calculations are given in Table 8.4.

As Table 8.4 shows, the turbulence coefficient within clouds is approximately 20% greater than in the cloudless sky.

Matveev, and also German [32, 33], determined the values of K in clouds of different types from airplane load data [see Eq. (5.3)]. The results of the calculations are given in Tables 8.5 and 8.6.

Litvinova and Silaeva calculated the values of K from the results of measurements of fluctuations of the vertical wind velocity on free ballons with the aid

TABLE 8.5. Frequency (%) of Different Values of Turbulence Coefficient in Clouds

Type of cloud	Turbulence coefficient (m^2/sec)								Number of cases
	≤ 10	10.1–20.0	20.1–30.0	30.1–40.0	40.1–50.0	50.1–75.0	75.1–100.0	100.1–125.0	
Warm half of year									
Cs	—	30.8	33.2	11.9	11.9	7.1	2.4	2.4	42
Ac	—	6.2	6.3	12.5	12.5	25.0	25.0	12.5	16
As	—	10.0	10.0	5.0	20.0	15.0	40.0	—	20
Ns	1.9	7.7	9.6	19.3	25.0	25.0	11.5	—	52
Sc	—	—	7.4	11.0	37.1	37.1	7.4	—	27
Cold half of year									
Ac	—	15.3	5.3	15.8	15.8	42.0	5.3	—	19
As	—	—	—	63.7	27.2	9.1	—	—	11
Ns	—	22.7	13.7	31.8	22.7	9.1	—	—	22
Sc	—	13.4	10.0	20.0	33.3	20.0	3.3	—	30

TABLE 8.6. Mean and Maximum Values of K (m^2/sec) inside and outside Clouds

	Inside clouds					Outside clouds at altitude (km)			Author
	Cs	Ac	As	Ns	Sc	⩽2	2–8	8–12	
				Warm half of year					
Mean	31	68	58	46	51	24	22	18	German
Maximum	101	109	89	92	88	35	42	34	German
				Cold half of year					
Mean		46	43	34	40	22	20	6	German
Maximum		79	62	66	35	38	55	14	German
Mean		55	27	56	59	–	–	–	Matveev

of the Hesselberg equation (5.4). The values of K that they obtained for St–Sc lay in the range 2–10 m^2/sec.

Thus, the values given for K in stratiform clouds of the lower layer by different authors do not agree with one another. The main reason for this is the great diversity of scales of atmospheric turbulence, which means that eddies of different sizes (L) are involved in the exchange; their specific contribution to this process is different. Since different methods of measuring w give data relating to different parts of the turbulence spectrum, the values of K calculated with their aid will not necessarily agree. Differences in measurement technique should also be taken into account. For instance, in the pilot-balloon method the measurements are made in the course of the ascent, i.e., on an approximately vertical line. Measurements on free balloons are conducted all the time in the same volume of air, i.e., they give the temporal, and not the spatial, fluctuations of K, and so on.

In the analysis of the values of K calculated from airplane loads it should be borne in mind that loads are produced only by relatively large eddies, whose size is comparable with the size of the airplane. In addition, the low threshold sensitivity of accelerographs leads to the filtering out of gusts with $w \leqslant 0.3$–0.5 m/sec. Finally, since airplanes experience loads only in zones of intense turbulence, the values of K obtained from them are not representative of the average level of turbulence for the whole cloud, but of its intensity in the turbulent zones within the cloud. We have mentioned above that the extent of such zones in stratiform clouds is less than half their length in any case and, hence, the mean turbulence coefficient for the whole cloud will be less than the values of K given in Tables 8.5 and 8.6.

We now consider the changes in K in a vertical direction within clouds. Examples of the distribution of K in stratiform clouds of the lower layer are given in Fig. 8.1. This figure shows that the value of K within clouds changes relatively rapidly with height, and it is greatest at the boundaries of the clouds. Particularly large values of K are encountered at the upper boundary of clouds with an un-

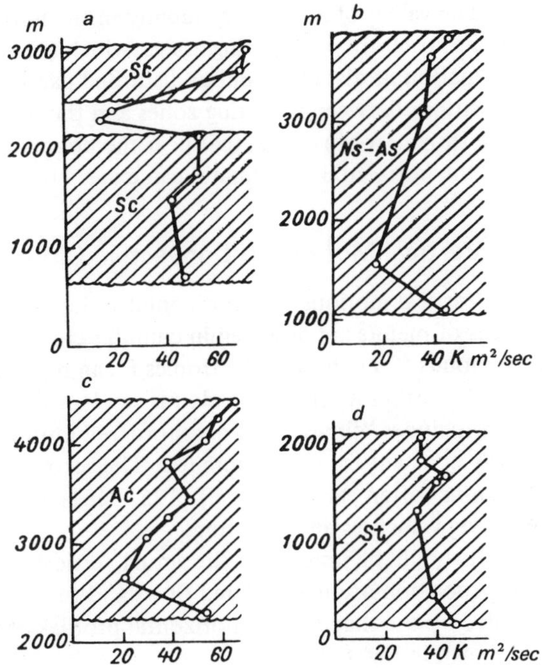

Figure 8.1. Vertical profiles of turbulence coefficient K. (a) Sc, February 20, 1961; (b) Ns–As, January 24, 1958; (c) Ac, January 23, 1958; (d) St, March 3, 1961.

dulating upper margin. In the case of stratified clouds, K decreases sharply in the cloud-free interlayers. This picture is typical also of internal and frontal cloud systems. The vertical distribution of turbulent exchange intensity in stratiform clouds is also illustrated by the vertical profiles of the spectral densities of turbulence. Examples of such profiles are given in Fig. 8.2 [42].

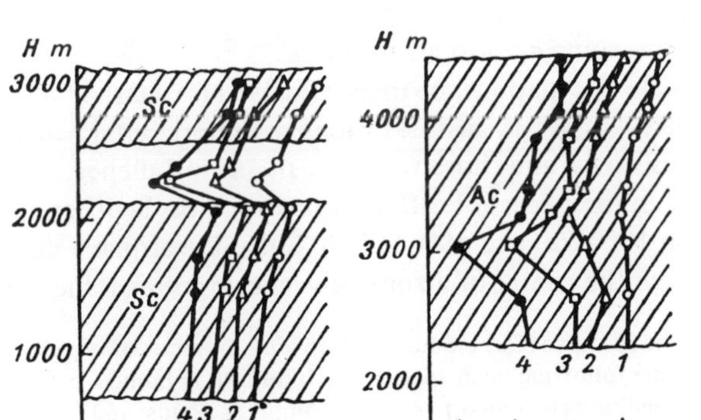

Figure 8.2. Vertical profiles of spectral densities of turbulence in Sc and Ac. (1) L = 250 m; (2) L = 500 m; (3) L = 750 m; (4) L = 1000 m.

The values of K also vary rapidly in the horizontal direction. This is revealed by the structure of buffeting zones, which are characterized by a continuous alternation of disturbed and undisturbed parts within the cloud and very sharp transitions between buffeting zones and the regions of calm flight. Measurements have shown, for instance, that near the boundaries of clouds $\partial K/\partial z$ and $\partial K/\partial s$ (s is the horizontal coordinate) can exceed 0.2–0.3 m/sec.

2. Vertical Motions in Convective Clouds

Vertical motions with horizontal scales L from centimeters to tens or hundreds of meters are observed in cumulus clouds of all species.† In piled-up cumulus clouds (Cu cong.) of the tropics L can reach 1–2 km.

Vertical motions within clouds can be subdivided into two classes. The first class includes "flows," i.e., quasi-ordered motions, whose direction and speed is fairly constant for several minutes. The second class includes *purely turbulent motions* (often called "gusts") [125, 148], which are characterized by rapidly varying velocities and direction.

It should be noted that the actual vertical motions in Cu are the result of superposition of quasi-ordered flows and turbulent gusts. This is distinctly revealed by the shape of the horizontal profiles of w in Cu, an example of which is shown in Fig. 8.3.

The vast majority of data on vertical motions in Cu has been obtained by research airplanes. Since the flight speed is tens or even hundreds of times greater than the characteristic velocities of air motions in clouds, it can be assumed that measurements of w in flights through clouds give an "instantaneous photograph" of the air motions. At the same time, the instantaneous character of such measurements does not allow determination of the change in w with time and, hence, they are unsuitable in principle (when a single airplane is used) for distinct separation of the motions into flows and turbulent fluctuations. Hence, such separation requires some *a priori* assumptions.

The small-scale fluctuations, or those which cause sharp loading of the airplane,‡ are usually classed as gusts. Motions with horizontal scales $L > L_{cr}$ are usually classed as flows, and those with $L < L_{cr}$ as turbulence. The value of the critical scale L_{cr} in the papers of different investigators is different, but it usually lies between 100–200 and 400–500 m. Since the airplane does not necessarily pass through the center of a zone of vertical motions, the frequency of

†The whole range of cumulus (Cu) clouds is subdivided, in order of increase in vertical dimensions, into the following main species: Cu humilis (Cu hum.)–cumulus clouds of fair weather; Cu mediocris (Cu med.)–medium cumulus clouds; and Cu congestus (Cu cong.)–piled up cumulus clouds (see [113]).

‡This assumption is not always justified, since loads can be caused not only by turbulence but also, for instance, by intersection of the side boundaries of even very large vertical flows.

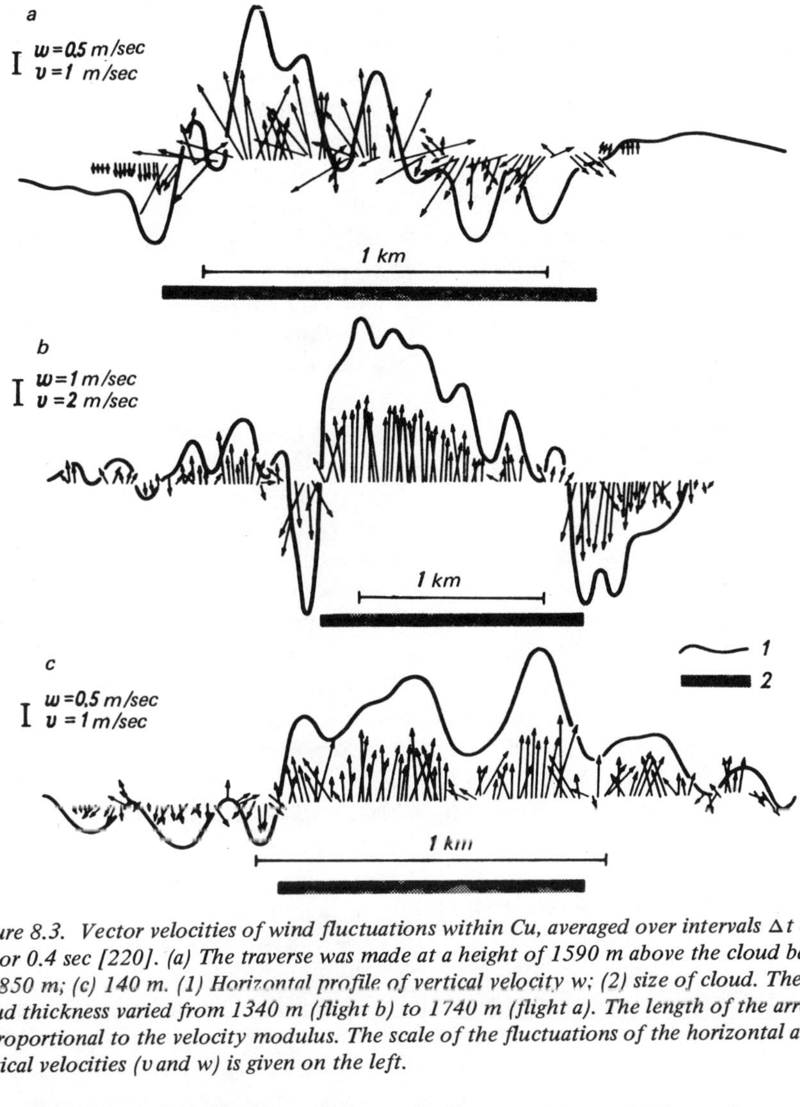

Figure 8.3. Vector velocities of wind fluctuations within Cu, averaged over intervals Δt of 0.2 or 0.4 sec [220]. (a) The traverse was made at a height of 1590 m above the cloud base; (b) 850 m; (c) 140 m. (1) Horizontal profile of vertical velocity w; (2) size of cloud. The cloud thickness varied from 1340 m (flight b) to 1740 m (flight a). The length of the arrows is proportional to the velocity modulus. The scale of the fluctuations of the horizontal and vertical velocities (v and w) is given on the left.

occurrence of small-scale fluctuations is usually overestimated. The underestimate of L can be very considerable.

In statistical studies the great range of motions in Cu is often not separated at all into quasi-ordered and turbulent motions, and they are lumped together.

2.1. Quasi-ordered Vertical Motions in Cumulus Clouds. The characteristics of vertical motions in Cu have been fairly well investigated (see, for instance, [24, 125, 221]).

In the lower two-thirds of growing Cu hum.—Cu med., ascending motions occupy 70–80% of the volume of the clouds and the air descends only at the sides. Inside Cu cong. ascending streams occupy about 40% of the cloud volume.

Within the top third of all varieties of Cu the ascending flows have smaller velocities and usually occupy not more than 30–40% of the cloud volume. The rest of the air within the cloud is usually in a highly turbulent state and, hence, it is impossible to determine the main direction of the flows. Quasi-ordered descending motions are often found in particular parts of the top of the cloud.

Within large Cu (Cu med. and, in particular, Cu cong.) there may be several ascending flows, which is clearly revealed by the multidomed structure of these clouds. The greater the horizontal dimensions of the vertical flows, the less frequently they are encountered. According to estimates given in [114], the probability of occurrence of vertical flows decreases exponentially with increase in L. For instance, flows with $L > 400$ m are observed in only 2% of the cases.

As measurements of w in Cu over the Ukraine, North Caucasus, and other regions of the USSR have shown [24, 50, 114], the values of w within cumulus clouds of all varieties do not exceed 3–5 m/sec in approximately 80% of the cases. The frequency of values of $w > 15$ m/sec is 0.5–1.0%. The maximum speeds of vertical flows in thick cumulus clouds can probably reach 20 m/sec.

Numerous measurements of w in Cu have been made in subtropical zones of the Pacific and Atlantic Oceans [189, 221]. As an example, we can cite the results of measurements over the eastern coast of Australia [220, 221]. Measurements made in 21 mature cumulus clouds in 68 traverses showed that the median (50%), 75%, and 90% quantiles for the velocities of ascending and descending flows were 4.9, 6.8, 10.0 and 3.5, 5.0, 7.5 m/sec, respectively. The maximum velocities of the ascending and descending flows were 12.7 m/sec and 9.2 m/sec, respectively. There was an increase in the standard deviations of the air velocities within clouds with reduction of stability in the surrounding atmosphere.

Figure 8.4 shows the distribution curve for w in Cu (irrespective of the horizontal dimensions of the vertical flows) in the layer from the cloud base to a height of several hundred meters above it [220]. In general, the values of w are normally distributed, since the plot in probability coordinates gives a straight line. The cumulative frequency of vertical gusts having effective velocities w_{eff} exceeding 5–6 m/sec within Cu from measurements over the southwestern U.S.A. [212, 213] was 10^2–10^3 times greater than in clear air (Fig. 8.5). This difference increased with increase in w_{eff}.† The standard deviations of the velocities of

†It should be noted that graphs of the type shown in Fig. 8.6, which are widely used in applied aerodynamics, do not always give the true probability distribution of w. This is because they are plotted from the peaks (local maxima) on records of variation of w with time or distance.

‡The fullest information about the distribution of w in Cu above the USSR is given in the article: I. P. Mazin and S. M. Shmeter, "Distortion of fields of meteorological elements in the region of cumulus clouds," in: *Questions of Cloud Physics* [in Russian], Gidrometeoizdat, Leningrad (1978), pp. 135–136.

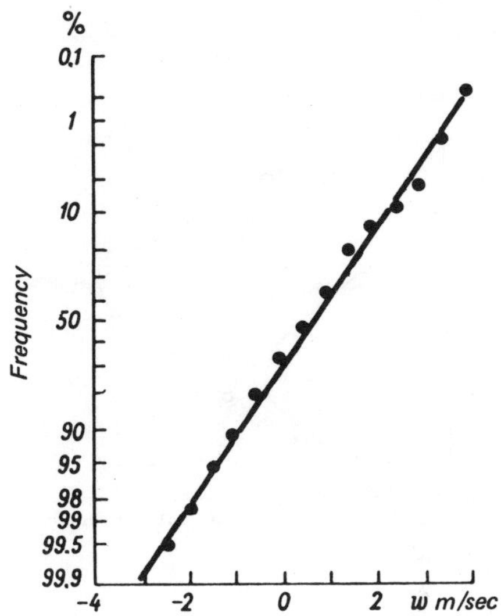

Figure 8.4. Frequency of velocities of ascending motions in layer several hundred meters thick above base of Cu.

vertical air gusts in Cu, according to [212], vary from 1.1 to 3.1 m/sec.‡

2.2. Variation of Velocities and Horizontal Dimensions of Ascending Motions with Height. Figure 8.6 shows mean vertical profiles of w in cumuliform clouds over the USSR. Table 8.7 illustrates the vertical distribution of the standard deviations σ_w of the vertical air velocity in Cu above Australia. The values of σ_w increased by 0.7 m/sec per km on the average.

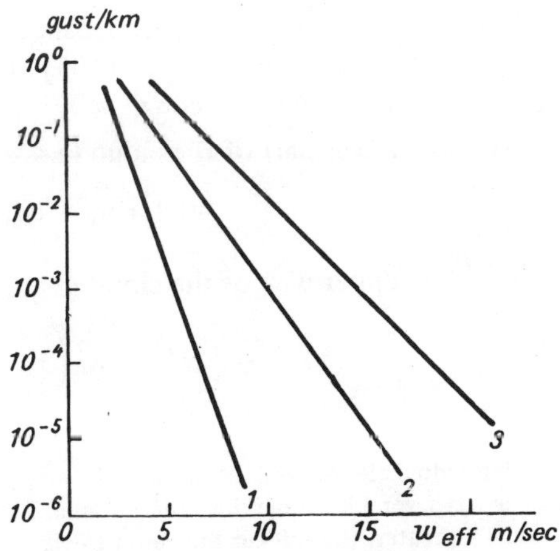

Figure 8.5. Probability of occurrence of velocities of vertical gusts exceeding a prescribed value. (1) Outside clouds; (2) in Cu; (3) in Cb.

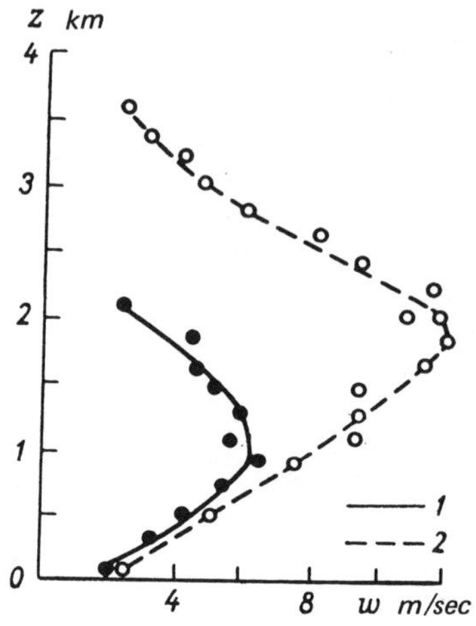

Figure 8.6. Vertical distribution of averaged vertical velocity in Cu med. (1) and Cu cong. and Cb (2).

According to the data of [221], the coefficient of correlation between σ_w and z is close to 0.6; in the active convection layer (i.e., where $\partial w/\partial z > 0$)

$$\sigma_w = A + Bz, \tag{8.2}$$

where $A = 1.1$–1.3 and $B = 0.7$–0.8.

The increase in σ_w with z in the active convection layer, which is clearly shown in Table 8.7, is probably observed† (though, perhaps, with different $\partial\sigma_w/\partial z$) both in "gusts" and in "flows." For instance, according to empirical data [29], for ascending motions whose average scale is close to 100 m we have the relation

$$w = 0.02z^{3/4}\left(1 - \frac{r^2}{R^2}\right)^{1/2}, \tag{8.3}$$

where R is the radius of the disturbance, and r is the distance from its center (all the quantities are in meters). In [105] the following empirical relations were given for flows within Cu cong.:

for the lower part of the cloud ($z \leqslant z_{max}$)

$$w(z) = w_0 + (w_{max} - w_0)\frac{z}{z_{max}}, \tag{8.4a}$$

for the upper part of the cloud ($z \geqslant z_{max}$)

$$w(z) = w_0\left(1 - \frac{z - z_{max}}{z_B - z_{max}}\right). \tag{8.4b}$$

†In individual clouds, w does not usually vary monotonically with z, and the vertical profile of w has several local maxima and minima. The raggedness of the vertical profile of w is usually greater, the thicker the cloud [50].

TABLE 8.7. Standard Deviations of Velocity of Vertical Motions in Cu for Different Heights z above Cloud Base

z (m)	σ_w (m/sec)	Number of cases
0– 300	1.35	8
300– 600	1.61	3
600– 900	1.98	6
900–1200	2.15	5
1200–1500	2.25	3
1500–1800	2.83	4
>1800	2.27	1

Here w_0 is the velocity of the ascending flow at the Cu base ($z = 0$), w_{max} is the maximum velocity of the ascending flow (at the level $z = z_{max}$), z_B is the height at which $w = 0$.

The relation between the horizontal dimensions ("diameter") of regions with ascending motions and height has not been well investigated. According to the data of [29], for comparatively small-scale ($L \leqslant 100$ m) ascending motions,

$$R = 27 + 0.015z\,, \tag{8.5}$$

where the flow "radius" R and height z are given in meters.

2.3. Vertical Motions in Vicinity of Cumulus Clouds. Quasi-ordered vertical flows are observed not only within Cu, but also in their vicinity. Measurements have shown that, on the average, air descends at the sides of Cu in 60–80% of the cases. In the other cases ordered vertical motions are absent.

The descending flows near Cu are of a compensatory nature and maintain the mass balance in the convection zone. Alongside disintegrating Cu, within which ascending motions die out, descending compensatory motions are either absent or very weak.

The width of the zone of descending flows near clouds can apparently vary from 0.5 to 2.0D (D is the cloud diameter). Because of the mutual superposition of the circulating flows near individual clouds, however, caused by the simultaneous existence of entire Cu fields, there are still no reliable data on the "radius of action" of the cloud on the field of vertical motions. Another characteristic feature is that the width of the zone of descending air alongside a particular cloud varies from level to level, and at certain heights such motions may be absent altogether.

The ratio of the area of descending motions near Cu cong. to the cross-sectional area of the cloud at the particular height is smaller than in less thick Cu [29]. This can obviously be attributed to the fact that Cu hum. (and often

Cu med.) grow as a unit whole, while Cu cong. grows "in parts," with the result that the zone of compensatory descending flows does not surround the entire cloud simultaneously.

The velocities of descending motions near clouds are 0.3–0.5 m/sec on the average, but in particular regions they can sometimes reach 2-3 m/sec. Particularly high air descent velocities are found close to the sides of Cu, where the descent of air continuously cooled by evaporation of the cloud elements is added to the compensatory descending flow.

The descent of air is observed also at the sides of the subcloud zone. Air also descends above the tops of Cu, where the descending flows are due to the compensatory processes accompanying the horizontal spread of the top of the convective flow within the cloud.

2.4. Vertical Motions within Cumulonimbus Clouds. In cumulonimbus clouds (Cb), just as in Cu, there are quasi-ordered and turbulent vertical and horizontal motions. The first have horizontal dimensions up to 10–12 km, which has led many investigators to call them mesoscale motions.

Such motions are branches of the convective circulation, and their intensity and spatial structure determine all the characteristics of the cloud–its dimensions, water content, nature of precipitation, etc. The major part of the volume of growing clouds is occupied by ascending motions, and the major part of dissipating clouds is occupied by descending motions [121]. Mature Cb sometimes contain a single closed circulation cell, consisting of ascending and descending flows. Vertical wind shears in the surrounding atmosphere play an important role in the formation of this "organized" circulation.

The most detailed data on vertical motions in Cb were obtained in the Thunderstorm Project investigations in the U.S.A. in 1946–1949 [148]. The vertical air velocities within clouds were measured by research airplanes equipped with accelerometers and highly sensitive flight-altitude recorders. In flying through the clouds the pilot intervened as little as possible in the control of the airplane.

The results of measurements made in intersections of 747 ascending and 339 descending flows showed that the greatest values of w were usually observed in the middle and upper parts of Cb.

In approximately 50% of the cases $w \leqslant 6$ m/sec. The maximum recorded velocity of ascending motions was 25 m/sec. The velocity of descending flows in most cases did not exceed 4.5 m/sec, and the maximum value was 24 m/sec.

The mean "diameter" of the ascending flows in Cb was $D \approx 1.5$ km, and that of descending flows was 1.2 km. The maximum diameter was 11.5 km for ascending flows and 6 km for descending flows. There was no distinct relation between D and altitude.

In 1957–1958 a large series of measurements of vertical air velocities within Cb associated with hurricanes was made in the U.S.A. [171]. Figures 8.7 and 8.8 illustrate the frequencies of occurrence of the velocities and dimensions of the vertical flows in Cb, while Table 8.8 gives the mean maximum velocities

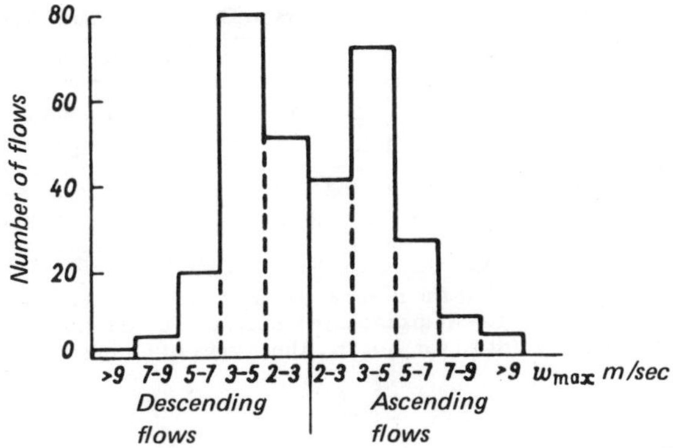

Figure 8.7. Histograms of maximum velocities of ascending and descending flows in Cb [171].

($\overline{w}_{max}$) and diameters ($\overline{D}$) of flows at altitudes of 4–5 km, measured in the flights of the Thunderstorm Project and in hurricane clouds.

Table 8.8 shows that the $\overline{w}_{max}$ and $\overline{D}$ data, obtained in internal cumulonimbus clouds (Thunderstorm Project) and in hurricane clouds, are close to one another. Other data on w, obtained in recent years in the U.S.A. and in the USSR [50, 105, 106, 122, 126, 212, 213], agree satisfactorily with the results of investigations of vertical air velocities in the Thunderstorm Project.

The largest values of w_{max} in ascending flows greatly exceed the values given above, reaching 63 m/sec in particularly thick Cb [212, 213]. The most typical values of w_{max} for Cb, however, from the data of all investigators are usually close to 30–40 m/sec (in mature, i.e., fully developed, clouds). The values $\overline{w}$, averaged over the horizontal cross section of the cloud, are usually not more than a few meters per second at any level.

The vertical profile of w and the spatial distribution of ascending and descending flows within Cb have been poorly investigated. Synchronous measurements of w at different levels within a cloud by several research airplanes have not yet been made. The results of such measurements, obtained by means of Doppler

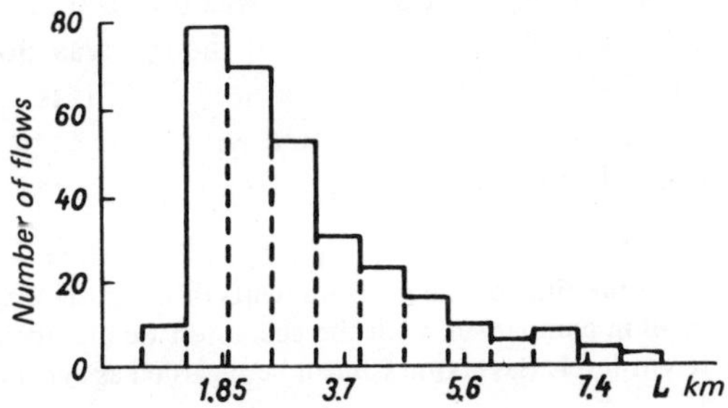

Figure 8.8. Histogram of horizontal dimensions of ascending and descending flows in Cb [171].

TABLE 8.8. Characteristics of Vertical Flows in Cumulonimbus Clouds

Nature of flow	Hurricanes			Thunderstorm Project		
	number of cases[a]	(m/sec)	$\overline{D}$(km)	number of cases[a]	(m/sec)	$\overline{D}$(km)
Ascending	155	3.9	2.4	206	4.2	2.4
Descending	158	3.6	2.8	95	3.6	1.8

[a]Gray [171] pointed out that the number of cases with vertical flows in measurements in hurricane–cloud systems included a certain (undetermined) number of measurements relating not to Cb, but to other types of clouds.

radar,† have still not been systematized, and their accuracy (and comparability with one another) is probably not very high, mainly due to ignorance of the size spectrum of the reflectors (cloud elements, including precipitation particles) and, hence, the spectrum of their falling velocities within the clouds in which w was measured.

According to measurements of w, obtained by radar tracking of sounding balloons with corner reflectors [105, 106], the average velocities of ascending flows within Cb increase with height from the base to the middle, or a little above the middle, of the cloud. The level at which $w = w_{max}$ is close to the upper boundary of the moist-labile layer of the troposphere.

Sometimes the level with $w = w_{max}$ is situated in the "anvil" of Cb [125], which can probably be attributed to the presence here of an additional buoyancy force due to the release of latent heat of crystallization of cloud droplets. Although the general increase in w with ascent of air from the base to the center or top half of the cloud has been reliably established, the details of the variation of w with height are still unknown. The change in w with height is probably not monotonic, so that there may be several local maxima of w at different heights in a cloud [50, 125].

The ascending flows in Cb are probably jetlike. This is indicated, for instance, by a unique experiment carried out by Vonnegut. He compared the data on the tension of the cable of a tethered balloon, situated within a Cb 500 m above its base, with the rate of growth of the top of the cloud. Although the latter was 3–7 km above the Cb base, it was found that with growth of a cloud with $w \geqslant 2$ m/sec the velocity of ascent of the top was closely correlated with the cable tension. This can easily be accounted for if it is assumed that both the balloon and the top of the Cb are situated within the same jet, which passes through the entire cloud.

†Since the time of the measurements of w by the Doppler method at each of the levels is small in comparison with the characteristic life of the cloud, such measurements at different levels (layers) in Cb can be regarded as synchronous in a first approximation.

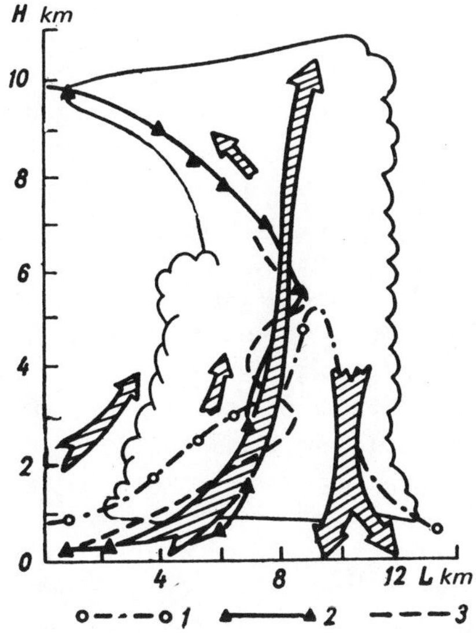

Figure 8.9. Kinematic scheme of Cb [173]. (1) August 17, 1967; (2) April 21, 1968; (3) postulated rotation.

Figure 8.9 gives a schematic diagram of flows within a Cb, plotted from the data obtained in the U.S.A. by launching sounding balloons with radar transponders into clouds [173]. According to these data, the ascending flow originates at a distance of about 40 km in front of the cloud.

Within the cloud the flow has the nature of a quasi-stationary laminar jet, whose central region rotates cyclonically around a vertical axis. Slight turbulence of the flow within mesoscale jets in Cb was reported in [105, 137].

The short-period fluctuations of w within mesoscale vertical jets in Cb usually do not exceed 5–10% of the mean value of the vertical velocity. Indirect evidence of relatively small short-period fluctuations of w in mesoscale jets within clouds is given by observations of the rate of growth of the tops of Cb. They show that the rate of growth of Cb and, hence, the vertical air velocity at the upper boundary of the cloud is subject to considerable low-frequency oscillations with period $T \approx 12$ min [125].

Mesoscale descending flows in Cb, beginning at the stage of maturity† of the cloud, are found at all heights, but their velocities are greatest in the precipitation zone [122]. In large quasi-stationary Cb the most powerful descending flow is localized in the bottom two-thirds of the cloud. Radar investigations of motions in Cb and, in particular, observations of the trajectories of dipole reflectors released in the vicinities of these clouds, suggest that the main descending flow consists of relatively dry air, with low potential temperature, coming from the

†Ascending motions predominate within "young," rapidly growing, Cb [118].

middle troposphere within the rear part of Cb. Its penetration into this region must be assisted by the large relative velocities of the air surrounding the middle and upper parts of the Cb, i.e., the "blow-through" of the cloud [119, 121].

Subsequently, owing to the additional cooling of this air caused by evaporation of the precipitation particles in it and its entrainment by the falling drops, the velocities of the descending motions increase.

The precipitation zones, however, do not always coincide exactly with the regions of the most intense descent of air. This indicates that there are other mechanisms of formation of descending motions, the most important of which is probably the spontaneous descent of air due to the instability of stratification in the cloud. In the upper part of the cloud the initial impetus may be, for instance, the local cooling of air within the tops of Cb due to entrainment of dry air from above the cloud.

2.5. Mesoscale Vertical Flows in Vicinity of Cb. According to [122], Cb is almost always accompanied by extensive zones in which air descends or rises and whose distance across may exceed the cloud "diameter." The velocities of the vertical flows $|w| \leqslant 1$-2 m/sec, and the frequencies of the ascending and descending flows are almost the same. Since the horizontal dimensions of the descending flows are usually larger, however, the air as a whole descends here. Alongside mature Cb the width of descending flows exceeds 11 km in almost a third of the cases.

Figure 8.10 shows the averaged spatial structure of vertical flows on different sides of Cb in relation to the wind shear vector. In pooling the individual measurements 0.1 of the diameter of Cb at the flight level was taken as the distance unit. Figure 8.10 shows that in front of the windward edge of the cloud there is a zone of relatively strong ($\overline{w} \approx 1$ m/sec) ascending motions. On the lee edge there is a distinct alternation of vertical velocities with wavelength close to the diameter of the cloud and amplitude decaying with increasing distance from it. Along the flanks of the cloud there is a narrow zone of descending motions, whose velocity decreases rapidly with increasing distance from the Cb. The frequency of the velocities of air flows above Cb is given in Table 8.9.

The general structure of the field of w alongside Cb resembles the distribution of vertical motions in the case of flow over hills, where the air ascends on the lee slope and a system of waves is formed downstream [122, 125]. The existence of such a flow was first postulated by C. W. Newton (see [195], for instance). The descending flows, located right at the boundaries of the Cb and often having velocities of more than 1 m/sec, probably owe their origin to cooling of the air here due to evaporation of the cloud elements.

Upward flow over clouds† usually occurs in cases where the atmosphere above their tops is not very stably stratified and, hence, the ascent and descent

†Flow past clouds can be observed not only near the tops of Cb, but also at the tops of Cu cong.

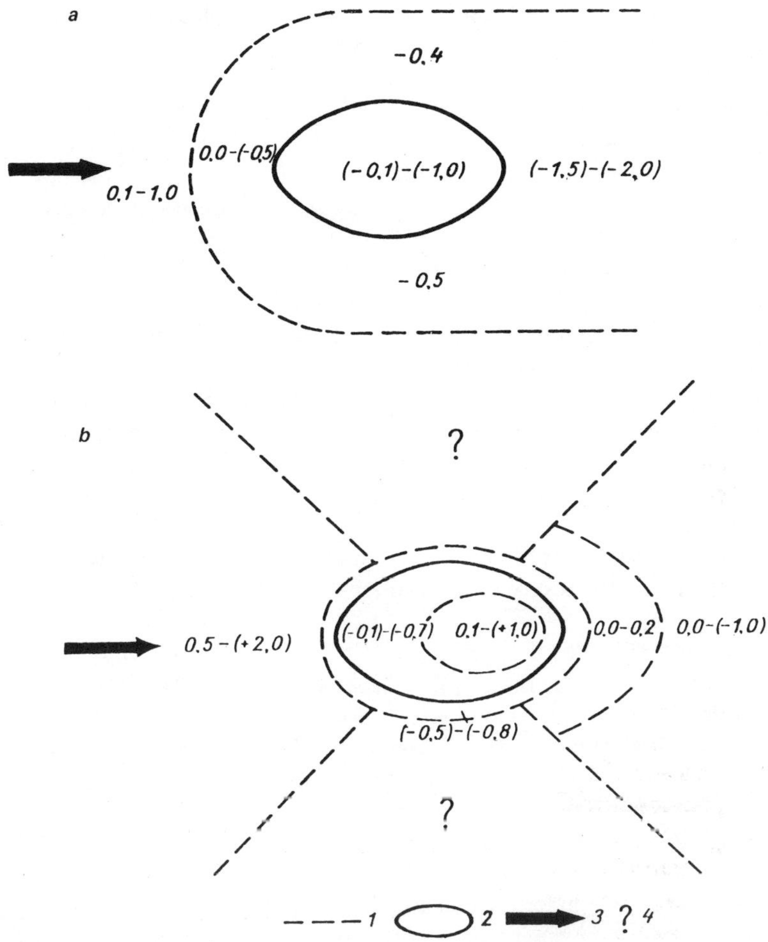

Figure 8.10. Averaged vertical velocities in region of cumulonimbus clouds. (a) Above Cb; (b) At sides of top and within top. (1) Lines demarcating regions of $\overline{w}$ with different sign; (2) cloud; (3) direction of motion of Cb; (4) regions for which there are no reliable data. The figures give the values of w in m/sec.

TABLE 8.9. Frequency (%) of Velocities of Mesoscale Vertical Flows in the 0- to 500-m Layer above Cb

Clouds	w (m/sec)							
	(−10.0)–(−8.0)	(−7.9)–(6.0)	(−5.9)–(−4.0)	(−3.9)–(−2.0)	(−1.9)–(0.0)	0.0 – 1.9	2.0 – 3.9	4.0 – 6.0
Growing	—	1	3	19	33	36	8	—
Mature	2	—	3	16	38	34	5	2

of air require no great expenditure of energy. If the cloud is situated under an isothermal or inversion (i.e., very stable) layer, the greater part of the flow goes along the flanks, causing a sharp increase in wind velocity alongside the cloud.

In recent years the presence of waves above convective clouds has been confirmed by observations in glider flights. Bradbury [142] suggested that the term "thermal waves" should be used to describe the structure of the flow over Cu cong.-Cb fields. According to the results of his investigations, the disintegration of thermal waves due to their loss of stability can lead to the formation of intense turbulence and to severe buffeting of aircraft above clouds.

Table 8.9 shows that above Cb, as well as alongside them, the velocities of the vertical motions rarely exceed ±2 m/sec, and the distribution of w over growing and mature clouds is almost the same. At the same time, above mature Cb the average width of descending flows is much greater than that of ascending flows, whereas above growing clouds the transverse dimensions of ascending flows are a little greater than those of descending flows (Table 8.10). Hence, as a whole, the transfer of air above mature clouds is downward, and above growing clouds it is upward.

The greatest velocities of descending flows above growing clouds are located above the edges of the top, and the smallest velocities are located above its center.

Above mature clouds the greatest velocities of descending motions are observed above the center of the Cb, and the least above the edges of the top. We can postulate that when the cloud attains maturity the descending flow is first formed over the axial part of the Cb and, hence, it is here that suppression of the ascending flows begins. At a distance from the sides of the top, descending flows are observed alongside both growing and mature Cb.

We turn now to the characteristics of the vertical motions immediately below the lower edge of Cb. Measurements have shown (see [125]) that a large part of the zone under the cloud is characterized by the presence of strong descending flows. As distinct from all other parts of the Cb zone, the velocities of the descending flows here sometimes greatly exceed the air ascent velocity.

TABLE 8.10. Frequency (%) of Horizontal Dimensions of Mesoscale Vertical Flows above Cb

Clouds	L, km						
	<1.0	1.0–3.0	3.1–5.0	5.1–7.0	7.1–9.0	9.1–11.0	>11.0
	Ascending flows						
Growing	20	42	20	10	4	–	4
Mature	19	29	29	15	–	2	6
	Descending flows						
Growing	39	25	14	14	4	4	–
Mature	6	19	25	15	10	6	19

In addition, the mean width of the descending flows is greater than that of the ascending flows.

The descending flows are located mainly in the region within which precipitation takes place. With approach to the ground, the velocity of descending flows increases to the level where the air descending from the cloud begins to spread to the sides, which slows down the descent. The dynamic interaction of the powerful descending flow with the underlying surface, which acts as a barrier to it, leads to a sharp intensification of the wind here, the production of squalls, a local increase in pressure (the "thunderstorm anticyclone"), and so on.

Ascending flows under the cloud base are due mainly to convection, i.e., the main source of cloud formation. They are also partially due to the dynamic ascent of air in front of moving Cb and the associated dome of cold air under the cloud.

The most detailed investigations of ascending subcloud flows were made by Auer and Sand [138], who in 1964–1965 made 229 measurements of w within ascending flows under Cu and Cb. As a single measurement they took the vertical air velocity averaged over a 1-min interval, which corresponded to a distance of about 3 km.

Table 8.11 gives the mean maximum values of w under clouds of different types separately under the left and right parts of the front sector of the cloud. The cloud sectors were taken relative to the direction of motion of the center of the radio echo from the cloud. The clouds were classified according to their intensity of precipitation.

As Table 8.11 shows, with increase in precipitation the value of w_{max} also increased. The greatest maximum velocities were observed under the right part of the front sector of the cloud. The same pattern was revealed (even more distinctly) by the mean vertical velocities w given in Table 8.12. The last column of this table gives the range of vertical air velocities.

The measurements revealed two cases (not included in Table 8.12) of extremely high velocities of ascending motions: 17.5 and 22.5 m/sec. In both cases the width of the flow within which these values of w were observed did not exceed 800 m. The larger the Cb, the broader were the ascending flows under them. Under clouds which gave moderate or heavy precipitation the width of the zone of ascending motions was 8–23 km. Usually turbulence was ob-

TABLE 8.11. Mean Maximum Velocities of Ascending Flows under Front Sector of Clouds Giving Precipitation of Different Intensity

Part of sector	No precipitation		Light precipitation		Moderate precipitation		Heavy precipitation	
	no. of cases	$\overline{w}_{max}$ (m/sec)	no. of cases	$\overline{w}_{max}$ (m/sec)	no. of cases	$\overline{w}_{max}$ (m/sec)	no. of cases	$\overline{w}_{max}$ (m/sec)
Left	4	1,3	1	3,0	6	3,0	—	—
Right	6	1.8	7	2.0	11	4.0	11	6,2

TABLE 8.12. Mean Vertical-Flow Velocities under Front Sector of Convective Clouds

Type	Number of cases	w (m/sec)	Range (m/sec)	
			from	to
No precipitation	71	0.8	−1.5	5.0
Light precipitation	42	1.0	−1.5	5.0
Moderate precipitation . . .	48	2.5	−3.0	9.5
Heavy precipitation	58	3.8	−1.0	11.5

served only close to the boundaries of these flows, while the flows themselves were very calm.

In addition, measurements of w in repeated traverses of large ascending flows usually gave similar results, which suggests that such flows are quasi-stationary.

3. Turbulence in Cumulus and Cumulonimbus Clouds

3.1. Statistical Characteristics of Turbulence within Cumulus Clouds. Statistical investigations of wind-velocity fluctuations in the Cu–Cb zone have just begun. The vast majority of them relate to vertical motions, and the investigations have resulted in several individual spectral density curves $S_w(\Omega)$, which have been used to determine the standard deviations of the vertical velocity fluctuations (σ_w) and the rates of dissipation of turbulent energy (ϵ). The calculation of $S_w(\Omega)$ for not very thick cumulus clouds entails serious difficulties due to two factors:

(1) the shortness of the regions of measurement of w (the width of Cu cong. usually does not exceed a few kilometers, and the width of Cu hum. is quite often less than 1 km);
(2) the considerable nonstationarity of dynamic processes in cumuliform clouds, especially during their growth.

The first of these circumstances prevents obtaining statistically accurate spectra in a sufficiently wide range of scales, since (see Chapter 2) the maximum value of L up to which the spectrum is statistically accurate is $\frac{1}{5}$ to $\frac{1}{10}$ of the length of the realization.

The nonstationarity of Cu convection means that the individual spectra are not sufficiently representative.

To increase the "effective" length of the realizations and to increase the statistical stability of the spectral estimates, artificial methods of some kind are sometimes used. For instance, several realizations relating to clouds with homogeneous characteristics are sometimes "lumped together" or a single realization is doubled.

3.2. Spectra of Vertical Velocity Fluctuations. Information about the spectra of w in Cu is at present meager.

Data on $S_w(\Omega)$ in Cu, obtained from measurements above the central part of the Ukraine [24], showed that for $L < 500$ m, $S_w(\Omega) \sim \Omega^{-5/3}$, and with further increase in L the slope of the spectral density curve decreases. The "saturation" of the spectrum $[\Omega \cdot S_w(\Omega)]$ at $L \approx 500$ m may be due to the fact that the energy-carrying disturbances ("ordered" vertical convective flows) have approximately these linear dimensions. These estimates are confirmed in general by the data in [221], which gave the spectral densities of vertical air gusts and their kinetic energy spectra in cumulus clouds observed above the Australian coast. The calculations were based on the results of 16 traverses of seven cumulus clouds. The values of $S_w(\Omega)$ were determined by using the autocorrelation function. The calculations were made up to $L \approx 0.2L^*$ (L^* is the length of the realization); the time interval was 0.1 sec.

Figure 8.11 shows examples of gust spectral density curves $S_w(N)$ at differ-

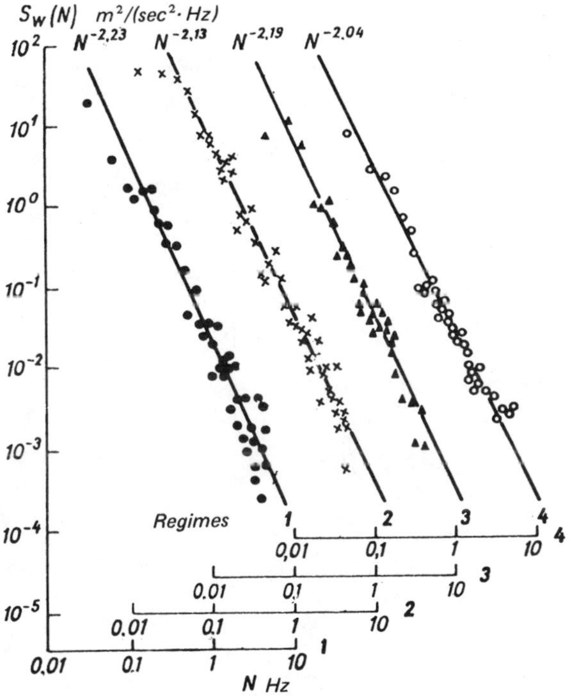

Figure 8.11. Spectral density of vertical gusts at different heights above Cu base in Australia [221]. Flights 1–4 were made at heights of 2.74, 2.13, 1.52, and 1.06 km, respectively; the standard deviations of the vertical velocities for these flights were 1.32, 1.95, 1.38, and 1.04 m/sec.

ent heights above the cloud base. It is apparent that in the wavelength range from tens of meters to 1-2 km the slope of these curves conforms to the law

$$S_w(N) \sim N^{-\alpha}, \tag{8.6}$$

where $\alpha = 2.04$-2.23, and N is the oscillation frequency in Hz.

The spectral distribution of the kinetic energy of vertical gusts at different wavelengths (frequencies) is shown in Fig. 8.12, where the values of $N \cdot S_w(N)$ for different N are plotted. Figure 8.12a shows the mean spectra for three traverses of Cu at different levels up to 300 m above the cloud base (curve 1) and down to 300 m from the upper edge of the cloud (curve 2). Figure 8.12b shows the spectrum obtained by averaging the two spectra shown in Fig. 8.12a (curve 3), and the mean spectra for layers 300-600 m (curve 4) and 600-900 m (curve 5) from the base or top of the cloud.

Figure 8.12 shows that the greatest amounts of kinetic energy are concentrated in the left-hand (long-wave) part of the spectrum. Another interesting feature is the local maximum of the energy spectrum at $L \approx 500$-600 m (in the layers 600-900 and 300-600 m) and near $L \approx 200$ m (in the layer 0-300 m from the Cu top or base).

3.3. Turbulence Coefficient in Cumulus Clouds. The intensity of vertical turbulent exchange in Cu can be estimated by means of the turbulence coefficient K calculated from Eq. (5.7). Such calculations, of course, include only

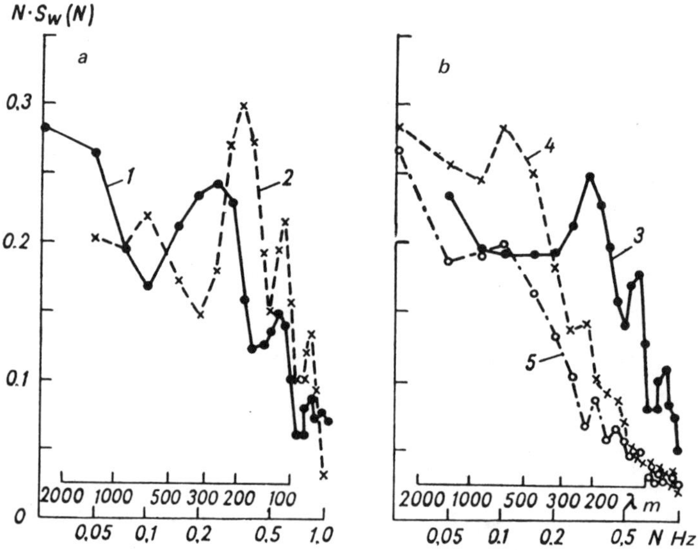

Figure 8.12. Spectral distribution of kinetic energy of vertical motions in Cu [221].

TABLE 8.13. Turbulence Coefficient $\overline{K}_{max}$ (m^2/sec) within Cu

Part of cloud	Cloud thickness Δz (km)		
	< 2.0	2.0–2.5	2.5–3.0
Top	178	198	356
Bottom	148	165	124

gusts of the scales which are measured by the apparatus used. Table 8.13 gives the mean maximum values $\overline{K}_{max}$ in Cu over the Ukraine.

It is apparent that in the upper part of Cu there is a rapid increase in $\overline{K}_{max}$ with increase in the thickness of the cloud. Inside the lower part of the cloud this increase continues to $\Delta z \approx 2.5$ km, and with further increase in Δz the value of K apparently decreases. This decrease might actually have been due to the fact that the airplanes did not enter the lower part of thick Cu before these regions dissipated (the flights usually started at the higher levels and gradually moved downward).

3.4. Turbulence in Vicinity of Cu. Turbulence becomes enhanced (in comparison with the background turbulence) at a distance of (0.5–1.0)D from a cloud: the closer to the cloud, the more intense the turbulence. Close to the sides of Cu it becomes fairly intense (the horizontal fluctuations become particularly large) and can cause moderate buffeting of aircraft. The dependence of the length and location of distributed regions on the wind has been investigated only in the case of cumulus clouds in the trade-wind region. The investigations showed that the turbulent zone is usually asymmetric and extends along the vector of the vertical wind shear. For instance, in 379 traverses of 78 clouds a symmetric disturbed zone was observed in only 27% of the cases.

According to Ackerman's data, turbulence is most intense in front of a cloud, whereas according to Malkus it is more intense behind a cloud (relative to the direction of the wind shear vector). This disagreement can probably be attributed to the fact that Malkus drew his conclusions for moderate and severe turbulence, whereas Ackerman considered turbulence irrespective of its intensity.

Ackerman's investigations showed that as clouds "age," the dimensions of their turbulent zone continuously increase (especially behind Cu). Even after the cloud completely disperses, the region of the atmosphere where it had been previously situated remains more disturbed than the surrounding air, i.e., the turbulent zones last longer than the clouds with which they are associated.

3.5. Turbulence inside Cb. Inside Cb the turbulence intensity is extremely high and in $\frac{1}{4}$ of all the turbulent zones within the clouds the mean vertical gust velocity $\overline{w}_T \geqslant 3$ m/sec [160]. The standard deviations of $w_T(\sigma_w)$ in both the lower and upper half of the cloud are equal to 2–5 m/sec, according to the results of different investigations, and the maximum values exceed 15 m/sec [122, 126, 148, 160]. It is characteristic that the fluctuations of the transverse and longitudinal flow velocity components usually differ from one another and from w by not more than 20-25% [212, 213].†

The maximum values of w are usually found in the central (relative to height) or subapical parts of a cloud. There may also be a local maximum of w_T near the level of the zero isotherm. Its appearance here may be due to the stratification instability induced by cooling of the air resulting from the thawing of ice particles. According to the results of measurements of w_T within Cb in the region of the Great Lakes (U.S.A.) the turbulence intensity generally increases from the cloud base to a level 1.5-2 km from its upper edge [170].

The horizontal distribution of w_T within Cb depends on the mesostructure of the cloud. In clouds consisting of a single convective cell the greatest values of w are observed near the axial part of Cb. In multicellular clouds the horizontal distribution of w_T has several specific maxima situated at the boundaries or within the mesoscale vertical flows. The main maximum is usually associated with the cell in which the mesoscale vertical velocities are greatest. Thus, there is a definite relation between mesoscale and turbulent vertical motions. According to the data of Byers and Braham [148], the coefficient of correlation between the velocities of such motions is approximately 0.42 ± 0.02. The presence of this relation is probably due to the fact that turbulent eddies originate close to the boundaries of mesoscale jets, i.e., where horizontal gradients of the vertical air velocity are particularly large. Subsequently, owing to their "own" horizontal velocities, the eddies can move into other parts of the cloud. Within disintegrating clouds the correlation between mesoscale and turbulent vertical motions is usually weaker. In addition, turbulence in them is intensified even when mesoscale convection completely ceases.

3.6. Statistical Characteristics of Turbulence in Cumulonimbus Clouds. Data on the statistical structure of turbulence in Cb are given in [122, 123, 212, 213, 216]. The most detailed information is contained in [212, 213], which are based on the results of measurements of u and w in 28 traverses of Cb by research airplanes at altitudes of 7.5 to 12 km above Oklahoma (U.S.A.).

Figure 8.13 shows the spectral densities of the velocities of vertical and horizontal gusts in Cb from measurements on five successive traverses of a mature cloud at an altitude of 11.7 km. The standard deviations σ (m/sec) of the gust velocity for each of the traverses are given. An interesting feature is the small variation of the turbulence spectral characteristics with time and the fact that they (like the velocity standard deviations) are almost the same for

†Results of investigations (not yet published in the periodical literature) that are inconsistent with the results cited above are given in [170]. It is reported that at "middle" heights in Cb horizontal gusts appear to be more frequent than vertical gusts, but the values of w_T greatly exceed the velocities of horizontal gusts.

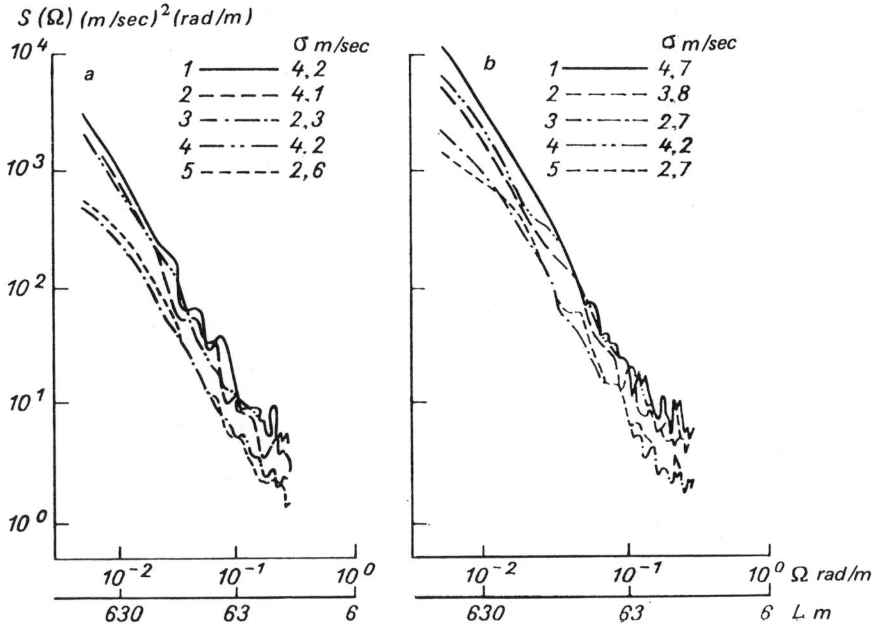

Figure 8.13. Variation of spectral density of vertical (a) and horizontal (b) turbulent gusts within Cb with time. (1–5) Numbers of successive traverses.

vertical and horizontal gusts. A similar picture was revealed by flights through Cb.

The actual variation of $S_u(\Omega)$, $S_w(\Omega)$, σ_u, and σ_w will, of course, be even less than that obtained, since the time between the successive traverses of Cb was several minutes and, in addition, the airplane could not follow the same path each time and, hence, the variation of u and w in space and time in Cb affected the results of the measurements.

The slight variation of the statistical characteristics of turbulence within mature Cb with time means that the latter can be regarded as quasi-stationary for periods of several minutes.

Figure 8.14 shows the averaged spectral density curves for turbulent gusts in different parts of the Cb zone from the results of measurements over the USSR [122, 123]. The first group includes spectra calculated from measurements made in flights through Cb at a level 2 km or more below the upper edge, while the second group was obtained from flights within the tops of Cb at a distance of a few tens to hundreds of meters below the upper boundary of the cloud.

The flights above Cb (third group) were carried out in the layer 0–200 m above Cb, and those under the cloud (fifth group) were made in the precipitation near the cloud base. The fourth group includes spectra for regions 0–500 m to the side of the side surface of the top half of Cb.

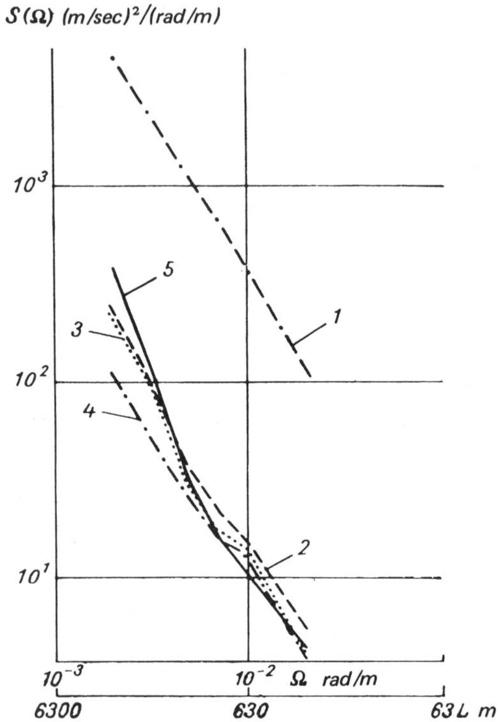

Figure 8.14. Averaged spectra of vertical velocities in region of cumulonimbus cloud:

Group number	1	2	3	4	5
Number of cases	4	17	18	18	10

Figure 8.14 and Table 8.14 show that the values of $S_w(\Omega)$ and their dependence on Ω (or L) in different parts of the Cb zone are different. The spectral density of vertical gusts within the central part of Cb is more than 10 times the value of $S_w(\Omega)$ in other parts. The disagreement between the values of $S_w(\Omega)$ for any of the other regions is slight. It is also apparent that the differences in the values of $S_w(\Omega)$ increase with increase in L.

The slope of the spectral density curves in general is close to $-\frac{5}{3}$, but there are deviations from the "minus five-thirds" law, which are particularly appreciable at $L > 600$ m. The data for $S_u(\Omega)$ and $S_w(\Omega)$ given in [212, 213] also indicate that the spectral density curves conform satisfactorily with the "minus five-thirds" law.

3.7. Turbulence Coefficient in Cumulonimbus Clouds. The mean turbulence coefficients in different parts of Cb are given below:

	Within top	On the edge	On side edge
$\overline{K}$, m²/sec.	215	146	95
Number of cases. . .	13	13	4

The value of K is greatest in mature Cb, where it can exceed 500 m^2/sec, and is least at the end of the cloud decay stage, when K is usually <150 m^2/sec.

An analysis of the results of calculations of K in individual clouds showed that in large multicellular Cb the field of K also has a cellular structure with different values of K in the different cells. Sometimes cells with $K = K_{max}$ are situated at the edge, and not at the center, of the cloud.

3.8. Turbulence in the Vicinity of Cb. Table 8.15 gives a characterization of vertical turbulent gusts above Cb from the results of 263 series of measurements. With ascent above the cloud the gust velocity decreases fairly rapidly, so that at a level of more than 200 m, the probability of $|w_T| > 2$ m/sec is less than 4%, and the probability of $|w_T| > 5$ m/sec is approximately 1%.

The increase in $|w_T|$ at heights of more than 400 m above the cloud is probably not due to the effect of Cb, but to the increase in turbulence under the tropopause.

TABLE 8.14. Values of $S_w(\Omega)$ [$(m/sec)^2/(rad/m)$] in Cb Zone for Different Linear Scales L

Region	S	L (m)			
		300	1000	2000	3000
Under Cb	$\bar{S}$	4	21	122	370
	S_{min}	3	14	53	90
	S_{max}	7	30	215	800
At side of Cb	$\bar{S}$	4	19	53	105
	S_{min}	1	7	27	34
	S_{max}	20	40	100	195
Above Cb	$\bar{S}$	4	21	105	215
	S_{min}	1	5	12	13
	S_{max}	21	77	360	850
Just under top of Cb	$\bar{S}$	6	26	100	220
	S_{min}	1	7	16	44
	S_{max}	10	44	230	730
Within central part of Cb	$\bar{S}$	119	800	2300	4300
	S_{min}	4	40	110	275
	S_{max}	280	1580	4150	7400

TABLE 8.15. Frequency (%) of Different Velocities $|w_T|$ of Vertical Turbulent Gusts above Cb

$\|w_T\|$ (m/sec)	Height above Cb (m)					
	0–10	11–100	101–200	201–300	301–400	> 400
< 0.5	34	55	66	83	100	86
0.5–2.0	44	27	34	13	0	10
2.1–5.0	21	18	0	0	0	4
> 5.0	1	0	0	4	0	0

The gust velocities are greatest over growing clouds. In mature Cb the turbulence characteristics above their dome are similar to those observed during growth of the cloud, but the turbulence is appreciably weaker above the projections of the anvil. Above dissipating clouds $|w_T|$ is usually <2 m/sec, but the thickness of the turbulent zone is the same as above mature Cb, and is sometimes even a little greater.

The turbulent regions in the layer 0-200 m above Cb are extended in the direction of the wind shear vector, and the frequency of occurrence of turbulence, like the lengths of the disturbed regions, is greatest behind the subshear edge of the top. This configuration of the turbulent zones is probably due to the flow of wind over the cloud, as mentioned above.

In Oklahoma (U.S.A.) in 1965 five flights over the tops of thunderstorm Cb at altitudes of 12-13.5 km were made. The greatest height of flight above the cloud varied from zero to 300 m. The measurements [160] showed that the relative extent of the turbulent zones was 11.4% of the total flight path of the airplane above the Cb (4100 km). The horizontal dimensions of the turbulent zones varied from 1.8 to 55 km; in 33% of the cases they were 5-9 km. The vertical gust velocities reached 6 m/sec, and exceeded 3 m/sec in 25% of the cases. It is characteristic that, as distinct from the results of investigations in the USSR [122], these measurements showed no correlation between height above the cloud and turbulence intensity. There was also no correlation between the length of the disturbed zone and the turbulence intensity within it.

It should be noted, however, that the conclusions of Dietz [160] were based on a relatively small amount of empirical data.

Table 8.16 gives the frequencies of the vertical velocities of turbulent gusts at a distance of not more than 1 km from the sides.

Table 8.16 shows that the frequency and intensity of turbulence alongside the upper half of Cb are greater than alongside the lower half. The velocities of turbulent gusts at all altitudes are greatest behind the lee side of the cloud.

The velocities of vertical and horizontal gusts at the side of Cb rapidly decrease with increase in distance from the cloud, and at a distance of a few tens (occasionally hundreds) of meters from it the velocity standard deviations σ_u,

TABLE 8.16. Frequency (%) of Different Values of w_T at Side of Cb

	w_T (m/sec)							
	<0.5	0.5–2.0	2.1–5.0	>5.0	<0.5	0.5–2.0	2.1–5.0	>5.0
	Lower half of Cb				Upper half of Cb			
Weather side	77	14	6	3	60	31	9	–
Lee side	88	9	3	–	53	35	10	2
Flanks	89	6	4	1	67	26	6	1

σ_v, and σ_w do not exceed 2-3 m/sec. According to some experimental data [216], on approach to Cb appreciable horizontal fluctuations are detected first, and after this the vertical fluctuations become appreciable.

9

Turbulence Affecting Flight of Aircraft

The flight of an airplane in a turbulent zone is accompanied by fluctuations of its velocity, angle of attack, roll angle, and other flight characteristics. This leads to additional alternating loads that are manifested in the "buffeting" of the airplane.

In investigations of the effect of atmospheric turbulence on aircraft (airplanes, rockets, etc.), two models of turbulence are usually used:

(a) turbulence is regarded as a set of individual (discrete) air gusts that are independent of one another;

(b) turbulence is regarded as a statistical ensemble of random gusts.

1. Methods of Assessing Intensity of Turbulence Affecting Flight of Aircraft

As was shown in Chapter 2, an airplane does not respond to all turbulent gusts, but only to a relatively narrow portion of the turbulence spectrum. When an airplane encounters long-period disturbances it adapts itself to the air motion and moves up or down along with it, without experiencing additional loads. High-frequency disturbances, on the other hand, lead to such small loads that they are practically unnoticed. The greater the flight speed, the longer the period of the air gusts that can impose considerable loads on a particular airplane. For instance, the flight of a plane at subsonic velocities is affected by disturbances 10–20 m to 3–4 km in length (up to 6–7 km for heavy airplanes), whereas the flight of supersonic aircraft at altitudes of 20 km or more is affected by disturbances up to 15–20 km in length.

It is clear from the above that in speaking of the intensity of turbulence causing buffeting we must have in mind not the intensity of turbulence as a whole, but the intensity in the part of the spectrum that affects flight conditions.

In other words, in this case the intensity of aircraft buffeting can serve as a

TABLE 9.1. Scale for Assessment of Intensity of Aircraft Buffeting[a]

Intensity of buffeting	Symbols: USSR	Symbols: ICAO	Range of loads and load increments in g	Behavior of airplane during buffeting	Recommendations to pilot	Comfort of passengers
Light	$b^{(1)}$		$0.8 \leqslant n \leqslant 1.2$ $\Delta n = \pm 0.2$	Spasmodic light shuddering of aircraft	Fly autopilot.	If it lasts for a long time some passengers find it unpleasant.
Moderate	$b^{(2)}$		$0.5 \leqslant n \leqslant 1.5$ $\Delta n = \pm 0.5$	Frequent jolts accompanied by rocking of airplane and changes in altitude, but no difficulty encountered in piloting of airplane. Noticeable oscillations of needles of variometer and speed indicator. The set flight conditions are maintained.	Autopilot flight permissible.	A large number of passengers experience unpleasant sensations, and walking in airplane is difficult.
Severe	$b^{(3)}$		$0 \leqslant n \leqslant 2.0$ $\Delta n = \pm 1.0$	Sharp shuddering motions and occasional bumps accompanied by frequent pronounced rolling and yawing. Changes in altitude and frequent and sudden loads make use of autopilot difficult. Unstable operation of variometer	Follow flying procedure laid down for each type of airplane in buffeting conditions.	Unsecured objects begin to move. Walking through airplane can lead to injury to passengers. Seat belts must be fastened. Most passengers are distressed and feel ill. Crew become fatigued.

				and speed indicator. Set altitude and course not maintained. Large negative loads cause weightlessness, and large positive loads push passengers into their seats.		
Storm or very severe	$b^{(4)}$		$\dfrac{-0.1 < n < 2.1}{\Delta n \geqslant \pm 1.1}$	Exceptionally sharp bumps accompanied by large loads that throw passengers violently against the belt and back into the seat. Altitude and course are greatly altered, set flight conditions are not maintained. Variometer and speed indicator readings highly distorted. Airplane altitude determined by average position of horizon bar. Airplane becomes difficult to handle; incorrect action by pilot can produce speeds that endanger the airplane.	Strictly observe the flying procedure to be followed in buffeting conditions.	Passengers can be thrown violently against seat belts, and back into their seats by positive loads. Passengers not restrained by seat belts can suffer severe head and other injuries. The sudden large loads make most passengers feel ill.

[a]The maintenance of flying speed and instrument checks are carried out in accordance with the flight manual for each type of airplane.

measure of the intensity of turbulence. This approach corresponds to the first of the turbulence models mentioned above.

1.1. Assessment of Intensity of Buffeting. Effective Velocity of a Discrete Vertical Gust. The intensity of buffeting, which is usually determined by the magnitude of the airplane load increments, depends not only on the magnitude of the turbulent fluctuations of the wind-velocity components and the gust scales, but also on the aerodynamic characteristics of the airplane. This is apparent, for instance, from the following formula, which provides an approximate estimate of the load increment of an inelastic airplane in horizontal flight when it encounters an isolated vertical air gust characterized by an effective velocity w_{eff}:

$$\Delta n = n - 1 = \pm \frac{\rho v w_{eff} c_y^\alpha}{2G/S}. \tag{9.1}$$

Here ρ is the air density, v is the flight velocity, c_y^α is the derivative of the lift force with respect to angle of attack α, G is the airplane flight weight, S is the airfoil surface, and n is the airplane load, equal to the ratio of its lift force to its weight. In calm horizontal flight $n = 1$. We have already discussed the relation between w_{eff} and the true gust velocity in Chapter 2.

The intensity of aircraft bufetting, assessed both from instrumental data and the "physiological sensations" of the crew, can be expressed on a four-point scale, as indicated in Table 9.1. Figures for the magnitude and frequency of occurrence of the effective velocity w_{eff} of vertical air gusts are mainly used in aircraft calculations. It should be noted that the idea of discreteness of the air gusts, as was seen in the previous chapters, does not, in general, correspond to the physical picture of atmospheric turbulence, which is usually a continuous random process.

The value of w_{eff} in the discrete-gust model is determined in practice by the velocity component w, the distance x in the direction of motion of the airplane at which the gust velocity increases from zero to w_{max}, and the postulated shape of the gust. It is assumed in Soviet investigations that the effective velocity w_{eff} of the vertical gust increases linearly (Fig. 9.1) over the length $x = h$ (gradient distance); as distinct from (9.1), it is calculated from the formula

$$w_{eff} = \frac{2\Delta n G/S}{k \rho_0 c_y^\alpha v}. \tag{9.2}$$

Here k is the "gust damping coefficient," equal to

$$k = 0.8\,\frac{1 - e^{-\lambda}}{\lambda}; \quad \lambda = \frac{c_y^\alpha g \rho_H h}{2}.$$

The gradient distance h of increase of the gust is assume to be 30 m.

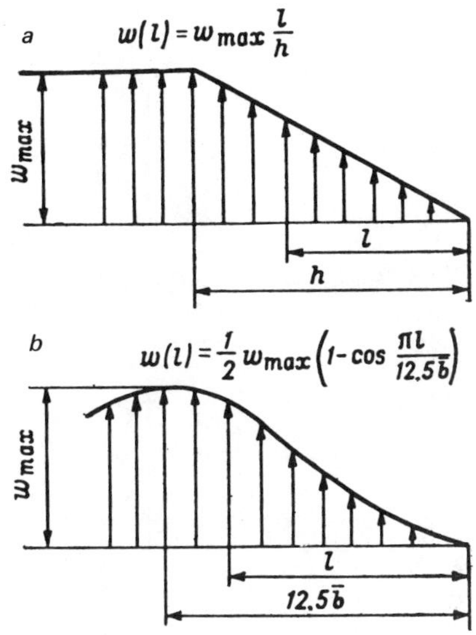

Figure 9.1. Diagrams of discrete gusts. (a) Rectangular; (b) "unity minus cosine."

In the U.S.A., and in several other foreign countries, the damping coefficient for calculations of the effective gust velocity, denoted here by k_g, is calculated on the assumption that the gust velocity increases in accordance with the "unity minus cosine" law (Fig. 9.1b) with gradient distance $h = 12.5\bar{b}$ ($\bar{b}$ is the standard mean chord of the airplane wing).

In this case the damping coefficient

$$k_g = \frac{0.88\mu_g}{5.3 + \mu_g},$$

where

$$\mu_g = \frac{2G/S}{\rho_0 c_y^\alpha \rho_H \bar{b}}$$

For comparison of Soviet and foreign data on the effective velocities of vertical gusts the u_{dE} values are converted to w_{eff} values by the formula

$$w_{\text{eff}} = \frac{k_g}{k} u_{dE}. \tag{9.3}$$

There is now a considerable amount of data on the effective velocities of vertical gusts in the troposphere, but there are not sufficient data for the stratosphere. Figure 9.2 shows the curves obtained by Arkhangel'skii *et al.* [2] for the integral frequency of gusts with velocity w_{eff} per kilometer of path in relation

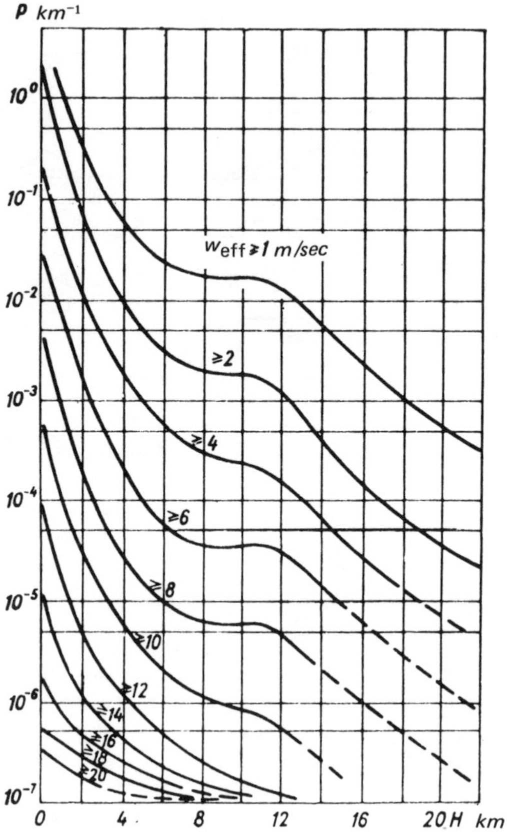

Figure 9.2. Integral frequency of w_{eff} *in relation to altitude.*

to altitude. The solid lines represent the generalized experimental data of Soviet and foreign investigations, and the dashed portions were obtained by extrapolation. It is apparent that high values of w_{eff} become less frequent with increase in altitude; an exception is altitudes of 8–12 km, where there is a maximum. This is the atmospheric layer where the tropopause and the tropospheric maximum of wind velocity are usually located.

1.2. Energy Spectrum of Atmospheric Turbulence Causing Buffeting of Aircraft. In several practical problems it is essential to take into account how flight is affected not only by the velocity of gusts, but also by their frequency characteristics. In this case one must resort to a statistical representation of turbulence (second model) and use the spectral density of turbulent gusts as a measure of turbulence intensity.

The fluctuating wind velocity is a random function of time and coordinates. If we consider a high-speed airplane as a turbulence indicator and bear in mind that the components of the fluctuating wind velocity $u'(x)$ are small in comparison with the flight velocity v, we can assume that the fluctuating velocity at any point in the airplane flight path is not significantly altered as long as the airplane flies in the vicinity of a point where the correlation between the wind-velocity fluctuations is high. In this case the field of fluctuating velocities of the wind vector for an airplane in flight can be regarded as a stationary homogeneous random field. For such a field the horizontal coordinate x and the time t are uniquely related to one another through the airplane flight velocity: $x = vt$. We can further assume that the fluctuating wind velocity is the same for all points on the airplane, i.e., within the dimensions of the airplane we can ignore the dependence of the wind-velocity fluctuations on the coordinates y and z. This assumption is probably more valid, the smaller the airplane. The assumptions made allow us to regard the wind-velocity fluctuations, measured by means of an airplane, as a function of only one variable—the coordinate x or the time $t = x/v$.

If the atmospheric turbulence is statistically homogeneous and isotropic, we require for its description only two correlation functions: $R_u(\Delta x)$ and $R_w(\Delta x)$; $R_u(\Delta x)$ is the correlation function of the fluctuations of the longitudinal wind-velocity component, and $R_w(\Delta x)$ is the correlation function of the transverse component. It is assumed in aerodynamic calculations that

$$R_u(\Delta x) = \sigma_u^2 e^{-\frac{\Delta x}{L}}, \tag{9.4}$$

where L is some average size of turbulent disturbances, defined as the distance in which the correlation function is reduced by a particular factor, usually to half its value, when $\Delta x = 0$.

It follows from the continuity equation [60] that for a three dimensional vector field the relation between the correlation functions for the longitudinal and transverse components of the wind-velocity fluctuations has the form

$$R_w(\Delta x) = R_u(\Delta x) + \frac{\Delta x}{2} R_u'(\Delta x), \tag{9.5}$$

and hence

$$R_w(\Delta x) = \sigma_w^2 \left[1 - \frac{\Delta x}{L}\right] e^{-\frac{\Delta x}{L}}. \tag{9.6}$$

For the correlation functions (9.4) and (9.6) we have the corresponding spectral densities

$$S_u(\Omega) = \sigma_u^2 \frac{L}{\pi(1 + \Omega^2 L^2)}, \tag{9.7}$$

$$S_w(\Omega) = \sigma_w^2 \frac{L(1 + 3\Omega^2 L^2)}{\pi(1 + \Omega^2 L^2)^2}, \tag{9.8}$$

where Ω is the wave number.

If we assume that the turbulence is isotropic, then, strictly speaking, the longitudinal correlation function is given by the expression

$$R_u(\Delta x) = \sigma_u^2 \left[1 - \frac{A}{2}(\Delta x)^{2/3}\right],$$

and not by expression (9.4). Here $A = c\epsilon^{2/3}/\sigma_u^2$, where ϵ is the rate of dissipation of turbulent energy, and c is a universal and dimensionless constant of the order of unity. However, the above assumption that the longitudinal correlation function is exponential fairly often provides a satisfactory approximation, from the practical viewpoint, of the experimental correlation functions obtained from an airplane.

When Eq. (9.8) is used for the energy spectrum of the fluctuations of the vertical wind-velocity component in various kinds of aviation calculations, the turbulence scale L is given by

$$L_u = \int_0^\infty R_u(\Delta x)\, d(\Delta x) \tag{9.9}$$

or, referring to (9.5),

$$L_u = 2\int_0^\infty R_w(\Delta x)\, d(\Delta x) = 2L_w; \tag{9.10}$$

L_u and L_w characterize the linear dimensions of the region within which there is a high correlation of the flow-velocity fluctuations and are called the longitudinal and transverse integral scales of this turbulent region. If the correlation function is exponential, the integral turbulence scale L is numerically equal to twice the value of Δx at which the correlation is reduced to 0.6.

The practical application of atmospheric turbulence data for the solution of some aviation problems requires not only the simple relationship of the form of (9.4), but also the functional form of the autocorrelation functions. Bulin (see [214]) proposed the following functional form for the family of normalized autocorrelation functions:

$$R_u(\Delta x) = \left[\left(\frac{\Delta x}{a}\right)^n / 2^{n-1}(n-1)!\right] K_n\left(\frac{\Delta x}{a}\right), \tag{9.11}$$

where a and n are parameters determining the shape and scale, K_n is a Bessel function of imaginary argument of the second kind, and $(n-1)!$ is a gamma function, where n is not a positive whole number.

Substituting (9.11) in (9.9) and (9.5), we obtain an expression for the integral turbulence scale

$$L = \frac{\sqrt{\pi}\left(n - \frac{1}{2}\right)!}{(n-1)!}a \tag{9.12}$$

and an expression for the transverse autocorrelation function

$$R_w(\Delta x) = \frac{\left(\frac{\Delta x}{a}\right)^n}{2^n (n-1)!}\left[2K_n\left(\frac{\Delta x}{a}\right) - \frac{\Delta x}{a}K_{n-1}\left(\frac{\Delta x}{a}\right)\right]. \tag{9.13}$$

Knowing $R_u(\Delta x)$ and $R_w(\Delta x)$, we can use the Fourier transformation to find expressions for the spectral density functions

$$S_u(\Omega) = \frac{4\sigma_u^2 L}{(1 + 4\pi a^2 \Omega^2)^{n+1/2}}, \tag{9.14}$$

$$S_w(\Omega) = \frac{2\sigma_w^2 L[1 + 8\pi^2 a^2 \Omega^2 (n+1)]}{(1 + 4\pi^2 a^2 \Omega^2)^{n+3/2}}, \tag{9.15}$$

where

$$a = \frac{L(n+1)!}{\sqrt{\pi}\,(n - 1/2)!}. \tag{9.16}$$

Thus, the turbulence spectrum is determined by only two quantities—the scale L and the parameter a. Two models of families of autocorrelation functions and spectral density functions are considered in aerodynamics: the Dryden model, for which $n = \frac{1}{2}$, and the von Karman model, for which $n = \frac{1}{3}$.

For the first model

$$L = a, \tag{9.17}$$

$$S_u(\Omega) = \frac{4\sigma_u^2 L}{1 + (2\pi\Omega L)^2}, \tag{9.18}$$

$$S_w(\Omega) = \frac{2\sigma_w^2 L[1 + 3(2\pi\Omega L)^2]}{[1 + (2\pi\Omega L)^2]^2} \tag{9.19}$$

For the second model, for which

$$a = L\frac{\left(-\frac{2}{3}\right)!}{\sqrt{\pi}\left(-\frac{1}{6}\right)!} = 1.339L, \quad L = 0.747a, \tag{9.20}$$

$$S_u(\Omega) = \frac{4\sigma_u^2 L}{[1 + \{2\pi(1.339L)\Omega\}^2]^{5/6}}, \tag{9.21}$$

$$S_w(\Omega) = \frac{2\sigma_w^2 L\left[1 + \frac{3}{8}\{2\pi(1.339L)\Omega\}^2\right]}{[1 + \{2\pi(1.339L)\Omega\}^2]^{11/6}}. \tag{9.22}$$

It should be noted that the autocorrelation functions given by (9.4) and (9.6) correspond to the Dryden model. The von Karman model, for which $n = \frac{1}{3}$, is the same as the model for which the "minus five-thirds" law is valid. Hence, (9.21) and (9.22) satisfactorily represent turbulent motions in the case of neutral thermal stratification of the atmosphere.

Equations (9.17)-(9.22) are widely used in aerodynamic calculations and are recommended for the standard model of the turbulent atmosphere [1]. It should be stressed, however, that they are approximate, of course, and cannot completely replace the actual spectra of atmospheric turbulence.

When expression (9.22), representing the spectral density of fluctuations of the vertical wind-velocity component, is used in various practical problems, it is essential to know the typical value of the turbulence scale L. Many investigators have determined this value by approximating the empirical spectral density functions by expression (9.22) and selecting an appropriate value of L for each $S_w(\Omega)$ curve.

As investigations have shown, the scale L for fluctuations of the vertical wind-velocity component in the atmospheric layer up to an altitude of 300-500 m generally increases with altitude. Above a smooth surface and in the case of neutral or unstable thermal stratification the scale L is approximately equal to the height above the surface, but over a hilly surface it is approximately twice the height; in both cases the standard deviation is about 25-30% of the mean value of L. This dependence of L on height above the ground cannot be extrapolated, as experimental data have shown, to heights exceeding 300-500 m. At greater heights the scale L varies in a wide range and depends in a complex manner on the thermal and wind stratification of the atmosphere.

Table 9.2 gives experimental data on the longitudinal integral turbulence scale L_u, defined as the integral of the normalized correlation function, for fluctuations of the horizontal wind-velocity component.

The value of $\overline{L}_u$ increases form 0.40 km in the 3-7 km layer to 1.0-1.2 km in the 12-18 km layer, while $L_{u\,\max}$ increases from 0.6 to 2.6-2.7 km. Table 9.2 shows that $L_{u\,\max} \approx 3\overline{L}_u$. In a first approximation, according to (9.10), the relation $L_w = L_u/2$ can be used.

According to the data of American investigators [157], in the lower stratosphere (13.5-21.0 km) above the U.S.A. the turbulence scale for fluctuations of

TABLE 9.2. Vertical Distribution of Integral Turbulence Scale L_u for Fluctuations of Vertical Wind-Velocity Component

H (km)	1–3	3–7	7–10	10–12	12–15	15–18
$\bar{L}_u$ (km)	0.464	0.40	0.54	0.63	1.21	0.992
$L_{u\max}$ (km)	0.9	0.6	1.5	1.9	2.7	2.6
$L_{u\min}$ (km)	0.1	0.1	0.2	0.1	0.3	0.3
Number of cases	11	8	34	23	26	25

TABLE 9.3. Standard Deviation σ_u of Velocities of Horizontal Wind-Velocity Fluctuations in Relation to Altitude

H (km)	< 1	1–3	3–7	7–10	10–12	12–15	15–18
$\bar{\sigma}_u$ (m/sec)	1.33	0.89	0.46	0.51	0.59	0.76	0.86
$\sigma_{u\max}$ (m/sec)	2.20	2.7	2.0	1.24	1.00	1.40	1.69
Number of cases	13	20	38	54	27	28	26

the vertical wind-velocity component, calculated from the spectral density by using the von Karman expression, is approximately 600 m.

In Chapter 5 we gave some empirical values of the wind-velocity fluctuations obtained by means of tethered balloons and radar observations. Table 9.3 gives the characteristics of the fluctuations of the horizontal wind-velocity component in the troposphere and stratosphere for spatial scales from a few tens of meters to 1-2 km.

The data in Table 9.3 show that, as was to be expected, $\bar{\sigma}_u$ decreases with altitude in the lower half of the troposphere, and increases at higher altitudes. The standard deviation of the velocity increases from 0.46 m/sec in the 3-7 km layer to 0.59 m/sec in the 10-12 km layer, and to 0.76-0.86 m/sec in the 12-18 km layer mainly above mountains. The maximum values of σ_u reach 1-2 m/sec or more.

An analysis of the experimental data showed that the distribution of wind-velocity fluctuations can be represented quite satisfactorily by a normal distribution. This is confirmed in [40]. Hence, we can assume that the maximum velocity of the fluctuations (gusts) can reach $3\sigma_{u\max}$.

The increase of $\bar{\sigma}_u$ with altitude in the lower stratosphere is due to the fact that most of the results that we used for the calculations related to flights above mountains. Looking ahead, we note that the dependence of the variance of the fluctuations of the wind-velocity components on the nature of the underlying surface and orography is confirmed by the results of flight investigations of turbulence by American scientists in the HICAT† program [186a]. The flights were conducted in clear skies at altitudes of 14-21 km. The values that they obtained for the standard deviations of the fluctuations of the longitudinal (σ_u),

†HICAT–High altitude clear-air turbulence.

transverse (σ_v), and vertical (σ_w) wind-velocity components in flights above different underlying surfaces and different orography are given in Table 9.4a. The variances of the fluctuations were calculated for disturbances with spatial scales $\lambda \leqslant 610$ m. The symbol h in the table means height above sea level.

The data of Table 9.4a show that above high mountains the standard deviations of the fluctuations of the wind-velocity components are 25–30% greater than the standard deviations of the same components above level ground.

The data for low altitudes are of particular interest for aviation. Experimental investigations by American scientists in the LO-LOCAT program in the lower 300-m layer of the atmosphere above deserts, level terrain, and on the lee side of low and high mountains showed that the values of u', v', and w' in clear air depend on the nature of the underlying surface and the thermal stratification of the atmosphere. The extrema of u', v', and w' are satisfactorily represented by a log-normal distribution. Table 9.4 gives the medians of the extreme fluctuations of all three wind-velocity components and their values for a cumulative probability of 99.0 and 99.9%. The table shows that in general the medians of the extreme fluctuations of the transverse component are a little larger than the medians of the extreme fluctuations of the other two components. The differences in the distribution curves of the extreme fluctuations of all three wind-velocity components are much smaller than the differences due to thermal stratification of the atmosphere and the nature of the underlying sur-

TABLE 9.4. Statistical Characteristics of Extreme Fluctuations of Wind-Velocity Components

Thermal stratification	Underlying surface	u' (m/sec)			v' (m/sec)			w' (m/sec)		
		median	99.0%	99.9%	median	99.0%	99.9%	median	99.0%	99.9%
Very stable	Desert	1.5	5.9	9.2	1.8	4.8	6.7	1.7	4.8	6.7
	Plain	1.8	6.4	9.8	2.1	6.2	8.9	2.1	5.7	7.9
	Low hills	2.6	7.9	11.4	2.9	8.6	12.2	2.6	7.1	9.7
	High mountains	3.0	8.7	12.4	3.3	9.7	13.9	3.2	8.8	12.3
Stable	Desert	2.5	7.3	10.3	2.9	7.9	10.4	3.0	7.5	10.1
	Plain	2.9	7.6	10.4	3.5	8.0	10.4	3.3	7.6	10.0
	Low hills	3.8	9.1	13.0	4.3	10.1	13.4	4.1	9.3	12.4
	High mountains	4.1	11.4	16.0	4.6	11.4	15.3	4.4	10.6	14.1
Neutral	Desert	3.0	6.7	8.6	3.7	7.6	9.6	3.4	6.9	8.7
	Plain	3.6	7.7	9.8	4.1	9.1	11.7	3.7	7.3	9.1
	Low hills	4.4	9.9	12.9	5.2	11.9	15.7	4.7	10.1	13.0
	High mountains	4.3	10.6	14.3	5.2	11.3	14.5	5.1	12.6	16.9
Unstable	Desert	4.2	7.7	9.4	4.6	9.3	11.7	4.6	7.5	8.8
	Plain	3.5	7.5	9.6	3.8	8.4	10.9	3.6	8.3	10.7
	Low hills	4.9	9.2	11.3	5.5	9.8	11.9	5.2	11.2	14.3
	High mountains	4.9	10.9	14.2	5.7	12.2	15.7	5.5	10.0	12.1

TABLE 9.4a. Standard Deviations of Fluctuations of Wind-Velocity Components (m/sec) at Altitudes of 14–21 km

Nature of surface	σ_u	σ_v	σ_w
Water	0.354	0.381	0.274
Level ground ($h < 900$ m)	0.348	0.375	0.268
Low hills (900 m $\leqslant h \leqslant$ 2100 m)	0.396	0.424	0.317
High mountains ($h > 2100$ m)	0.442	0.469	0.363

face. Thermal stratification is the most significant meteorological condition affecting clear-air turbulence intensity at low altitudes.

Experimental investigations of turbulence on TU-104 and IL-18 airplanes led to the determination of tentative spectral density ranges corresponding to aircraft buffeting of different intensity in the gradations indicated in Table 9.1. These ranges of $S(\omega)$ were determined for the high-frequency part of the spectrum, in which the "minus five-thirds" law is probably always valid, i.e., for the wave-number region $\omega \geqslant 2 \cdot 10^{-3}$ m^{-1}. In this case it is sufficient for assessment of aircraft buffeting to indicate the range of spectral densities for some one wave number, particularly for $\omega = 2 \cdot 10^{-3}$ m^{-1} (Table 9.5).

2. Characteristics of Turbulence in Troposphere and Lower Stratosphere

In the USSR, U.S.A., Great Britain, Canada, and other countries in recent years great attention has been given to the study of turbulence affecting the flight of aircraft. The use of aircraft as turbulence indicators in turn has allowed the most thorough investigation of disordered motions of air in the atmosphere.

The most valuable, of course, are instrumental observations, particularly when the results of measurements are continuously recorded. The principal drawback of visual observations is the subjectivity of assessment of turbulence intensity, which depends on the physiology and flying experience of the aircrew. This is apparent from Table 9.6, which gives load increments recorded by the

TABLE 9.5. Intensity of Aircraft Buffeting in Relation to $S(\omega)$ for $\omega = 2 \cdot 10^{-3}$ m^{-1}

Intensity of aircraft buffeting	$S(\omega)$ (m^3/sec^2)
Light, $b^{(1)}$	$\sim 10^2 - 5 \cdot 10^2$
Moderate, $b^{(2)}$	$5 \cdot 10^2 - 10^3$
Severe, $b^{(3)}$	$10^3 - 3 \cdot 10^3$
Very severe, $b^{(4)}$	$\geqslant 3 \cdot 10^3$

TABLE 9.6. Relation between Turbulence Intensity Assessed by Pilots and from Aircraft Loads

Load increments in g	Turbulence intensity assessed by pilots		
	light, $b^{(1)}$	moderate, $b^{(2)}$	severe, $b^{(3)}$
<0.2	71	24	5
0.2	53	40	7
0.2–0.5	–	78	22
$\geqslant 0.5$	–	–	100

accelerograms, and the frequency (in percent) of visual estimates of intensity of buffeting by the aircrew.

At airplane load increments not exceeding $0.2g$, i.e., the case of light buffeting (see Table 9.1), the pilots often assessed the turbulence as moderate (25% of the cases), or even as severe. Atmospheric turbulence producing load increments of 0.2-$0.5g$, i.e., moderate turbulence, was assessed by the pilots as severe in 22% of the cases. Visual observations usually lead to an overestimate of the frequency of moderate and severe turbulence. Similar results were obtained in [163] from a comparison of visual (subjective) and instrumental data for turbulence in the stratosphere in flights on the supersonic XB-70 airplane: The pilots reported more intense turbulence than was shown by the airplane instrument.

We now consider the special features of the vertical distribution of turbulence and some parameters of turbulent zones.

2.1. Frequency of Turbulence at Different Altitudes. According to the nature of the turbulence, its frequency, etc., the troposphere can be divided into three layers: lower (0-2 km), middle (2-6 km), and upper (6-10 km). We can also consider the troposphere and stratosphere separately.

Figure 9.3, plotted from the data of aircraft measurements, shows that turbulence is most frequent in the lower layer of the troposphere and least frequent in the middle troposphere. Its frequency then increases with approach to the tropopause or to the level with maximum wind velocity. The frequency of buffeting is greatest at altitudes of 8-12 km, where it reaches 5-20%.

It should be noted that the results of determination of the frequency of aircraft buffeting depend on the nature of the flights. For instance, aircraft used on short routes and flying at relatively low altitudes are subject to more frequent buffeting than aircraft performing long-distance flights at high altitudes.

Transport aircraft flying at an altitude of 5 km experience load increments exceeding $0.5g$ only about one-tenth as frequently as aircraft flying at altitudes of 1-2 km.

The horizontal extent of turbulent zones is many times greater than their thickness. Hence, the entry of an airplane into a turbulent zone in horizontal flight is less probable than in vertical probing of the atmosphere (including the climbing and descent of airline aircraft). In addition, it should be taken into account that the dynamic stability of an airplane during ascent and descent is somewhat lower than in horizontal flight. For instance, according to Barakhtin's data [4], the frequency of buffeting of regular TU-104 airplanes is 2-3 times greater during ascent, and especially during descent, than in horizontal portions of the flight.

We return to a consideration of Fig. 9.3. Curve 1, plotted from the data of vertical probing of the atmosphere, characterizes the frequency of turbulence in southern latitudes of the USSR. In temperate latitudes of the USSR the frequency of turbulence at altitudes of 8–12 km is 11.7% on the average [94]. It is less than the frequency of turbulence at the same altitudes in southern regions of the USSR, which is 16–18% on the average. This estimate is close to Hyde's data for the London-North Africa and London-Far East routes via Rome, Cairo, Karachi, and Delhi. An interesting feature is the low frequency of aircraft buffeting at altitudes of 2–8 km above the North Atlantic, obtained by Crossly from the data of more than 15,000 reports of crews of transport aicraft: At altitudes of 3–4 km it was 4.7% on the average in January and 2.1% in July, and, at altitudes of 7–8 km it was 3.1% and 1.0%, respectively.

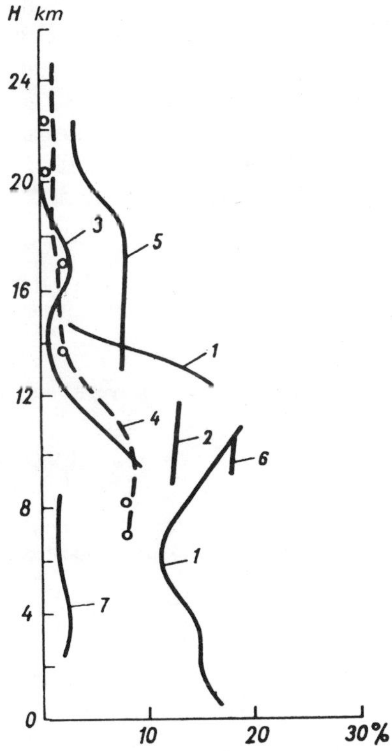

Figure 9.3. Frequency of turbulence at different altitudes. (1) USSR, southern latitudes; (2) USSR, temperate latitudes; (3) European USSR and West Siberia; (4) U.S.A. (U-2 data); (5) U.S.A.; (6) London–Far East and London–North Africa routes; (7) North Atlantic.

It should be noted that curve 1 in Fig. 9.3 was plotted mainly from flight observations. The vertical distribution of the frequency of turbulence for winter may differ from this curve owing to seasonal differences in the temperature and wind profiles. Of great interest in this connection are Barakhtin's data [4] for the seasonal variation of the frequency of buffeting on the Moscow-Khabarovsk route, obtained from accelerograms of more than 10,000 flights of TU-104 airplanes (Table 9.7). The calculations included only measurements made on horizontal portions of the flights.

Table 9.7 shows that over the whole route turbulence was most frequent in summer, and least frequent in winter, except in the Moscow-Omsk leg, where it was least frequent in spring. In the upper troposphere, where the flights mainly occurred, the frequency of buffeting in summer was 2.4–3.6 times greater than in winter.

Information about turbulence in the stratosphere is of great importance for flights of supersonic passenger aircraft. Such data are sparse and very contradictory at present. This is apparent from Table 9.8, which gives the frequency of aircraft buffeting over the USSR (European USSR and West Siberia, the Main Caucasus Range) and the U.S.A. In the 11–20 km layer the frequency of turbulence was about 1% on the average. In certain experiments, however, especially above mountains, turbulence was several times more frequent.

The great difference in frequency of turbulence over the U.S.A., especially from the data of [211] and [184], is due not only to weather conditions, but also to the fact that the results given in [211] were obtained from the data of flights of the subsonic U-2 airplane with instruments having an initial (threshold) sensitivity of $0.1g$, whereas the results given in [184] were obtained from data of the flights of the supersonic XB-70 airplane with instruments having an initial sensitivity of $0.06g$. In addition, the XB-70 is sensitive to disturbances of greater scale than the U-2.

Curves 1, 3, 4, and 5 in Fig. 9.3 give a picture of the frequency of turbulence at different altitudes in the lower stratosphere. It is apparent that the frequency of turbulence in the lower stratosphere decreases fairly rapidly with altitude: At altitudes of 14–15 km over the southern USSR it does not exceed 3–4%, and from the data of U-2 flights over the U.S.A. it is about 2.5%, and subsequently decreases to 0.5–0.1% at altitudes of 18–23 km [211]. From the data of flights

TABLE 9.7. Frequency (%) of Turbulence of Moscow-Khabarovsk Route in Relation to Season

Leg of route	Winter	Spring	Sum.	Fall	Average
Moscow–Omsk	3.2	2.1	7.6	4.1	4.2
Omsk–Irkutsk	2.4	3.8	8.5	4.0	4.7
Irkutsk–Khabarovsk	5.2	3.1	12.4	7.6	3.3.

TABLE 9.8. Frequency of Turbulence in Lower Stratosphere

Flight altitude (km)	Total flight length	Flight in turbulent zones (10^3 km)	Flight in turbulent zones %	Reference	Flight altitude (km)	Total flight length	Flight in turbulent zones (10^3 km)	Flight in turbulent zones %	Reference
USSR					U.S.A.				
11–20	3.128	24.97	0.8	[103]	12–23	1.248	11.83	0.95	[211]
11–20	3.768	20.76	0.6	[104]	13–21	664	30.00	4.51	[157]
9–18	26	1.84	7.0	[90]	12–23	192	7.78	6.4	[184]
					13–20	198	0.79	0.4	[192]

of the supersonic XB-70 the frequency of turbulence at altitudes of 14-20 km is 7.1-7.4%, and only at high altitudes does it decrease with height. Above the European USSR and West Siberia [104] the frequency of turbulence in general decreases with increase in altitude—from 8.9% at 10 km to 0.6% at 20 km; a small maximum of frequency is observed at altitudes of 16-18 km. A similar maximum is observed at approximately the same altitudes above the U.S.A. [157].

Measurements carried out in the USSR and the U.S.A. have shown that in the stratosphere turbulence is most frequent in the winter.

2.2. Intensity of Turbulence. The intensity of turbulence is assessed approximately from the load increments on aircraft (i.e., from the effective velocity of vertical gusts). Outside thunder clouds the load increments rarely exceed $0.9g$. However, in a flight of a Comet in a clear-air turbulent zone, Hislop observed load increments up to $\pm 1.5g$. Bindon described the case of an experimental flight in a clear-air turbulent zone over Canada at an altitude of 7.3 km, where $\Delta n \approx 3g$. In this flight many of the measuring instruments were put out of order. There was a case recorded in the U.S.A. where a Boeing B-52 that entered a zone of clear-air turbulence above a mountain region had its fin and elevator damaged.

When aircraft encounter intense clear-air turbulence they usually lose height very rapidly. An analysis of such cases has shown that pilots cannot always bring the wheel back and pull the airplane out of the dive. As an example, Fig. 9.4 shows an oscillogram of the flight of a Boeing-720B in a clear-air turbulent zone over the western U.S.A. [146]. The airplane was in the turbulent zone for about 4 min, including approximately 30 sec in very severe turbulence, when the load increments reached +2.0, $-2.8g$; the airspeed varied in the range 505-575 km/h. The airplane lost about 2 km of altitude.

At altitudes of 8-12 km in temperate latitudes of the USSR there were no cases where the value of Δn in flights of a TU-104 airplane in a cloudless atmosphere was greater than $\pm 1g$ [94]. Even load increments reaching $\pm 1g$ were observed in only a few isolated cases.

On the recommendation of the International Civil Aviation Organization

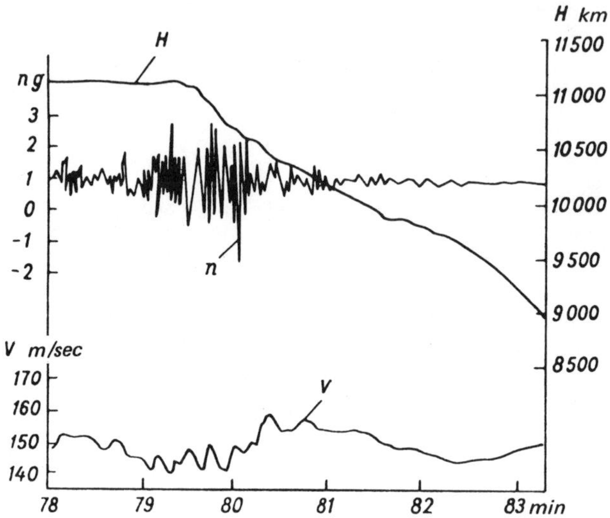

Figure 9.4. Oscillogram of flight of Boeing-720B in a clear-air turbulent zone.

(ICAO) and the World Meteorological Organization (WMO) information about turbulence at altitudes of 6-7 to 10-12 km was collected from aircrews of airline aircraft in many countries in 1964-1965. The information was collected in the following periods: December 9-13, 1964, March 10-14, June 9-13, and September 8-12, 1965. The collected information about the airplane flight conditions was processed by a standard method, which allowed comparison of the results of investigations carried out in different countries.

Table 9.9 gives data for the frequency of turbulence of different intensity in the upper troposphere above different geographical regions [94, 144, 152, 167]. The table shows that the greatest frequency of turbulence, including moderate, severe, and even very severe turbulence, was observed above Japan and the adjacent seas. Turbulence was most frequent in the upper troposphere in the zone 15-45° lat. N. This is due to the fact that intense jet streams are

TABLE 9.9. Frequency (%) of Turbulence of Different Intensity in Different Geographical Regions

$b^{(i)}$	Greenland, Canada	U.S.A.	Atlantic		Western Europe	USSR, temperate latitudes	Japan
			25–45° lat. N	45–70° lat. N			
$b^{(0)}$	84.0	69.0	81.0	88.0	92.0	88.3	53.0
$b^{(1)}$	10.4	21.3	12.0	7.5	5.0	5.7	34.5
$b^{(2)}$	5.2	9.3	7.0	4.3	2.8	5.6	11.8
$b^{(3)}$	0.4	0.4	0.0	0.2	0.3	0.4	0.6
$b^{(4)}$	—	—	—	—	—	—	0.1

TABLE 9.10. Frequency (%) of Turbulence in Relation to Height of Relief *h* in Southern Europe [144]

Flight altitude (km)	*h* (m)					
	<1000		⩾1000		⩾2000	
	$\sum_1^3 b^{(i)}$	$\sum_2^3 b^{(i)}$	$\sum_1^3 b^{(i)}$	$\sum_2^3 b^{(i)}$	$\sum_1^3 b^{(i)}$	$\sum_2^3 b^{(i)}$
6–9	9.2	3.2	9.4	3.9	10.5	4.9
9–11	7.8	2.5	6.8	2.7	9.1	3.5

most frequently observed here. Moderate and severe turbulence was usually recorded at altitudes of 9–13 km in an even narrower band: 15–30° lat. N.

An analysis of the data collected in Europe and America showed that the frequency of turbulence above continents depends strongly on the nature and height of the relief: Above mountain regions the frequency of turbulence causing aircraft buffeting is significantly greater than above flat regions (Table 9.10). The difference in the frequency of moderate and severe buffeting are particularly pronounced.

The data of U-2 airplane flights carried out in the U.S.A. as part of a program of research on clear-air turbulence at high altitudes [157] showed that in the stratosphere the frequency and intensity of turbulence were greatest above mountain regions. The results of these investigations are given in Table 9.11.

The data in Table 9.11 were obtained from 50 U-2 flights in turbulent zones over water, 176 flights over level terrain, and 207 flights in turbulent zones above mountain regions. It is apparent that the calmest flights were above expanses of water, where severe buffeting was never observed. In flights above level terrain, buffeting was moderate and sometimes severe in a third of the cases, and in flights above mountain regions buffeting was often not only moderate, but also severe.

These results are confirmed by the data of investigations on a high-altitude probing plane over a plain and spurs of the Main Caucasus Range [23]. Table 9.12 gives the mean maximum positive (in numerator) and negative (in denominator)

TABLE 9.11. Frequency (%) of Turbulence of Different Intensity

$b^{(i)}$	Above water	Above mainland	
		$l \leqslant 750$ m	$l > 750$ m
$b^{(1)}$	80.0	67.0	49.0
$b^{(2)}$	20.0	30.0	32.0
$b^{(3)}$	0.0	2.0	19.0

TABLE 9.12. Intensity of Turbulence (in *g*) in Lower Stratosphere

Flight altitude (km)	Above plain		Above mountains		Flight altitude (km)	Above plain		Above mountains	
	$\overline{\Delta n}$	Δn_{max}	$\overline{\Delta n}$	Δn_{max}		$\overline{\Delta n}$	Δn_{max}	$\overline{\Delta n}$	Δn_{max}
9–12	0.09 / 0.25	0.10 / 0.29	0.36 / 0.43	1.25 / 1.14	14–16	0.09 / 0.16	0.18 / 0.38	0.24 / 0.36	0.81 / 1.08
12–14	0.07 / 0.09	0.16 / 0.14	0.33 / 0.44	0.54 / 0.85	16–18	0.03 / 0.09	0.07 / 0.14	0.31 / 0.48	0.97 / 1.23

load increments and absolute maximum increments obtained in summer and autumn above mountain (1966) and plain (1967) regions.

The intensity of turbulence depends not only on the altitude, the nature of the underlying surface, and the relief, but also, of course, on weather conditions.

2.3. Spatial Dimensions of Turbulent Zones. Turbulent atmospheric layers in which aircraft buffeting occurs have relatively sharp boundaries. This can probably be attributed to the fact that at the boundaries of disturbed layers the turbulent conductance [30] is small, which is characteristic of undisturbed or almost undisturbed regions surrounding a turbulent zone in which aircraft buffeting occurs. The presence of sharp boundaries allows a fairly accurate determination of the vertical and horizontal extent of turbulent zones in the atmosphere by means of accelerograms. The greatest amount of data on the spatial dimensions of turbulent zones relates mainly to the middle and upper troposphere.

Figure 9.5 shows curves of the integral frequency of the thickness of turbulent zones for various latitudes in the Soviet Union. It is apparent that reduction of latitude leads to an increase in the frequency of relatively thick turbulent

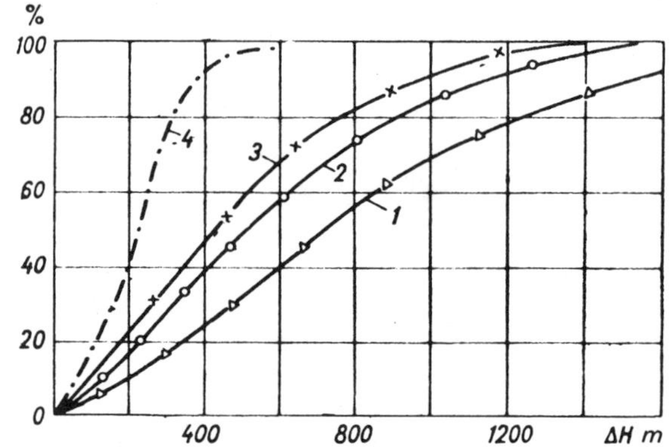

Figure 9.5. Integral frequency of thickness ΔH of turbulent zones in upper troposphere (1–3) and stratosphere (4). (1) Southern latitudes; (2, 4) temperate latitudes; (3) northern latitudes.

layers. In southern regions of the USSR a thickness of less than 1000 m was observed in approximately 70% of the cases, whereas in middle and high latitudes the figure was 85–90%.

Curve 4 in Fig. 9.5 characterizes the integral frequency of the thickness of turbulent layers observed at altitudes of 11–20 km above the European USSR and West Siberia [103]. It was obtained from the results of determination of 429 zones of buffeting of high-altitude aircraft.

The curve shows that in 80% of the cases the thickness of the turbulent layers was less than 300–350 m.

According to American data, the thickness of clear-air turbulent layers at altitudes of 3–15 km is usually less than 300 m. Later investigations conducted on a U-2 airplane established that the thickness of turbulent layers in the stratosphere above the U.S.A. is 500–1000 m on the average. A comparison of the curves in Fig. 9.5 shows that on the average the thickness of turbulent zones in the lower stratosphere is less than in the upper troposphere.

The determination of the horizontal extent of clear-air turbulent zones involves certain difficulties, since the airplane can pass through them in different directions and at arbitrary distances from the center of the horizontal cross section of these zones. In addition, the shape of the horizontal cross section of turbulent zones is still unknown. Some authors indicate that in jet streams the turbulent zones are extended in the direction of the air flows: They are longer than broad.

Figure 9.6 gives the integral frequency of the horizontal extent of turbulent zones. It is apparent that the length of turbulent zones in the upper troposphere

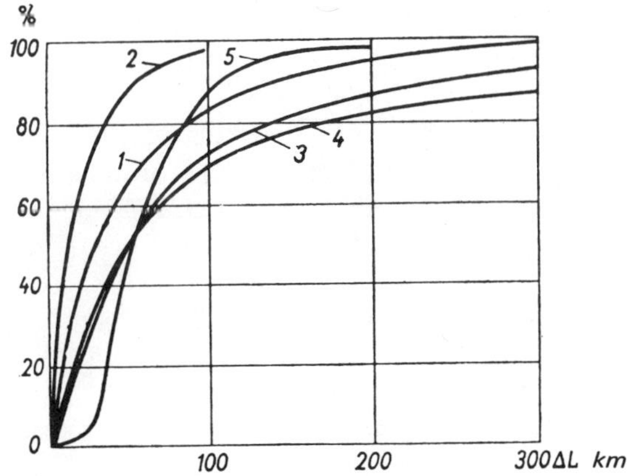

Figure 9.6. Integral frequency of horizontal dimensions ΔL of turbulent zones. (1) U.S.A. (upper troposphere); (2) U.S.A. (stratosphere); (3) USSR (upper troposphere, temperate latitudes); (4) USSR (upper troposphere, southern latitudes); (5) USSR (stratosphere, temperate latitudes).

varies from a few kilometers to 400-500 km or more, although it usually does not exceed 60-80 km. In temperate latitudes of the USSR the length of the turbulent zones was less than 100 km in approximately 72% of the cases and more than 400 km in 4% of the cases; for southern latitudes the corresponding figures were 68 and 10%, respectively. Thus, the horizontal extent, like the thickness of the turbulent zones, increases with reduction of latitude.

Colson [152] investigated the relation between the duration of flight in turbulent zones and the buffeting intensity. If the mean duration of flight with strong buffeting is taken as unity, the mean relative duration of flight with moderate buffeting is 1.7, and that with light buffeting is 2.4.

In addition, the duration of flight with severe buffeting is distinguished by relatively small dispersion ($\sigma_t/\bar{t} = 0.58$).

King [182] thoroughly investigated 36 flights in clear air in the upper troposphere with severe and very severe turbulence.

The maximum values of Δn in these flights were

$\Delta n(g)$	0.75-0.8	0.8-0.9	0.9-1.0	1.0-1.1	1.1-1.2	1.2-1.45
Number of cases . .	8	12	7	5	3	1

The duration of the flight in such conditions usually did not exceed 15-30 sec, and it was only in isolated cases that it reached 2-3 min. In King's opinion, a small horizontal extent of the zones is an important characteristic of intense turbulence.

The horizontal dimensions of zones observed in the lower stratosphere are smaller on the average than in the upper troposphere; in temperate latitudes of the Soviet Union the horizontal dimensions of these zones were less than 80 km in 80% of the cases. For comparison we point out that in the lower stratosphere above the U.S.A. the horizontal dimensions of turbulent zones were less than 40 km, and those in the upper troposphere less than 85 km, in 80% of the cases.

The relation between the horizontal dimensions of turbulent zones and the altitude is confirmed by the results of flights of a high-altitude research airplane over the spurs of the Main Caucasus Range [90]. At altitudes of 10-12 km in 80% of the cases ΔL was less than 35 km, and at altitudes of 16-18 km it was less than 18 km in 80% of the cases. This reduction of size of turbulent zones appears to be a specific feature of turbulence above mountain regions.

To illustrate what we have said, Fig. 9.7 shows the flight path of a supersonic XB-70 airplane in the stratosphere above the Western U.S.A. The positions of turbulent zones encountered by the airplane are marked on the path. Data on the horizontal dimensions ΔL of the turbulent zones and the maximum load increments Δn_{max} of the airplane are given in Table 9.13.

Thus, clear-air turbulent zones are flat ("pancake"-shaped) and exhibit pronounced spatial anisotropy. The coefficient of spatial anisotropy (the ratio of the thickness of the turbulent zone to its horizontal extent), calculated for 80%

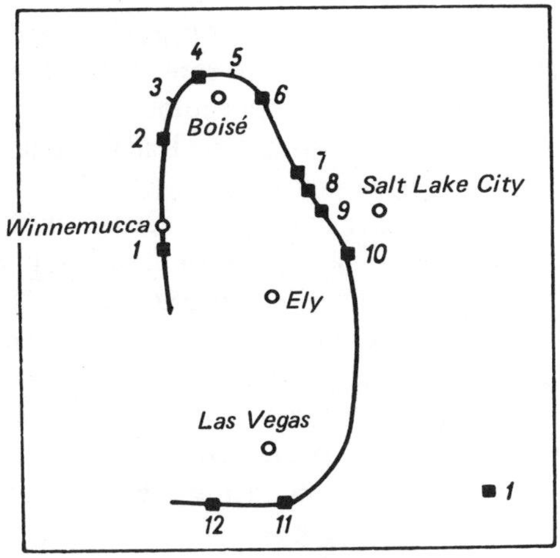

Figure 9.7. Turbulent zones (1) encountered by XB-70 airplane. (1–12) Numbers of turbulent zones.

integral frequency, is $7 \cdot 10^{-3}$ for the upper troposphere in moderate latitudes, and $4 \cdot 10^{-3}$ for the lower stratosphere.

2.4. Some Structural Features of Turbulent Zones. An analysis of accelerograms shows that two types of turbulent zones—continuous and discontinuous—are encountered in the free atmosphere. Discontinuous zones consist of several continuous turbulent regions between which flights are calm. At altitudes of 8–12 km the number of local turbulent regions varies from two to seven, but in 79.5% of the cases it does not exceed three. In 86.6% of the cases the extent of these zones is less than 60 km and in only 6.6% of the cases is it more than

TABLE 9.13. Characteristics of Turbulent Zones Encountered by XB-70 Airplane on October 11, 1967

Number of turbulent zone	Greenwich Mean Time	Flight altitude (km)	Δn_{max} (g)	ΔL (km)
1	16 38	15.76	0.127	63.9
2	16 43	13.40	0.095	23.0
3	16 48	16.61	0.111	3.4
4	16 49	16.76	0.127	15.4
5	16 50	16.76	0.095	6.8
6	16 52	16.89	0.254	9.5
7	16 56	16.95	0.095	10.9
8	16 57	16.83	0.095	26.2
9	16 58	16.95	0.127	19.5
10	17 02	16.89	0.143	8.2
11	17 20	15.48	0.111	8.7
12	17 24	13.11	0.127	12.4

100 km. The extent of calm regions in discontinuous turbulent zones is less than 100 km in 80% of the cases. Thus, the calm regions are usually a little larger than the turbulent regions.

In temperate latitudes the frequency of discontinuous turbulent zones is half as high as that of continuous zones.

The presence of two types of turbulent zones is confirmed by the results of investigations of Japanese scientists [167]. As Table 9.14 shows, the intensity of turbulence in continuous zones is usually greater than in discontinuous zones.

The conditions leading to the formation of discontinuous turbulent zones are still not clear. These zones may be due to loss of stability of atmospheric waves or, what is probably more likely, they reflect the intermittency of decay of turbulence in turbulent zones.

In the stratosphere, as Fig. 9.6 shows, the extent of turbulent zones is less than in the upper troposphere. The results of 814 flights in turbulent zones in the stratosphere [104] showed that continuous zones were present in 38% of the cases, and discontinuous zones in only 2% of the cases. In 60% of the cases the effect of the turbulence consisted of separate sharp jolts (bumps) of the airplane.

An analysis of accelerograms revealed that in the upper troposphere there are two modes of increase in turbulence intensity (aircraft buffeting) at the boundaries of turbulent zones [94]:

(1) when the intensity of buffeting at the boundary of the zone is at least moderate;
(2) when the intensity of buffeting increases more or less gradually from very light to light, or moderate, from the boundary into the zone.

In the first case the entry of the airplane into the zone of atmospheric turbulence in a clear sky is not only unexpected, but also sharp, and the crew must decide what measures to take to ensure the safety of the flight; in the second case the crew has more time to prepare for the encounter with the moderate or severe turbulence.

In temperate latitudes the frequency of sudden entry into turbulent zones with moderate or severe turbulence at the boundary is 3.6%, but in some regions it reaches 5-6%.

TABLE 9.14. Frequency (%) of Turbulence of Different Intensity in Discontinuous and Continuous Turbulent Zones

Type of zone	Intensity			
	$b^{(1)}$	$b^{(2)}$	$b^{(3)}$	$b^{(4)}$
Discontinuous	78.6	21.0	0.4	—
Continuous	71.8	26.6	1.3	0.3

2.5. Persistence of Turbulent Zones. An important characteristic of turbulent zones is their lifetime. Experimental studies of the lifetimes of these zones require continuous investigation of the evolution of turbulence within them. Such investigation is hampered by the fact that turbulent zones outside clouds cannot be discerned visually and, in addition, move along with the flow.

In [94] repeated flights between geographical checkpoints were carried out to determine the lifetime of turbulent zones. This allowed the determination of the lifetime of turbulence in the air flow between the points, which probably depends both on the variation of the turbulence and on the size of the turbulent regions. These data are very useful for assessment of the validity of inertial prediction of turbulence from weather reconnaissance reports.

Figure 9.8 gives the results of observations of atmospheric turbulence between Nizhnetambovskoe and Elabuga in the Far East on May 14, 1960. The direction of the wind was almost perpendicular to the line connecting the indicated points. The observations, at an altitude of 7200 m, lasted 1 h 40 min. Since the wind speed at this height was 85 km/h, then a flow 145 km "long" passed between the points during the observations. During this time, as Fig. 9.8 shows, the nature of the turbulence and its distribution between the control points were almost constant. The observations at an altitude of 8300 m were separated by an interval of 2 h 49 min. The mean wind speed at an altitude of 8300 m was 120 km/h; hence, a flow 366 km in "length" passed between the checkpoints. It is apparent that the nature of the turbulence was greatly altered.

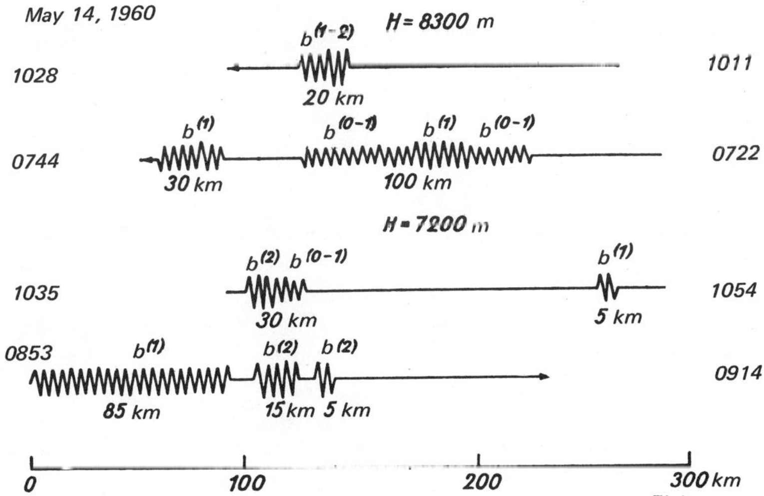

Figure 9.8. Nature of turbulence in repeated flights of an airplane between two points.

According to Barakhtin's experimental data [3], turbulence at altitudes of 8–10 km sometimes lasts for a day or more.

An idea of the persistence of turbulence above points is given by observations with radiosondes equipped with load attachments. Belyaev *et al.* conducted several series of radiosonde launches with intervals of 1 to 3 h between the launches. They found that the probability of the turbulent state of the atmosphere in the 8–12 km layer lasting for 1 h was 80–90%. In the middle and upper troposphere the probability of turbulent conditions lasting for 1.5–6.0 h was 60–75%, and at higher altitudes (12–20 km) it did not exceed 50%. For intervals greater than 6 h at almost all altitudes this probability again did not exceed 50%. Near the tropopause and the level of the tropospheric maximum of wind velocity the probability of persistence of the atmospheric state was 70%.

The data presented show that $t = 5$ h can be taken as a characteristic time scale for duration of the turbulent regime for the upper troposphere with a probability of about 75%.

3. Turbulence in Jet Streams and in Region of Tropopause

If we leave aside flights in clouds, the highest frequency of buffeting and maximum loads in the upper troposphere are observed in jet streams and near the tropopause. This is where clear-air turbulence is usually found; encounters with such turbulence are most hazardous because of their unexpectedness.

We will discuss buffeting in jet streams first. Since it is different in different sectors of these streams, we will give a brief account of current ideas on the aerological structure of a jet stream.

3.1. Frontal Structure of Jet Streams. Results of investigations of the aerological structure of jet streams, based mainly on the data of standard aerological observations, have been summed up by Vorob'ev [25], Reiter [203], and Georgii and Petrenko [39].

Special flight investigations of jet streams have been carried out in several countries. Their main results are given by Endlich and McLean [165], Brundidge and Clodman [145], Pinus and Shmeter [94], and Briggs and Roach [143]. In particular, in Project Jet Stream in the U.S.A., experimental investigations were made on two airplanes equipped with instruments for recording temperature, pressure, and aircraft loads, and also with Doppler devices for measuring the velocity and direction of the wind.

The experimental data showed that jet streams at different latitudes have a frontal structure. The front in the upper troposphere, called the jet-stream front, separates warm air from colder air. It can easily be followed downwards from the tropopause to an altitude of 5–3 km. The jet-stream axis is located in warm air, in the sector between the frontal surface and the tropical tropopause.

The horizontal and vertical wind shears on the cold (cyclonic) side of the jet stream are greater than on the warm side. In particular, the horizontal wind shears here are approximately 1.5 times greater than on the warm (anticyclonic) side of the stream. They decrease with increasing distance from the core of the jet stream.

A wide range of experimental investigations of jet streams was conducted in the USSR [94] by means of TU-104 and IL-18 airplanes equipped with instruments for recording the velocity and direction of the wind (Doppler devices), air temperature and pressure, airplane loads, and pitch and roll angles.

The flights were made over the European USSR, the Far East, and Central Asia. They were usually made across the jet stream, and each experiment involved four to six horizontal crossings of the jet stream, each one 300–500 km long.

The investigations showed that jet streams of temperate latitudes can be typified by two models. In the first model the jet stream is associated with frontal activity affecting the entire thickness of the troposphere, i.e., from the tropopause to the ground. In the second model, as distinct from the first, the jet stream is associated with frontal activity that is clearly manifested only in the upper half of the troposphere. In the upper troposphere in the Far East the transitional zone between the warm and cold air masses is a pronounced front with large horizontal and small vertical temperature gradients. The upper part of this front adjoins the tropopause in a region where it is discontinuous or has a steep slope, while the lower part can usually be traced to the 400- to 500-mb level. The front associated with the jet stream corresponding to the second model can be inclined to the horizon at different angles. In the Far East the inclination of the jet-stream fronts to the horizon is $\frac{1}{225}$ to $\frac{1}{150}$. It is of interest that at small angles of inclination the front is drawn out in such a way that its lower part forms an almost horizontal layer with temperature inversion.

Figure 9.9 shows an averaged vertical cross section of a jet stream. It is apparent that even on the average the horizontal temperature differences in the region of the front associated with the jet stream reach 2.5°C per 100 km. The isolines of relative velocity, as Fig. 9.9 shows, are also closest together in the region of the front. The wind velocity decreases by approximately 50% in the layer 3–4 km above the jet-stream axis. The horizontal wind-velocity differences on the cold side of the jet stream are greater than on the warm side, where they are generally small.

The empirical data showed that the changes in wind velocity with altitude below the maximum wind-velocity level and above it could be represented by the exponential function

$$u_z = u_0 e^{-\alpha|z - z_0|}, \tag{9.23}$$

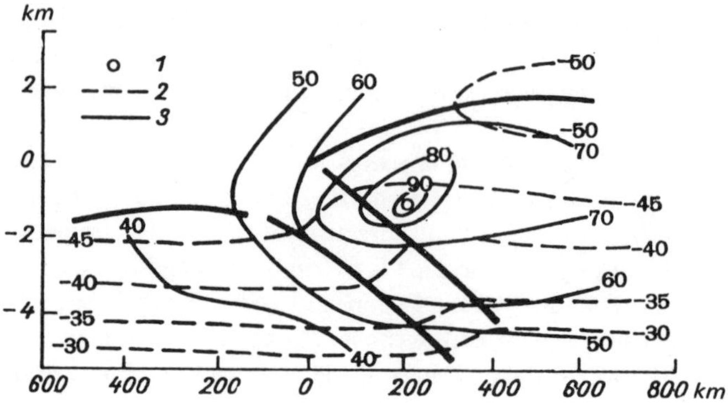

Figure 9.9. Vertical cross section of jet stream. (1) Stream axis; (2) isotherms (°C); (3) relative isotachs (%).

where z_0 is the altitude of the maximum wind velocity u_0. The wind velocity in a horizontal plane relative to the jet-stream axis is given by a similar relation:

$$u_y = u_0 e^{-\beta |\Delta y|} \tag{9.24}$$

It should be noted that in many cases, particularly on the anticyclonic side of jet streams, the wind velocity decreases approximately linearly with increasing distance from the axis or the maximum wind level. This does not destroy the generality of the approximation, however, since expression (9.24) can be written in the form

$$u_y = u_0 (1 - \beta |\Delta y|), \tag{9.25}$$

if we confine ourselves to the first two terms of the expansion of exp $(-\beta|\Delta y|)$ in a series. This approximation introduces an error that does not exceed 10% on the average for $|\Delta y|$ not exceeding 500 km. The values of α and β are given in Table 9.15. The value of β depends on the altitude, increasing as it approaches the jet axis from below and decreasing above the jet stream.

The jet streams observed in the Far East correspond with the model proposed by Endlich and McLean [165].

TABLE 9.15. Average Values of α and β

Region of stream	$\bar{\alpha}$ (km^{-1})	$\bar{\beta}$ (km^{-1})
Under jet axis	0.1314	—
Above jet axis	0.2388	—
On cyclonic side of jet	—	0.0015
On anticyclonic side of jet	—	0.0008

Twin jet streams are very often observed in low latitudes. The fronts associated with these jets are clearly manifested in the temperature field. Jet-stream complexes of this kind, which are often observed above Central Asia, have been described in detail by Georgii.

3.2. Experimental Data on Turbulence Causing Aircraft Buffeting in Jet-Stream Region. An investigation of the frequency and mean intensity of turbulence causing aircraft buffeting was carried out by Pinus and Shmeter [94], who took into account the models of the cross section of the jet stream. The frequency of turbulence was calculated as the ratio of the path length in turbulent zones to the total path of the airplane at the particular level, and the mean intensity of turbulence in the disturbed zones was determined from the formula

$$\bar{b} = \frac{\sum_{i=1}^{4} n_i b^{(i)}}{\sum_{i=1}^{4} n_i}, \tag{9.26}$$

where $b^{(i)}$ is the intensity of aircraft buffeting on the point scale and n_i is the number of cases of buffeting of a given intensity; $b^{(1)}$ was taken as unity, $b^{(2)}$ as 2, and so on.

Figure 9.10 illustrates the distribution of the frequency of turbulence in different parts of the jet stream. The isoline for 50% frequency encloses a region between the front and the polar tropopause in the cold part of the jet stream.

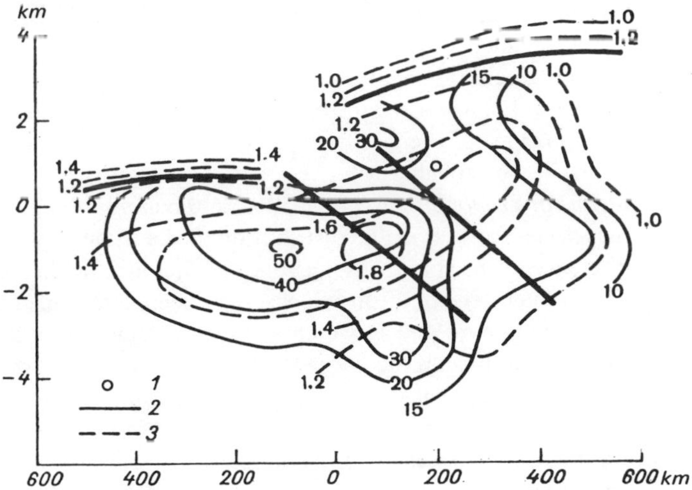

Figure 9.10. Frequency of turbulence and mean intensity of turbulence in jet stream (cross section). (1) Stream axis; (2) frequency (%); (3) mean intensity (point scale).

The frequency of turbulence decreases most rapidly on the warm side of the jet stream. At a distance of 200-300 km from the front it does not exceed 15%. An increased frequency of turbulence (up to 30% or more) is observed in the warm part of the jet above the front.

The region of most intense turbulence, where $b = 1.6$-1.8, is located below the jet stream axis near the front.

It should be noted that in the warm part of the jet stream the intensity of turbulence decreases less sharply than the frequency in a horizontal direction. This means that although the frequency of turbulence in the warm part of the jet stream is low as a whole, turbulence of considerable intensity can be encountered here. According to the results of airplane investigations over England [143], the relative number of cases of severe buffeting is greatest below the axis in the cyclonic part of the jet stream, and above the axis in the anticyclonic part. Barakhtin obtained similar results for Siberia and the Far East [4].

According to the investigations of Bannon, Jones, and Chambers, aircraft buffeting in 75% of the cases was observed mainly on the cold side of the jet stream.

Davydov points out that significant turbulence is encountered in the region of the largest vertical gradients of mean wind velocity under the tropopause and above the front on the northern periphery of the jet.

The described distribution of frequency and mean intensity of turbulence in different regions of the jet-stream cross section may be altered by mountain relief. According to the data of Helmholtz [31], above Kazakhstan the most turbulent part of jet streams is that located above mountains.

An analysis of the distribution of turbulence in jet streams where their warm part lay above mountains and the cold part above level terrain showed the presence of increased turbulence in the warm part of the jet stream, but it was still less intense than on the cold side.

It is obvious that in the lower stratosphere at levels above the jet-stream axis, where the anticyclonic part of the jet is also its cold part, the opposite distribution of turbulence is to be expected.

3.3. Turbulence in Tropopause Region and near Maximum-Wind Level. We consider first the special features of the distribution of the frequency and mean intensity of turbulence in the tropopause region. The presence or absence of turbulence, and also the intensity of turbulence, have been determined separately for layers situated below and above the polar and tropic tropopause in horizontal sections 25 km long in the layers 0-500, 500-1000, and 1000-1500 m below and above the tropopause. Figure 9.11 shows the distribution of frequency and average intensity of turbulence in the region of the polar and tropic tropopause from the results of experimental investigations in the Far East [94]. The distribution curves for these types of tropopause are different.

The frequency of turbulence in the 1500-1000 m layer under the polar tropopause is 78%, in the 1000-500 m layer it is 67%, and in the 500-0 m layer it

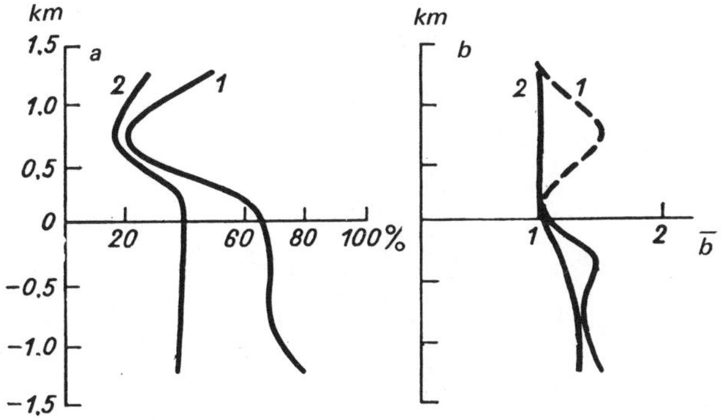

Figure 9.11. Distribution of frequency (a) and mean intensity (b) of turbulence in region of polar (1) and tropic (2) tropopause.

is 69%. Above the lower boundary of the polar tropopause the frequency decreases to 57% in the 0-500 m layer, then to 20% in the 500-1000 m layer, after which it shows a tendency to increase, becoming equal to 50% in the 1000-1500 m layer. The mean intensity of turbulence under the tropopause is greatest in the 500-0 m layer, where $\bar{b} = 1.45$; above the lower boundary of the polar tropopause in the 0-500 m layer the intensity of turbulence is 1.0, and in the 500-1000 m layer it increases to 1.5, after which it drops to 1.0 in the 1000-1500 m layer. Reshchikova [101] found approximately the same distribution of turbulence relative to the polar tropopause for the European USSR.

We will now consider the frequency of turbulence in the region of the tropic tropopause. Figure 9.11 shows that in the 1500-0 m layer under the tropopause the frequency of turbulence is constant and equal to 38%. Above the lower boundary of the tropopause it decreases to 17% in the 500-1000 m layer, and subsequently increases a little. The mean intensity of turbulence under the tropopause is 1.2-1.3, and above its lower boundary $\bar{b} = 1.0$.

Turbulence in the polar tropopause above Kazakhstan decreases in general with altitude. Turbulence is sharply reduced in the tropic tropopause. On its lower boundary and at a distance of approximately 500 m above and below it there are maxima of frequency and intensity of turbulence [31]. As will become clear from what follows, this difference in the distribution of frequency and mean intensity of turbulence for the polar and tropic tropopause can be attributed to the particular nature of the vertical distribution of temperature and mean wind velocity, particularly to the difference in the relative heights of the tropopause and the maximum wind-velocity level in temperate and tropic latitudes.

We now consider the distribution of turbulence in relation to the level of maximum wind velocity. The distribution for the Far East region is shown in Fig. 9.12. Curve 1 in this figure characterizes the frequency of turbulence caus-

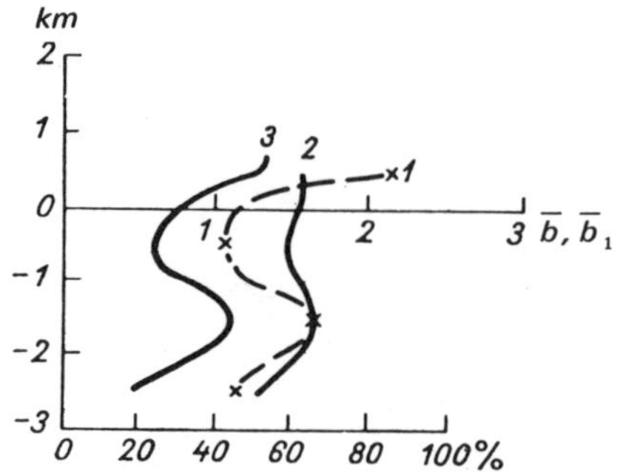

Figure 9.12. Distribution of frequency (1) and mean intensity (2 and 3) of turbulence in region of maximum wind velocity.

ing aircraft buffeting, while curve 2 characterizes the mean intensity of turbulence, including calm portions of flights, calculated from the equation

$$\bar{b}_1 = \frac{\sum_{i=0}^{4} n_i b^{(i)}}{\sum_{0}^{4} n_i}. \tag{9.27}$$

Here $b^{(0)}$ is calm flight and n_i is the number of flights. Curve 3 characterizes the mean intensity of turbulence $\bar{b}$.

It is apparent that there are two atmospheric layers (one under the maximum-wind level, and the other above it) in which turbulence has the greatest frequency and the greatest mean intensity. If it is assumed that flight conditions are appreciably affected only by turbulence with intensity $\bar{b}_1 \geqslant 1.0$, the mean thickness of the turbulent layer, which we can call the effective thickness, below and above the maximum-wind level can be estimated. Figure 9.12 shows that for the Far East region the effective thickness under the maximum-wind level is approximately 1.0 km.

The experimental data available at present allow a fairly thorough investigation of turbulence only in the layer under the maximum-wind level. Not enough experimental data on turbulence above this level have been amassed so far. The data show that turbulence above the maximum-wind level can sometimes be very intense.

Investigations of turbulence in the lower stratosphere by a high-altitude probing airplane equipped with special measuring apparatus showed that the level of minimum wind velocity plays the same role in the formation of turbulent zones as the tropospheric maximum-wind level [23]. As an example, Fig. 9.13 shows the results of a flight in the region of the Main Caucasus Range, where the maximum-wind level was under the tropopause, and the wind velocity, after attainment of the minimum above the tropopause, then increased sharply with altitude. At an altitude of 18 km there were two short turbulent zones.

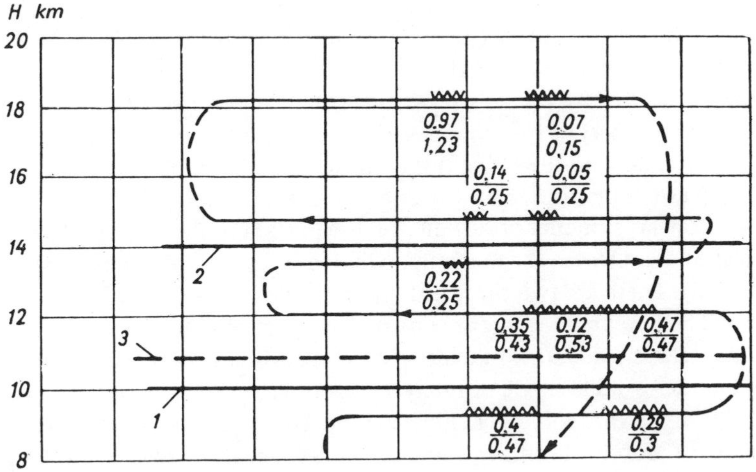

Figure 9.13. Distribution of turbulent zones relative to levels of maximum (1) and minimum (2) wind, and to tropopause (3). September 14, 1966; 1530–1710 h. The numerator gives the positive values of Δn, and the denominator gives the negative values.

In one zone the airplane was subjected to light buffeting ($\Delta n = 0.07\text{–}0.15g$), and in the other zone, situated approximately 20 km from the first, there was severe and even very severe buffeting ($\Delta n = 0.97\text{–}1.23g$). It is of interest that a similar distribution of turbulent zones was observed in a flight that took place 2.5 h earlier.

The data on the distribution of the frequency and mean intensity of turbulence in the region of the tropopause and the maximum-wind level lead to three models of turbulence distribution: the one-layer, two-layer, and three-layer models (Fig. 9.14).

The one-layer model of the turbulence frequency distribution is typical of cases of light winds in the tropopause region, when the tropospheric wind-velocity maximum is indistinct. In this case the layer with maximum frequency and intensity of turbulence is entirely under the tropopause (Fig. 9.14a).

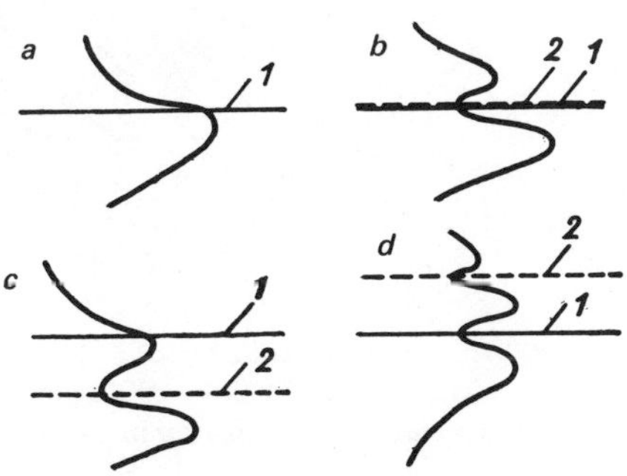

Figure 9.14. Models of turbulence distribution in region of tropopause and maximum-wind level.

The two-layer model characterizes the situation where the tropopause (1) and the maximum-wind level (2) almost coincide, or the tropopause lies above the maximum-wind level (Fig. 9.14c). In the first case we can have a layer of turbulence, intensified by the jet stream, under the tropopause, and a second turbulent layer at the level of the tropopause due to the effect of the jet stream.

In the second case (Fig. 9.14c) the turbulent layer immediately under the tropopause is intensified by the effect of the jet stream.

The three-layer model of turbulence distribution in the tropopause region is typical of cases where the maximum-wind level lies above the tropopause (Fig. 9.14d). There are two turbulent layers above the tropopause and one under it. The two upper layers are due to the effect of the jet stream.

Hence, statistical data on the difference in altitudes Δz of the tropopause and maximum-wind level are of interest. Figure 9.15 shows curves of the integral frequency of Δz for Moscow, Leningrad, and Tbilisi. This figure shows that in the region of Moscow in spring, summer, and autumn, cases where the tropopause lies above the maximum-wind level predominate, while in winter it lies below this level. Some numerical characteristics are given in Table 9.16. Above Moscow the tropopause lies below the maximum-wind level in spring and summer in 33–34%, in autumn in 39%, and in winter in 54% of the cases; i.e., in autumn and, in particular, in winter above Moscow, the distribution corresponding to the three-layer model is frequently found.

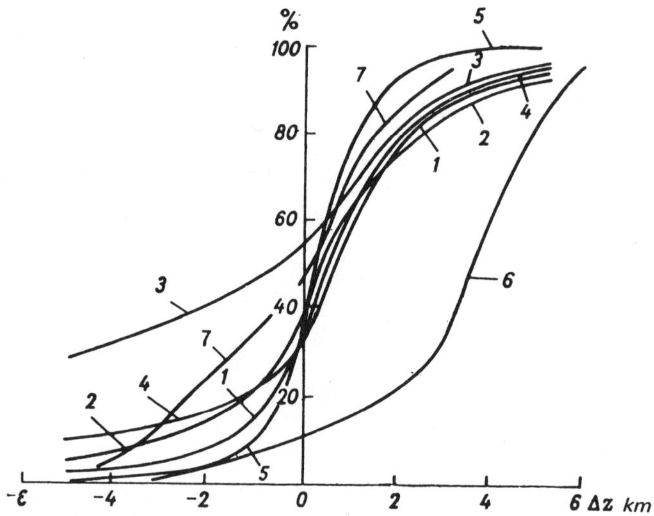

Figure 9.15. Integral frequency of difference between altitudes of tropopause and maximum-wind level. Moscow: (1) Summer; (2) autumn; (3) winter; (4) spring. Leningrad: (5) yearly average. Tbilisi: (6) summer; (7) winter.

TABLE 9.16. Frequency (%) of Difference between Altitudes of Tropopause and Maximum-Wind Level (Moscow)

Season	$\Delta z>0$	$\Delta z<0$	$0<\Delta z\leqslant 3$ km	$-3\text{km}\leqslant\Delta z<0$	$-3\text{km}\leqslant\Delta z\leqslant 3$ km
Spring	67	33	54	20	74
Summer	66	34	53	29	82
Autumn	61	39	48	28	76
Winter	46	54	36	18	54

According to [25], the level of maximum wind velocity above Leningrad on the average for the year lies below the tropopause in 68% of the cases, the two levels coincide in 10% of the cases, and the tropopause lies below the maximum-wind level in 22% of the cases. Thus, the frequency of cases where the maximum-wind level lies below the tropopause for Moscow is close to the corresponding figure for Leningrad in all seasons, except winter.

In those cases where the maximum-wind level lies below the tropopause the difference in their altitudes for the Moscow Region in spring, summer, and autumn does not exceed 3 km in 48–54% of the cases, and when this level lies above the tropopause the corresponding figure is only 20–29%. On the average, the maximum-wind level in summer and in the intermediate seasons lies in the layer ±3 km relative to the tropopause in 74–82% of the cases, and in winter in 54% of the cases. According to [25], above Leningrad in 83% of the cases the maximum-wind level lies in a layer from 2 km under the tropopause to 1 km above it.

Above Tbilisi in winter the maximum-wind level lies below the lower boundary of the tropopause in 47% of the cases, and in summer it is found here in only 10% of the cases [51]. A difference in altitudes of more than 3 km is found in 67% of the cases in summer, and in 16% of the cases in winter.

The data presented on the sign and magnitude of Δz for different geographical regions provide a picture of the frequency of occurrence of the different models of vertical distribution of turbulence.

4. Relation between Turbulence Causing Aircraft Buffeting and Thermal and Dynamic Conditions in Atmosphere

4.1. Possible Use of Richardson Number as a Criterion of Turbulence Causing Aircraft Buffeting. In recent years there have appeared many papers devoted to the investigation of turbulence causing aircraft buffeting in relation to various characteristics of the temperature and wind velocity fields. In particular, it has been discovered that aircraft buffeting depends on the vertical and horizontal gradients of the mean wind velocity, the rotation of the wind vector with altitude, horizontal temperature gradients, Richardson number, etc. According to Hislop's data, buffeting is encountered at vertical gradients of mean wind velocity exceeding 2 m/sec per 100 m of altitude and horizontal temperature gradients greater than 5°C per 100 km. Similar results were obtained by Klemin and Pinus. As regards horizontal wind-velocity gradients, turbulence occurs at a gradient of 5 m/sec or more per 100 km according to Hislop's data, and at 8-12 m/sec or more per 100 km according to Jones' data. According to Reshchikova's data [101], at gradients greater than 17 m/sec per 100 km the frequency of aircraft buffeting reaches 75%, whereas at lower gradients it decreases sharply and is only about 5%.

The Richardson number Ri has been used in many investigations to assess atmospheric turbulence responsible for aircraft buffeting. Klemin and Pinus found that light aircraft buffeting is observed when $0.5 < \mathrm{Ri} < 4$, and moderate and severe buffeting when $\mathrm{Ri} \leqslant 0.5$. Many foreign investigators have arrived at approximately similar criteria.

Briggs investigated 105 cases of severe clear-air turbulence in the upper troposphere above England and found that in 70% of the cases turbulence was recorded when $\mathrm{Ri} < 0.5$. In 80% of the cases of strong turbulence these small values of Ri were accompanied by horizontal gradients of mean wind velocity exceeding 8 m/sec per 100 km. The investigations of Endlich and Mancuso showed that the vertical gradient of mean wind velocity and Ri are the best indices of turbulence causing aircraft buffeting. As turbulence indices these two parameters give approximately the same results when they are compared with aircraft-buffeting data.

This can probably be attributed to the fact that in the absence of clouds in the upper troposphere the vertical temperature gradient lies in a relatively narrow range and, hence, the value of Ri is determined mainly by the vertical wind shear. Colson [151] and Vasil'ev [12, 13] indicate a satisfactory correlation between clear-air turbulence and the values of Ri and the vertical gradients of mean wind velocity.

Radar and simultaneous aircraft measurements by Brochet *et al.* [144] revealed that clear-air turbulence at altitudes of 6–9 km is found in stably stratified layers with vertical gradients of mean wind velocity exceeding $2 \cdot 10^{-2}$ $\sec^{-1}$.

On the other hand, some investigators have failed to find a satisfactory correlation between Ri and turbulence causing aircraft buffeting, especially in clear air in the upper troposphere.

The main reason for the disagreements in the assessment of the role of Ri is probably that turbulence, as shown in Chapter 1, is not correlated uniquely with this number. In addition, the calculation of Ri from aerological network data entails large errors due to errors of measurement of the wind velocity and direction, especially in the upper troposphere and lower stratosphere.

Theoretical investigations have shown that the local critical value of Ri is 0.25. The detection of such values of Ri in the free atmosphere is very difficult.

The rawinsonde method cannot provide data about the wind at fixed levels; the wind velocity is averaged over layers several hundred meters thick and relates to the center of the layer, which leads to coarse averaging of the vertical wind-velocity gradients. To investigate turbulence and use Ri for assessment of turbulence conditions it is essential to measure the wind with good height resolution, at least for layers not less than 50 m thick.

According to Roach [209], values of Ri are greatly overestimated due to the artificial underestimation of the vertical gradient of mean wind velocity. He suggested that for calculations of Ri the wind gradient should be measured not from the modulus of the difference in wind velocity vectors at the boundaries of the considered layer, but from the integral length of the hodograph of the wind vector in this layer, since the vertical wind profile within the layer may be distinguished by bends (peaks) with velocity maxima and minima.

Erroneous values of Ri are very often produced by an incorrect choice of layer thickness for which the vertical gradients of temperature and mean wind velocity are calculated. In addition, in assessing the results of investigations of the frequency and intensity of buffeting in relation to Ri it should also be borne in mind that investigators did not always have synchronous aerological and aircraft-buffeting data.

The spatial and temporal variation of temperature and wind affects very appreciably the results of calculation of Ri. The effect is pronounced even in "ideal" cases where the turbulence, temperature, and wind are measured on the same airplane. In this case the measurements of T and u at different altitudes are asynchronous.

Pinus and Shmeter [94] compared averaged data on aircraft buffeting in jet streams with the mean values of Ri. In the calculations they used the dry-adiabatic vertical temperature gradient for cloudless layers and the moist-adiabatic gradient for cloudy layers. In the majority of calculations they used the horizontal temperature gradient instead of the vertical gradient of mean wind velocity. They did not calculate Ri for the tropopause region.

Figure 9.16 shows the distribution of averaged values of Ri on a vertical cross section of the jet stream. This figure and Fig. 9.10 show that in the zone where $\bar{b} \geqslant 1.8$, i.e., the intensity of aircraft buffeting is close to moderate on the average, $\mathrm{Ri} \leqslant 1$. In the cold part of the jet stream, where $\bar{b} < 1.8$, $1 < \mathrm{Ri} < 5$. On the right of the jet stream axis, where $\bar{b} < 1.4$, the values of Ri increase fairly rapidly.

Figure 9.16 shows that the Ri can be used to obtain the general distribution of turbulence in the jet-stream region, but not the details of this distribution, which are clearly seen in Fig. 9.10.

Flight investigations have established the presence in general of a high correlation between clear-air turbulence and the horizontal temperature variations in a turbulent region of the atmosphere. This is consistent in principle with the high correlation between turbulence and the vertical gradients of mean wind

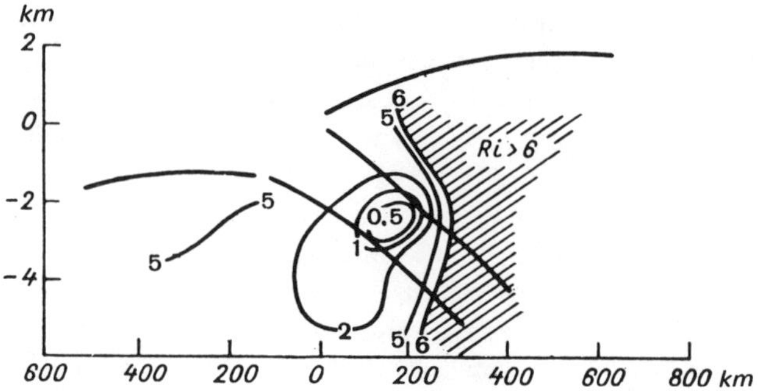

Figure 9.16. Distribution of averaged values of Ri in vertical cross section of jet stream.

velocity, since in the free atmosphere the horizontal temperature gradients and vertical wind gradients are related to one another. The data given below were obtained in special flights in the upper troposphere [101]:

Horizontal temperature gradient (°C/100 km)	0–2	2–4	4–6	6–8	8–10	10–12	>12
Frequency (%)	9.0	46.8	65.0	87.5	75.0	100.0	80.0

It is apparent that at horizontal temperature gradients exceeding 6–8°C/100 km the probability of turbulence is more than 75%, although in several cases and at gradients greater even than 12°C/100 km, aircraft buffeting was not observed. The investigations of MacPherson and Morissey [192] showed that in the stratosphere the horizontal changes of temperature in turbulent regions usually exceeded 2.5°C per 30 sec of flight. The highest correlation was obtained for the data of flights on the lee side of high mountain ranges perpendicular to strong tropospheric jet streams.

The correlation between clear-air turbulence and horizontal temperature differences is confirmed by the data of flights of the Concorde supersonic airliner [161a]. As an example, we give the results of one of the experimental flights of this airplane at an altitude of 15.5 km above the south Atlantic. During this flight the tropopause was located at an altitude of about 12 km (temperature −58°C), and the maximum-wind (58 m/sec) level at an altitude of about 13 km. Over a distance of 2350 km 14 load increments in the range 0.05–0.15g, nine in the range 0.15–0.25g, and one increment of 0.35g were recorded. An idea of the nature of the changes in the Concorde loads in the region of large air temperature changes at an altitude of 15.5 km is given by Fig. 9.16a.

The largest airplane load increment (0.35g) was observed in the part of the flight where there was the sharpest change in air temperature.

Using the data of aircraft measurements of turbulence in the lower stratosphere up to altitudes of 18-20 km, Buldovskii [8, 9] investigated the role of various atmospheric parameters which might indicate the presence of turbulence

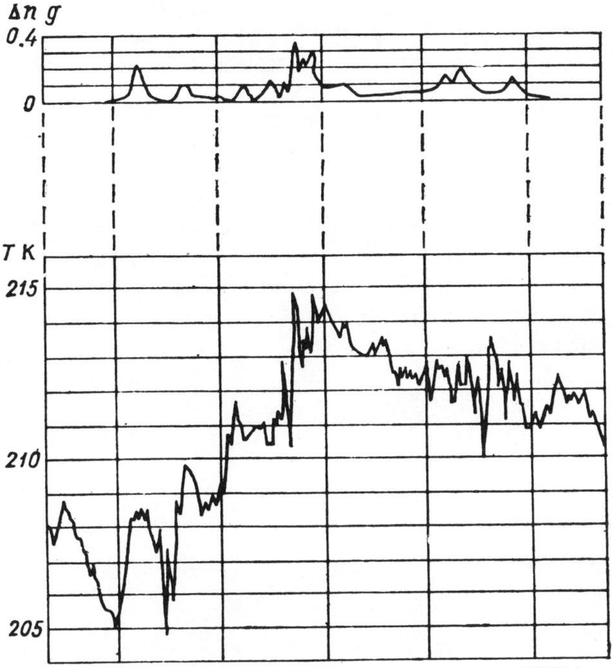

Figure 9.16a. Air temperature and load increments of Concorde airplane in a flight at an altitude of 15.5 km.

causing aircraft buffeting. He found that the best parameters for the stratosphere were the Richardson number and the vertical gradient of mean wind velocity.

These parameters are equally reliable for the diagnosis of aircraft buffeting. For the stratosphere this can be attributed to the small variation of the vertical temperature gradient [97], which is contained in the first degree in Ri. The diagnosis is improved if the change in thermal stratification with altitude is taken into account. Buldovskii proposed a criterion that takes both these parameters into account:

$$k = \frac{1}{\beta} - 8q^2 \left| \begin{array}{l} < 1.2 - \text{presence of buffeting} \\ > 1.2 - \text{absence of buffeting}\,. \end{array} \right. \quad (9.28)$$

Here β is the vertical wind-velocity gradient and q is the difference in vertical temperature gradients in the lower and upper layer. The values $\beta = 8 \cdot 10^{-3}$ sec^{-1} and $q = 0.4 \cdot 10^{-2}$°C/m are critical.

Colson and Panofsky [153] suggested that clear-air turbulence should be assessed with the aid of a special turbulence index. They regarded the energy of the fluctuations of the vertical wind-velocity component ($E' = \overline{w'^2}/2$) as the main quantitative characteristic of the intensity of turbulence. The expression for E' was found from dimensional considerations on the assumption that E' is uniquely determined by the vertical eddy scale λ and the energy influx rate

$$\Phi_* = K\left(\frac{d\bar{u}}{dz}\right)^2 + \alpha K_T \frac{g}{\theta}\frac{d\theta}{dz}, \tag{9.29}$$

where α is a quantity a little greater than unity, characterizing the fact that hydrostatic forces act directly on fluctuations of the vertical wind-velocity component, whereas shear generates fluctuations of the longitudinal wind-velocity component, whose energy is then redistributed among all the components.

A further assumption is that $K \approx K_T \approx \lambda\sqrt{E'}$. The thickness Δz of the turbulent layer is used as λ, and $\partial\bar{u}/\partial z$ is replaced by a ratio of finite differences. The authors thus obtained the expression

$$I \equiv (\Delta u)^2\left(1 - \frac{\mathrm{Ri}}{\mathrm{Ri}_{\mathrm{cr}}}\right) \approx E', \tag{9.30}$$

where Δu is the difference in wind velocity in the layer.

From an assessment of empirical data it is assumed that

$$\mathrm{Ri}_c = \frac{1}{\alpha\dfrac{K}{K_T}} \approx 0.5.$$

The quantity $I \equiv E'$ is taken as the index of clear-air turbulence. Colson and Panofsky found typical values of I for turbulence of different intensity above the U.S.A., which agreed satisfactorily with experimental data for the frequency of aircraft buffeting.

Laikhtman and Al'ter-Zalik [61] devised a method of using aerological data to assess turbulence causing aircraft buffeting in the free atmosphere. They assumed that the turbulent flow is stationary and horizontally homogeneous. The mean spatial scale of turbulence was estimated from the "mixing length" in the Prandtl–von Karman sense, and the role of thermal stratification of the atmosphere was regarded as similar to that of the vertical wind-velocity gradient. The well-known von Karman formula for the turbulence scale

$$l \approx 2\varkappa\left|\frac{\left(\dfrac{d\bar{u}}{dz}\right)^2}{\dfrac{d}{dz}\left(\dfrac{d\bar{u}}{dz}\right)^2}\right|, \tag{9.31}$$

where $\kappa \approx 0.4$ is the von Karman constant, can then be extended to the case of a thermally stratified atmosphere by replacing $(d\bar{u}/dz)^2$ in (9.31) by the function

$$\Phi = \left(\frac{d\bar{u}}{dz}\right)^2 + \left(\frac{dv}{dz}\right)^2 - \alpha\,\frac{g}{\theta}\,\frac{d\bar{\theta}}{dz}. \tag{9.32}$$

Here α is the ratio of the turbulent mixing coefficients for heat and momentum.

If diffusion of turbulence is neglected, the relation between the kinetic energy of turbulence, the mean turbulence scale, and the vertical distribution of wind velocity and air temperature is given by the following equations:

$$K\Phi - \varepsilon = 0, \tag{9.33}$$

$$K = l(E')^{1/2}, \tag{9.34}$$

$$\varepsilon = c(E')^{3/2} l^{-1}, \tag{9.35}$$

$$l = -\tilde{\varkappa}\left[\frac{d}{dz}\ln\Phi\right]^{-1}, \tag{9.36}$$

where K is the coefficient of turbulent mixing, E' is the kinetic energy of turbulence, ϵ is the rate of dissipation of turbulent energy as heat, and c and $\tilde{\kappa}$ are dimensionless constants equal to 0.046 and 0.37, respectively. The constant $\tilde{\kappa}$ is connected with the von Karman constant by the relation $\tilde{\kappa} = 2\kappa c^{1/4}$

If the vertical distribution of temperature and mean wind velocity is known, the main characteristics of turbulence can be obtained:

$$E' = 0.75\cot^2\varphi, \qquad l = 0.185\Phi\cot\varphi, \tag{9.37}$$

whoro

$$\cot\varphi = \left[\frac{d}{dz}\Phi^{1/2}\right]^{-1} \tag{9.38}$$

Calculations from Eqs. (9.37) are confirmed by the data of experimental investigations.

Shnaidman [127], using the equations of motion, balance of turbulence kinetic energy, and semiempirical relations for the turbulence coefficient and the characteristic dimension of the eddies, obtained a closed system of equations whose solution allows calculation of the characteristics of turbulent layers in the free atmosphere from the vertical profiles of the geostrophic wind (or horizontal temperature gradient) and temperature.

The system of hydrodynamic equations was written in the following way:

$$\frac{d^2\varphi}{dz^2} + \frac{2\omega_z}{K}\psi = 2\omega_z(v_g - v_a), \tag{9.39}$$

$$\frac{d^2\psi}{dz^2} - \frac{2\omega_z}{K}\varphi = -2\omega_z(u_g - u_a), \tag{9.40}$$

$$\varphi^2 + \psi^2 - \alpha_T K^2 \frac{g}{T}\left(\gamma_a + \frac{dT}{dz}\right) + \alpha_b K \frac{d}{dz} K \frac{db}{dz} - \alpha_\varepsilon b^2 = 0, \tag{9.41}$$

$$l = mz\left[1 - \frac{g}{T}\,\frac{K^2\left(\gamma_a + \dfrac{dT}{dz}\right)}{(\varphi^2 + \psi^2)\,\mathrm{Ri}_{\mathrm{cr}}}\right]\left(1 - \frac{z}{2h}\right), \tag{9.42}$$

$$K = b^{1/2} l. \tag{9.43}$$

Here $\varphi = K(\partial u/\partial z)$ and $\psi = K(\partial v/\partial z)$ are the components of the vector of the turbulent friction tangential stress, u_a and v_a are nonlinear terms in the equations of motion (they are assumed to be known and are determined by means of the geostrophic approximation), K and b are the turbulence coefficient and the turbulence intensity, l is the characteristic dimension of the eddy, $2h$ is the thickness of the turbulent layer, and u_g and v_g are the geostrophic-wind components. The constants contained in the equation were taken as $\alpha_T = 1, \alpha_b = 0.73, \alpha_\epsilon = 0.046, m = 0.185$.

The problem was solved numerically by the method of successive approximations in conjunction with matrix and ordinary pivoting on the assumption that at the boundaries of the turbulent layer the components of the tangential stress vector and the turbulence intensity become zero. Shnaidman used two calculation procedures:

(1) Vertical profiles of wind velocity and temperature were assigned, and the intensity of turbulence, the turbulence coefficient, and characteristic dimension of the eddy were determined;
(2) the field of geopotential and temperature on the isobaric surfaces 500, 300, and 200 mb was assigned and the complete system of equations was solved.

Calculations by the first procedure showed that the characteristic dimension of the eddies was 70 m on the average and did not exceed 200 m; the turbulence coefficient could reach several hundred m^2/sec for cases of severe aircraft buffeting, and the standard deviations of the velocity fluctuations could be as high as 10 m/sec.

In the second procedure the problem was solved only for the case of neutral stratification. He found that when the turbulent layer was about 2 km thick and the horizontal temperature gradient was 5°C/100 km the turbulence coefficient reached 200 m^2/sec, and the turbulence intensity approximately 5 m^2/sec^2. These results should be regarded as maximum estimates for turbulent layers of the free atmosphere. Furthermore, fairly large horizontal temperature gradients

(5–10°C/100 km) are required to maintain the intensity of turbulent mixing. Such gradients can exist in local regions of small horizontal dimensions.

4.2. Relation between Turbulence and Divergence of Horizontal Wind Velocity. In the theoretical treatment of turbulence in the free atmosphere it is usually assumed that the flow is horizontally homogeneous. As we saw, however, the horizontal wind-velocity gradients in a jet stream are large and very variable. The reason why many investigators have neglected the role of horizontal wind-velocity gradients is because they are at least an order of magnitude less than the vertical gradients of mean wind velocity.

Kao and Sizoo [177] investigated the frequency of turbulence in relation to the sign and modulus of the divergence of the horizontal wind velocity (the "plane divergence").

For calculations of the plane divergence they used the expression

$$\nabla_z \cdot \vec{v} = \frac{\Delta v_s}{\Delta s} + \frac{\Delta v_n}{\Delta n}, \tag{9.44}$$

where z is the altitude, v_s is the wind velocity parallel to the jet axis, and v_n is the wind velocity normal to the jet axis.

The value of Δv_n was obtained from the data of an airplane Doppler navigation system, since the flights were made across the axis of the jet stream; Δv_s was calculated from the data of network rawinsonde observations; the values of Δs and Δn were 555 and 111 km, respectively.

They found that the frequency of turbulence (P, %) in relation to the plane divergence of the wind velocity could be represented by the exponential function

$$P = ae^{-b\nabla_z \cdot v} \tag{9.45}$$

The parameters a and b depend significantly on the intensity of turbulence (Table 9.17).

With increase in intensity of turbulence the parameter a is reduced, whereas the parameter b is increased. In addition, with increase in the convergence of the wind velocity ($\nabla_z \cdot \vec{v} < 0$) the intensity of the turbulence affecting aircraft flight increases.

TABLE 9.17. Parameters a and b and Range of Variation of Plane Divergence $\nabla_z \cdot \vec{v}$ in Relation to Intensity of Turbulence

Intensity of turbulence	a	b (sec)	$\Delta_z \cdot \vec{v} \cdot 10^{-5}$ (sec)	
			from	to
Light	8.00	$2.16 \cdot 10^3$	−16	+28
Moderate	6.83	$7.70 \cdot 10^3$	−18	+ 8
Severe	2.05	$1.20 \cdot 10^4$	−30	+10
On average	19.00	$4.33 \cdot 10^3$	−30	+28

4.3. Some Synoptic Conditions Promoting Development of Turbulence Affecting Flight of Aircraft. Constant-pressure charts have been used to determine the special features of the pressure field or wind-velocity field which make the development of turbulence most probable. The aircraft-buffeting data obtained from aircrews reveal the regions in which aircraft buffeting is most frequent.

Pchelko [99] believes that favorable conditions for the development of turbulence in the upper troposphere are found in the cyclonic part of the jet stream, and mainly on the left of the stream axis, where, in his opinion, the dominant role is played by horizontal wind shears across the flow, and also in the zone of divergence of the streamlines, accompanied by anticyclonic curvature of the streamlines, where the horizontal wind shears are directed along the flow. Moderate and severe turbulence is observed if the horizontal gradient of the mean wind velocity over short distances is greater than 50 km/h per 100 km.

According to [100], in the cyclonic part of jet streams moderate and severe turbulence is usually associated with comparatively small horizontal wind shears (less than 10 m/sec per 100 km), but with relatively large vertical shears (more than 0.8 m/sec per 100 m), and much more rarely with large horizontal shears (less than 0.8 m/sec per 100 m). According to [100], in the anticyclonic part of jet streams moderate and severe turbulence is associated with small horizontal and vertical wind shears (less than 10 m/sec per 100 km, and 0.8 m/sec per 100 m, respectively). Yet, according to Reshchikova's data [101], the frequency of turbulence at wind shears less than 17 m/sec per 100 km does not exceed 9–13%, whereas at larger shears it reaches 75%.

One should note the contradictory nature of the data of many investigators for the frequency of turbulence affecting aircraft flight in relation to the curvature of the isohypses, the stream-line divergence (convergence) in the jet streams, and the characteristic value of the horizontal wind shear. This is largely due to differences in the technique of analysis of constant-pressure charts and comparison with aircraft-buffeting data. In addition, some authors used wind data obtained from a relatively sparse aerological network, with the result that the averaging in calculations of the horizontal wind shear was coarse, whereas other authors used aircraft measurements that allow calculation of the wind shear for relatively short distances, comparable with the horizontal scales of the turbulent zones.

4.4. Typical Values of Parameters of Clear-Air Turbulence. We now generalize the quantitative values of the physical characteristics of clear-air turbulence, which may be useful for theoretical and practical calculations of various kinds. Pinus [90] gives typical values of turbulence parameters for the upper troposphere and lower stratosphere in temperate latitudes. This paper contained no data for lower altitudes, since clear-air turbulence, in particular, in the lower half of the troposphere strongly depends on the nature of the underlying surface and the thermal instability of the atmosphere that is often observed here [87].

In addition, for the lower and middle troposphere there is very little information at present about the spatial dimensions and structure of clear-air turbulent zones.

Quantitative values of the physical characteristics of clear-air turbulence for the upper troposphere and lower stratosphere are given in Table 9.18.

For the solution of many practical problems it is not only the averaged spectra of the fluctuations of the wind-velocity components that are of interest, but also information about the statistical distribution of spectra with different levels of turbulent energy [93]. Figure 9.17 (solid lines) gives the functions $S(\omega)$ of fluctuations of the horizontal wind-velocity component for the atmospheric layer 7–18 km, plotted for 50, 80, and 95% cumulative frequency of the spectral density values. The calculations were made from the results of flight investigations by the Laboratory of Atmospheric Dynamics (TsAO) at different times of

TABLE 9.18. Parameters of Clear-Air Turbulence in Atmosphere [90]

Physical characteristics of clear-air turbulence	Altitude (km)	Typical numerical values
Frequency of occurrence (relative to total length of flight in km), %	7–10	10
	10–20	1
Horizontal extent of turbulent zones (for 80% cumulative frequency), km	7–10	140
	10–20	80
Vertical extent of turbulent zones (for 80% cumulative frequency), km	7–10	0.9
	10–20	0.3
Coefficient of spatial anisotropy	7–10	$7 \cdot 10^{-3}$
	10–20	$4 \cdot 10^{-3}$
Typical lifetime of turbulent zones, h	7–10	5
Hydrodynamic and thermodynamic characteristics		
Vertical gradient of mean wind velocity, sec^{-1}	7–20	$\geqslant 8 \cdot 10^{-3}$
Horizontal gradient of mean wind velocity, sec^{-1}	7–20	$\geqslant 10^{-4}$
Horizontal temperature gradient, °C/m	7–20	$\geqslant 6 \cdot 10^{-5}$
Interlayer differences in vertical temperature gradient, °C/m	7–20	$\geqslant 4 \cdot 10^{-3}$
Spectral density of fluctuations of wind-velocity components $S(\omega)$, m^3/sec^2:	7–20	
$\omega \geqslant 10^{-3}$ m^{-1}	7–20	$\sim\omega^{-5/3}$
$\omega \leqslant 10^{-3}$ m^{-1}	7–20	$\sim\omega^{-2.7}$
$S(\omega)$ for $\omega = 2 \cdot 10^{-3}$ m^{-1}:	7–20	
light turbulence, $b^{(1)}$	7–20	$\sim 10^2$–$5 \cdot 10^2$
moderate turbulence, $b^{(2)}$	7–20	$5 \cdot 10^2$–10^3
severe turbulence, $b^{(3)}$	7–20	10^3–$3 \cdot 10^3$
very severe (storm) turbulence, $b^{(4)}$	7–20	$\geqslant 3 \cdot 10^3$

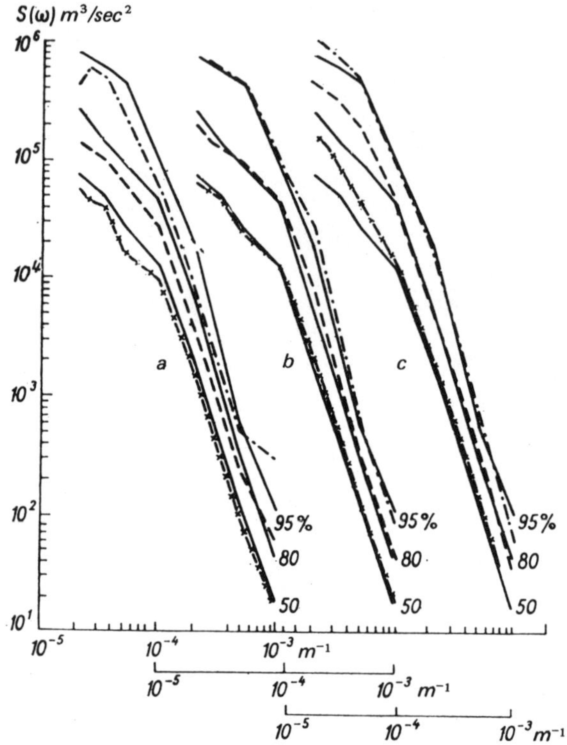

Figure 9.17. Curves of spectral density of wind-velocity fluctuations.

the year and in different geographic regions of the Soviet Union. The spectra for altitudes 12–18 km were obtained mainly from the results of flights above mountain regions [23, 91]. An idea of the number of individial spectra of wind-velocity fluctuations contained in the statistics is given by the following data:

Altitude (km)........	7–20	10–12	12–18	7–18
Number of spectra....	274	110	128	512

As models for the upper troposphere and lower stratosphere we can take functions $S(\omega)$ with cumulative frequency 50, 80, and 95%, obtained for the atmospheric layer from 7 to 20 km. In Fig. 9.17 the spectra of the fluctuations of the horizontal wind-velocity component for the same cumulative frequencies for the atmospheric layers 7–10 km (a), 10–12 km (b), and 12–20 km (c) are shown as dashed lines against the background of the model spectra.

Table 9.19 gives the standard deviations of the wind-velocity fluctuations for the wave-number region 10^{-3} to $2 \cdot 10^{-5}$ m^{-1} and intermediate intervals in relation to the altitude and cumulative probability. Since the spectral densities

TABLE 9.19. Values of σ_u (m/sec) for Cumulative Frequency of 50, 80, and 95% as Function of Altitude and Different Regions of Wave Numbers

$\Delta\omega$ (m^{-1})	7–10 km			10–12 km			12–18 km		
	cumulative frequency (%)								
	50	80	95	50	80	95	50	80	95
$10^{-3}-5\cdot10^{-4}$	0.24	0.28	0.62	0.25	0.38	1.45	0.23	0.37	1,34
$5\cdot10^{-4}-2\cdot10^{-4}$	0.57	0.93	1.40	0.70	1.31	2.86	0.66	1.10	2.33
$2\cdot10^{-4}-10^{-4}$	0.87	1.52	2.30	1.10	2.05	3.27	1.04	1.95	2.91
$10^{-4}-5\cdot10^{-5}$	1.13	2.00	3.91	1.34	2.58	4.77	1.52	3.10	4.95
$5\cdot10^{-5}-2\cdot10^{-5}$	1.44	2.40	5.00	1.64	2.80	5.89	2.21	4.20	6.70
$10^{-3}-10^{-4}$	1.06	1.80	2.72	1.33	2.46	4.54	1.25	2.26	3.96
$10^{-3}-2\cdot10^{-5}$	2.10	3.60	6.72	2.50	4.53	8.86	2.97	5.66	9.20

given in Fig. 9.17 were calculated from the formula

$$S(\omega) = \frac{1}{\pi} \int_0^\infty R(r) e^{-i\omega r} dr,$$

where $R(r)$ is the correlation function of the fluctuations, the variance was calculated from the $S(\omega)$ data from the formula

$$\sigma_u^2 = 2 \int_{\omega_1}^{\omega_2} S(\omega)\, d\omega.$$

Table 9.19 shows that for the wave-number region 10^{-3} to $2\cdot10^{-5}$ m^{-1} the standard deviation σ_u of the fluctuations in 95% of the cases does not exceed 6.72 m/sec in the 7–10 km layer, 8.86 m/sec in the 10–12 km layer, and 9.2 m/sec in the 12–18 km layer. The data of most importance for aviation applications are those for the wave-number region 10^{-3} to 10^{-4} m^{-1}, for which σ_u in 95% of the cases does not exceed 2.72 m/sec in the 7–10 km layer, 4.54 m/sec in the 10–12 km layer, and 3.96 m/sec in the 12–18 km layer.

5. Prediction of Clear-Air Turbulence

At present there are still no reliable methods of predicting clear-air turbulence in the atmosphere. The main reason for this is the absence of a well-developed theory of this atmospheric phenomenon. The methods used in practice, usually of an experimental nature, for the prediction of turbulence causing aircraft buffeting are usually based on empirically established statistical relations between the presence of clear-air turbulence according to aircrew reports and some atmospheric parameters.

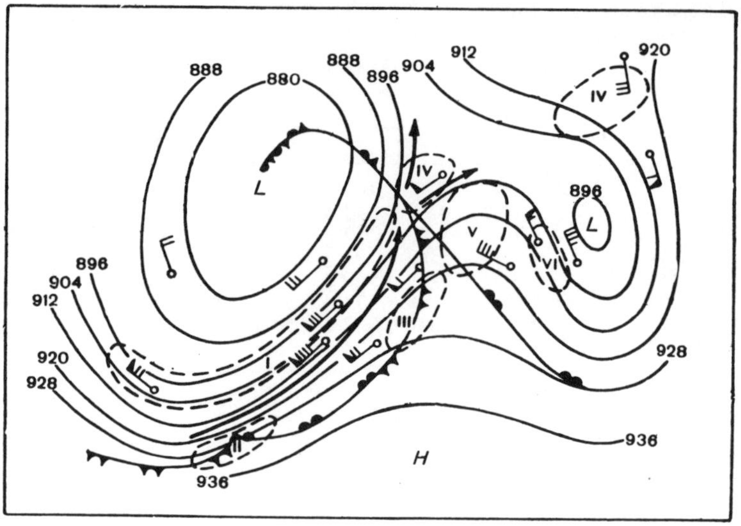

Figure 9.18. Diagram of synoptic situation characterizing position of zones most favorable for the onset of clear-air turbulence (according to Pchelko).

In most of the forecasting agencies of the Soviet Union and foreign countries the predictions of turbulence causing aircraft buffeting are based mainly on the synoptic method, which takes into account the pressure field at different altitudes, the wind-velocity field, and the vertical profiles of wind velocity and temperature. Figure 9.18 shows the unified scheme, proposed by Pchelko [72], characterizing the synoptic conditions that most favor the development of turbulence causing aircraft buffeting in the upper troposphere. In this figure regions where turbulence is most probable are bounded by dashed lines and indicated by Roman numerals. In this scheme an important role in the formation of turbulent zones is attributed to the divergence of high-altitude air flows with anticyclonic curvature of the streamlines. In the prediction of turbulence the vertical wind shear or Ri, as the best criteria of turbulence causing aircraft buffeting, must be taken into account.

Davydov *et al.* [37] proposed a synoptic-statistical method for the diagnosis and prediction of clear-air turbulence. In the development of this method information about aircraft buffeting obtained from pilots of airline planes during the period of the already-mentioned four five-day series of observations conducted in 1964–1965 in the ICAO and WMO programs was used. All the information on the occurrence or absence of aircraft buffeting over flat terrain in the Soviet Union was distributed in squares with dimensions on the ground of 300 × 300 km. In the statistical treatment the observed cases of aircraft buffeting were divided into two groups: The first group included all cases of buffeting, irrespective of intensity, while the second group included only cases of moderate and severe aircraft buffeting. On the average the probability of occurrence of buffeting in the square was 0.41 for the first group (I), and 0.16 for the second group (II). Table 9.20 gives data on the probability of aircraft buffeting in relation to some synoptic features s_i, selected by the Davydov *et al.* [37] for the isohypse field at the 400- and 300-mb levels.

Table 9.20 shows that the characteristics for the first and second groups behave in the same ways for cases of jet streams ($c > 90$ km/h) and for other cases ($c < 90$ km/h), but for jet streams they give a more reliable indication of the probability of aircraft buffeting.

The probability of aircraft buffeting is greatest on the cyclonic side of the jet and in its crest; it is 50–51% in the anticyclonic part of the jet, in the trough and linear part of the jet, and in the region of convergence and divergence of the isohypses.

The data of Table 9.20 were used to devise a probabilistic method of predicting aircraft buffeting. The method is essentially as follows. Information about the initial situation is given by an n-dimensional predictor-vector D_j with component features s_i:

$$D_j = (s_1, s_2, \ldots, s_i, \ldots, s_n)_j \qquad j = \overline{1, N}. \tag{9.46}$$

For each predictor-vector there is a particular probability of buffeting (b) or absence of buffeting ($b^{(0)}$). Assuming that the features s_i in the set are independent, the conditional probability of aircraft buffeting on the basis of the hypoth-

TABLE 9.20. Probability of Aircraft Buffeting in Relation to Feature s_i

Number of feature i	Name of feature s_i	I	II
		$P(b/s_i)$	
	A. For velocities $c \geqslant 90$ km/h	0.54	0.26
1	Cyclonic part of jet	0.61	0.33
2	Anticyclonic part of jet	0.51	0.23
3	Axial part of jet	0.28	0.07
4	Crest in jet	0.75	0.41
5	Trough in jet	0.51	0.26
6	Linear jet	0.50	0.23
	B. For velocities $c < 90$ km/h	0.35	0.12
7	Crest	0.39	0.11
8	Trough	0.43	0.20
9	Linear flow	0.31	0.10
	C. For all velocities	0.41	0.16
10	Convergence or divergence of isohypses	0.51	0.21
11	Absence of convergence or divergence of isohypses	0.37	0.15

esis probability formula (Bayes formula) can be written, after normalization, in the form

$$P(b/D_j) = \frac{P(b)\,k_1\,\ldots\,k_n}{P(b)\,k_1\,\ldots\,k_n + P(b^{(0)})\,\bar{k}_1\,\ldots\,\bar{k}_n}, \tag{9.47}$$

where

$$k_i = \frac{P(s_i/b)}{P(s_i)}, \qquad \bar{k}_i = \frac{P(s_i/b^{(0)})}{P(s_i)}$$

Thus, the numerical measure of each predictive feature is given by the value of k_i, called the compatibility coefficient.

The data in Table 9.20 were used to obtain complexes of features D_j (Table 9.21) and to select as predictors those which have the highest probability. As we see, complexes D_j of features s_i, corresponding to the values $i = 1, 4, 10; 2, 4, 10; 1, 4, 11$, have a reliability exceeding 80% for the first group (buffeting irrespective of intensity) and 60–70% for the second group (moderate and severe aircraft buffeting).

Lomonosov and Lunin [65] also used the Bayes formula on the assumption of independence of the predictive features for the diagnosis and prediction of turbulent zones in which aircraft buffeting occurs. As simple features s_i they selected hydrodynamic factors whose values can be obtained from constant-pressure charts by means of a computer. These authors selected the 14 features s_i given in Table 9.22. In this table u_T and v_T are components of the thermal wind velocity.

The transformations M_1, M_2, and M_3 were calculated for $\mu_T = 1$. The basic data for calculations of the features s_i are, as we see, the values of z, $\partial z/\partial t$, and

TABLE 9.21. Probability (%) of Aircraft Buffeting in Relation to D_j

j	I		II		j	I		II	
	D_j	$P(b/D_j)$	D_j	$P(b/D_j)$		D_j	$P(b/D_j)$	D_j	$P(b/D_j)$
1	1, 4, 10	90	1, 4, 10	71	13	2, 5, 11	54	3, 4, 10	28
2	2, 4, 10	87	1, 4, 11	62	14	8, 10	53	8, 10	25
3	1, 4, 11	85	2, 4, 10	59	15	2, 6, 11	49	3, 5, 10	23
4	2, 4, 11	79	1, 5, 10	58	16	7, 10	48	3, 4, 11	20
5	1, 5, 10	77	1, 6, 10	51	17	3, 6, 10	45	8, 11	17
6	1, 6, 10	77	2, 4, 11	50	18	3, 5, 10	41	7, 10	15
7	3, 4, 10	72	1, 5, 11	49	19	9, 10	41	3, 5, 11	15
8	2, 5, 10	69	2, 5, 10	48	20	8, 11	37	3, 6, 10	14
9	2, 6, 10	69	1, 6, 11	41	21	7, 11	34	9, 10	13
10	1, 5, 11	67	2, 5, 11	40	22	3, 5, 11	33	7, 11	10
11	1, 6, 11	66	2, 6, 10	39	23	3, 6, 11	32	9, 11	9
12	3, 4, 11	59	2, 6, 11	30	24	9, 11	27	3, 6, 11	9

TABLE 9.22. Predictive Features s_i from [65]

i	Features s_i	Analytical expression for feature
1	Richardson number Ri	$\frac{g}{T} \cdot \frac{\gamma_a - \gamma}{\left(\frac{\partial \bar{c}}{\partial z}\right)^2}$
2	Variation of Ri in horizontal plane	$\sqrt{\left(\frac{\partial \mathrm{Ri}}{\partial x}\right)^2 + \left(\frac{\partial \mathrm{Ri}}{\partial y}\right)^2}$
3	Transformation of kinetic energy of mean motion to turbulent energy (M_1)	$\mu\left[\left(\frac{\partial u}{\partial z}\right)^2 + \left(\frac{\partial v}{\partial z}\right)^2\right]$
4	Transformation M_2	$\mu\left(\frac{\partial v}{\partial x} + \frac{\partial u}{\partial y}\right)^2$
5	Transformation M_3	$2\mu\left[\left(\frac{\partial u}{\partial x}\right)^2 + \left(\frac{\partial v}{\partial y}\right)^2\right]$
6	Transformation $M_2 + M_3$	
7	Wind velocity c	
8	Horizontal gradient of wind velocity normal to flow	$\frac{\partial c}{\partial n}$
9	Horizontal gradient of wind velocity along flow	$\frac{\partial c}{\partial s}$
10	Vertical component of vorticity $\Omega + l$	$\frac{\partial v}{\partial x} - \frac{\partial u}{\partial y} + l$
11	Thermal wind velocity c_T	$\sqrt{u_T^2 + v_T^2}$
12	Vertical component of thermal wind vorticity	$\frac{\partial v_T}{\partial x} - \frac{\partial u_T}{\partial y}$
13	Temperature advection $(\partial T/\partial t)_a$	$-u\frac{\partial T}{\partial x} - v\frac{\partial T}{\partial y}$
14	Modulus of trend of altitude of isobaric surface z	$\left\lvert\frac{\partial z}{\partial t}\right\rvert$

T. In [65] the initial calculation data were taken for the isobaric surfaces 400 and 300 mb; the step of the grid for the computer calculations was 300 km. The range of values of the features s_i was then divided into eight equal intervals ($m = 1, 2, \ldots, 8$). The following characteristics were calculated for each interval:

(a) the probability of appearance of gradation in the whole range of values of the feature s_i

$$P_m(s_i) = \frac{N_i^{(m)}}{N};$$

(b) the conditional probability of appearance of gradation in the presence of buffeting;

$$P_m(s_i/b) = \frac{L_i^{(m)}}{L};$$

(c) the compatibility coefficient for the ith gradation

$$k_i^{(m)} = \frac{P_m(s_i/b)}{P_m(s_i)} = \frac{1}{P(b)} \frac{L_i^{(m)}}{N_i^{(m)}}.$$

Here N is the total number of squares considered, L is the number of squares with buffeting, and $P(b) = L/N$ is the *a priori* probability of occurrence of buffeting.

The criterion used to assess the success of the prediction is the coefficient of accuracy of alternative predictions proposed by Obukhov:

$$Q = 1 - (\alpha + \beta), \tag{9.48}$$

where α is the error of the first series, characterizing the relative number of unsuccessful predictions of absence of buffeting, and β is the error of the second series, characterizing the relative number of unsuccessful predictions of buffeting. A measure of the usefulness of the feature s_i is the value of Q_i in the layer:

$$Q_i = P(s_i/b) - P(s_i/b^{(0)}). \tag{9.49}$$

If $Q_i > 0$, expectation of s_i means expectation of aircraft buffeting in the corresponding square.

To assess the role of the features s_i, Lomonosov and Lunin [65] used the reports of airline crews for the periods in 1964–1965 recommended by the ICAO and WMO, i.e., the same data as were used by Davydov *et al.* in [37]. Table 9.23 gives values of the success index Q_i of diagnosis of aircraft buffeting for the features s_i.

TABLE 9.23. Values of $Q_i \cdot 10^2$ for Features s_i

s_i	Winter	Spring and autumn	Summer	s_i	Winter	Spring and autumn	Summer
1	8.5	6.9	8.6	8	8.2	9.1	5.5
2	3.1	6.3	4.7	9	17.2	9.4	4.9
3	8.9	7.2	6.4	10	10.7	8.7	6.8
4	9.3	2.4	2.8	11	3.3	8.3	7.1
5	3.3	4.0	1.5	12	5.4	7.2	5.3
6	10.5	4.9	6.3	13	6.6	7.4	4.4
7	8.0	12.6	13.6	14	8.2	13.4	9.5

It is apparent that the values of Q_i depend significantly on the time of year. The maximum and most stable values of Q_i in the course of the year were obtained for features s_i with numbers 1, 3, 7, 10, and 14, i.e., Ri, the vertical wind-velocity gradient, the wind velocity, the vertical vorticity component, and the modulus of the trend of altitude of the isobaric surface. Table 9.24 shows the success indices Q for complexes s_i. This table shows that obtaining $Q = 0.28$–0.33 in summer requires the use of up to 13–14 features s_i, whereas complexes consisting of seven features are sufficient for other seasons.

TABLE 9.24. Values of Q for Complexes of Features s_i

Winter complexes s_i	$Q \cdot 10^2$	Spring and autumn complexes s_i	$Q \cdot 10^2$
9	17.2	14	13.4
9, 6	22.2	14, 7	19.1
9, 6, 14	25.6	14, 7, 3	22.0
9, 6, 14, 3	28.5	14, 7, 3, 8	22.6
9, 6, 14, 3, 1	29.9	14, 7, 3, 8, 9	24.2
9, 6, 14, 3, 1, 13	31.7	14, 7, 3, 8, 9, 12	26.1
9, 6, 14, 3, 1, 13, 12	32.6	14, 7, 3, 8, 9, 12, 10	26.8
9, 6, 14, 3, 1, 13, 12, 2	33.0	14, 7, 3, 8, 9, 12, 10, 13	27,1
		14, 7, 3, 8, 9, 12, 10, 13, 2	27.9

Summer complexes s_i	$Q \cdot 10^2$
7	13.6
7, 12	18.3
7, 12, 8	20.8
7, 12, 8, 3	21.4
7, 12, 8, 3, 10	22.5
7, 12, 8, 3, 10, 1	24.7
7, 12, 8, 3, 10, 1, 2	25.7
7, 12, 8, 3, 10, 1, 2, 5	25.8
7, 12, 8, 3, 10, 1, 2, 5, 6	26.0
7, 12, 8, 3, 10, 1, 2, 5, 6, 14	26.6
7, 12, 8, 3, 10, 1, 2, 5, 6, 14, 11	26.8
7, 12, 8, 3, 10, 1, 2, 5, 6, 14, 11, 4	27.5
7, 12, 8, 3, 10, 1, 2, 5, 6, 14, 11, 4, 13	27.6

TABLE 9.25 Average Seasonal Values of Parameters

Parameter	Winter	Spring and Fall	Summer
α	0.30	0.38	0.30
β	0.37	0.34	0.42
Q	0.33	0.28	0.28
P	0.62	0.46	0.39

For the diagnosis (prediction) of aircraft buffeting the most significant feature is selected from Table 9.23. It is "multiplied" in turn by the others, and the pair of features s_i leading to the highest resultant value of Q is then selected. This pair of features s_i is combined in turn with the other features, and from the obtained set of three-dimensional vectors the vector with the highest value of Q is selected. The formation of the n-dimensional predictor-vector on the computer continues until the resultant success index Q ceases to increase. The features corresponding to the maximum value of Q are used to predict the probability of aircraft buffeting.

A possible average seasonal measure of the success of the numerical method of predicting the probability of aircraft buffeting is illustrated by the data of Table 9.25, which show once again that the predictor-vectors for different seasons have different degrees of significance.

We have discussed the works of [37, 65] in detail, since synoptic features were mainly considered in the former, and hydrodynamic features, including Ri, in the latter. The proposed methods of diagnosis and prediction of aircraft buffeting were tested on the same aircraft buffeting data obtained from flying personnel of Aeroflot USSR. The main point, in our opinion, is that the statistical data given in [37] and [65] provide an idea of the role of these features and their complexes as possible predictors of turbulence causing aircraft buffeting.

Yarovskaya [134] proposed a fundamentally similar method of diagnosis and prediction of turbulence in the stratosphere. The main predictors were the vertical gradient of mean wind velocity and its variation with altitude, and also the vertical temperature gradient.

10

Mesostructure of Wind Field

1. Mesoinhomogeneities of Wind Field

Experimental investigations of the mesostructure of the wind field have shown that local wind-velocity maxima are superimposed on a general transfer of air at some mean velocity. These local zones take the form of flattened jets of varying size. In a number of cases the local zones, bordered by a common isotach in the horizontal plane or in a vertical section of the atmosphere, consist in turn of a series of small zones with wind-velocity maxima (Fig. 10.1). Aircraft passing through these inhomogeneities usually experience buffeting.

It should be noted that local spatial changes in wind velocity are not always correlated with spatial changes in air temperature and that, consequently, the appearance of disturbances with a scale of several tens of kilometers in the wind field cannot always be accounted for by means of the thermal wind concept.

Flight experiments have shown that such inhomogeneities in the wind field are usually observed in the region of divergence of air flows and in low-gradient pressure fields.

As is known from aerodynamics, even a slight constriction of the flow has a stabilizing effect on it and, other conditions being equal, the critical Reynolds number Re_{cr} is greater than for unconstricted flow. On the other hand, when the walls diverge slightly, turbulent flow begins at smaller Re_{cr}.

With increase in the angle of divergence of the wind-tunnel walls, there is a back flow and the flow separates from the walls, though not simultaneously from both sides.

Laboratory investigations have shown that instability of diffuser flow is greater, the smoother the surface of the diffuser walls. In application to the free atmosphere, this means that regions with flow divergence can have several alternating velocity maxima and minima on the vertical and horizontal wind profiles. The presence of bends in the wind-velocity profile is an important factor leading to instability of the flow and the onset of turbulence.

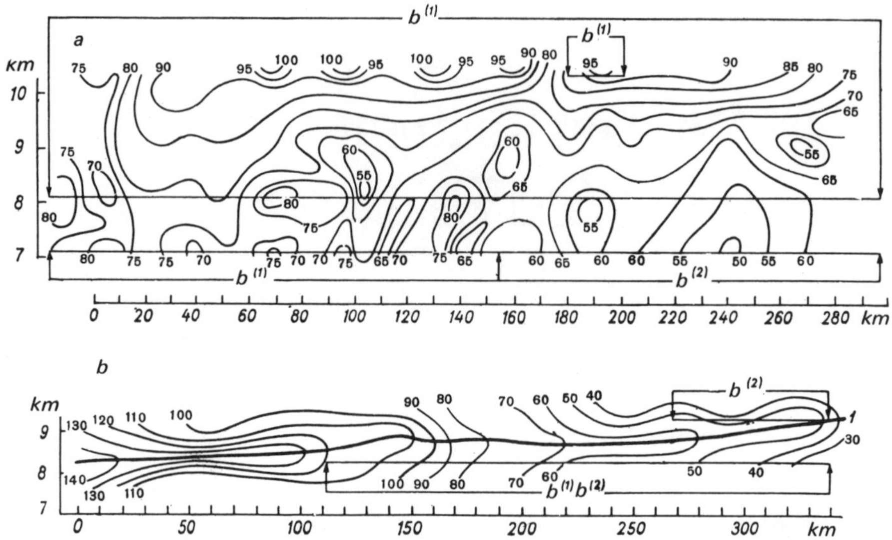

Figure 10.1. Vertical section of wind field (km/h) in zone of: (a) divergence; (b) convergence of the air flows. Light ($b^{(1)}$) and moderate ($b^{(2)}$) buffeting of TU-104. (1) Maximum wind level.

2. Energy Spectra of Mesoscale Turbulence

Measurements of the wind velocity and direction at different heights in the free atmosphere by means of Doppler navigation systems have revealed turbulent mesoinhomogeneities of the wind field and have provided information about the statistical characteristics of mesoscale turbulence [83, 84]. So far such investigations have been made only in the upper troposphere and lower stratosphere.

In [83, 84] empirical structure functions of the wind-velocity fluctuations were calculated from the results of wind-velocity measurements on horizontal areas about 200-300 km long. Areas of this length provided a sufficient number of wind-velocity measurements for statistical calculations in the case of a flight experiment of the minimum duration for which the velocity field can be regarded as approximately stationary.

The structure function curves showed that an incease in the scale Δx is accompanied by an increase in $\sigma^2_{\Delta u}$. The values obtained for the exponent of the structure function in this case [the empirical data were approximated by the expression $\sigma^2_{\Delta u} = A(\Delta x)^n$] have a large spread—from 0.2 to 1.5. Figure 10.2 shows curves of the integral frequency of values of the exponent n from the results of observations above regions of the European USSR and the Far East. In approximately 25% of the cases, n was less than 0.5, i.e., the increase in $\sigma^2_{\Delta u}$ with increase in Δx was relatively slow, and in 20-30% of the cases n was greater

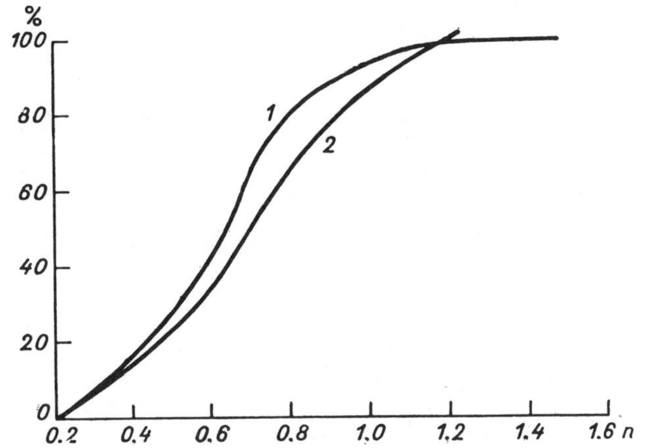

Figure 10.2. Integral frequency of exponent of structure function. (1) 1960; (2) 1962.

than 0.8. On the average, as can be seen from Table 10.1, which gives the mean values $\bar{n}$ at different heights, the relation between $\sigma^2_{\Delta u}$ and Δx is close to the "two-thirds" law.

The increase in $\sigma^2_{\Delta u}$ with increase in Δx usually proceeds only up to a certain characteristic value $(\Delta x)_{\text{char}}$. When $\Delta x > (\Delta x)_{\text{char}}$ the value of $\sigma^2_{\Delta u}$ either does not increase, or its increase is slowed down and it oscillates around a mean value $\overline{\sigma^2_{\Delta u}}$. Figure 10.3 shows the curve of integral frequency of $(\Delta_1 x)_{\text{char}}$, while the mean values $\overline{(\Delta_1 x)_{\text{char}}}$ at different heights are given in Table 10.1. Figure 10.3 shows that in approximately 70% of the cases the characteristic distance was less than 20 km, including 47% cases where it was less than 15 km; in certain cases $(\Delta_1 x)_{\text{char}}$ reached 45–50 km.

It is of interest to compare the characteristic scale for fluctuations of the horizontal wind-velocity component with the characteristic scale $\Delta_2 x$ for the vertical component. For this purpose we plot on Fig. 10.3 the curve of integral frequency of $(\Delta_2 x)_{\text{char}}$, obtained by Yurgenson from airplane loads in flights in a turbulent atmosphere. It is apparent that the curves of integral frequency of $(\Delta_1 x)_{\text{char}}$ and $(\Delta_2 x)_{\text{char}}$ have the same shape, but $(\Delta_2 x)_{\text{char}}$ is an order of magnitude less than $(\Delta_1 x)_{\text{char}}$.

The fact that the exponent of the empirical structure function corresponds on the average to the "two-thirds" law provides evidence that on the average the

TABLE 10.1. Parameters of Structure Functions of Fluctuations of Horizontal Wind-Velocity Component

H (km)	7±0.5	8±0.5	9±0.5	10±0.5	11±0.5
$\bar{n}$	0.685	0.646	0.646	0.611	0.611
$\overline{(\Delta_1 x)}_{\text{char}}$ (km)	16.9	17.4	18.3	19.2	18.6

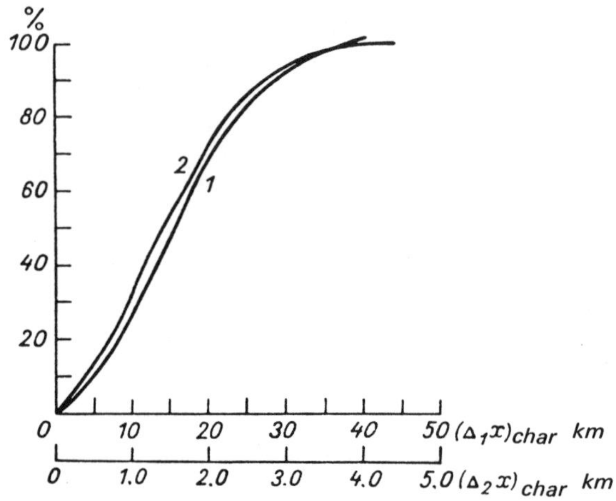

Figure 10.3. Integral frequency of $(\Delta x)_{char}$. (1) For horizontal wind-velocity component; (2) for vertical component (according to Yurgenson).

spectral density is represented by the "minus five-thirds" law. As investigations [86] showed, deviations from this law are related to the nature of the thermal and wind stratification of the atmosphere. A theoretical treatment of the relation between the fluctuating wind field and the thermal stratification is the subject of a paper by Panchev [78]. These structure functions will, of course, be reflected in the spectral density functions of the wind-velocity fluctuations.

The experimental energy spectra plotted in logarithmic coordinates can have bends leading to different slope of the spectrum. These bends indicate (see Chapter 4) the presence of a sink of turbulent energy in the particular region of scales of turbulent motions (say, on overcoming negative buoyancy forces) or to an influx of energy from some external source. At low altitudes, for instance, a source of turbulent energy in the high-frequency region is the disturbing effect of the underlying surface and unstable thermal stratification on the flow, whereas in the free atmosphere the main source is the loss of stability of wavelike motions and the nonlinear interactions of turbulent motions of different scales, and also of the turbulent motions with the mean flow. In view of this the term WIT (wave-induced turbulence) has recently been used in addition to the term CAT.

Figure 10.4 shows in logarithmic coordinates the averaged spectral density $S(\omega)$ curves for fluctuations of the horizontal wind-velocity component in the troposphere [92]. The averaged curves were derived from spectra obtained at different times of the year and in different geographic regions. For altitudes of 0.5–1.0 km the averaging included 11 individual spectra; for 1–3 km, 21; for 3–7 km, 51; and for 7–10 km, 194 spectra. For comparison the figure shows the lines corresponding to the density $S(\omega)$ proportional to $\omega^{-5/3}$ and ω^{-3}.

The spectrum for the atmospheric layer 0.5–1.0 km has two bends, forming special flat portions, with the result that the regions of the spectrum for high frequency lie in a region of higher energies than if the spectrum, based on the hypothesis of cascade energy transfer across the spectrum, was linearly extrapolated into regions of the same higher frequencies. The characteristic linear dimensions of the disturbances, in the region of which there is an influx of external turbulent energy, are approximately 2–5 and 10–20 km, respectively.

Figure 10.4 also shows that the spectra for atmospheric layers 1–3, 3–7, and 7–10 km have flat portions for turbulence scales from 10 to 20 km. Thus, in the free atmosphere the spectra of meso- and microscale turbulence can contain at least two or three different inertial subranges with different rates of dissipation of turbulent energy.

Theoretical investigations [53] of stably stratified jet streams, which are characteristic of the upper troposphere, have shown that in a linear approxima-

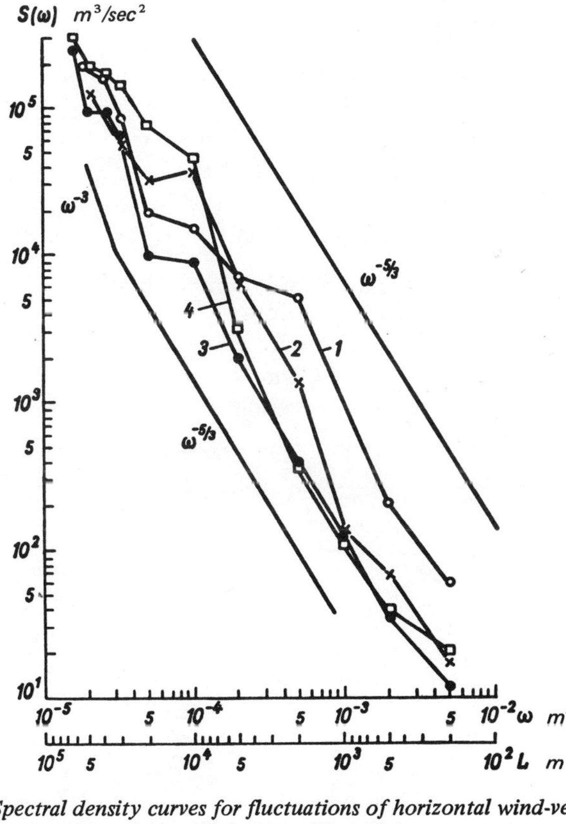

Figure 10.4. Spectral density curves for fluctuations of horizontal wind-velocity component. (1) 0.5–1.0 km; (2) 1–3 km; (3) 3–7 km; (4) 7–10 km.

tion wavelike disturbance with lengths of approximately 8–30 km are unstable at very low Ri ($<\frac{1}{4}$). This means physically that in a moving flow whose stratification is hydrodynamically unstable ($\text{Ri} < \frac{1}{4}$) there arise waves of increasing amplitude, which at a certain instant break, and then the flow becomes turbulent. In application to the upper troposphere with its characteristic vertical temperature gradient $\gamma = 0.007°\text{C/m}$ this means that there is a critical vertical gradient of mean wind velocity β at which waves of a particular length are unstable. An increase in β increases the range of lengths (or frequencies) of unstable waves (toward shorter or longer waves). This relation is illustrated in Fig. 10.5; the hatched region characterizes the range of hydrodynamically unstable wavelike disturbances. According to Fig. 10.5, when $\beta = 3.2 \cdot 10^{-2}$ sec^{-1} (Ri = 0.1) waves of length 8.5–16.0 km are unstable. The energy of these disturbances can be converted to turbulence energy and produce (or intensify) the turbulent regime. The solid curve bounding the region of instability in Fig. 10.5 corresponds to neutral (neither increasing nor decreasing in amplitude) waves.

Figure 10.5, plotted on the basis of linear theory, can be used to determine the conditions leading to instability of the wavelike disturbances, but it cannot be used to characterize the onset, far less the development, of turbulence, since loss of stability *per se* is still not the same as conversion of the wave regime to a turbulent regime. The theoretical description of such a transition entails consideration of the nonlinear interaction of the disturbance and the main flow, and also wave disturbances of different lengths. An account of some results of investigations of the nonlinear effects of mesoscale disturbances and the mechanism of turbulence in the jet stream is given in [183]. It was found that the

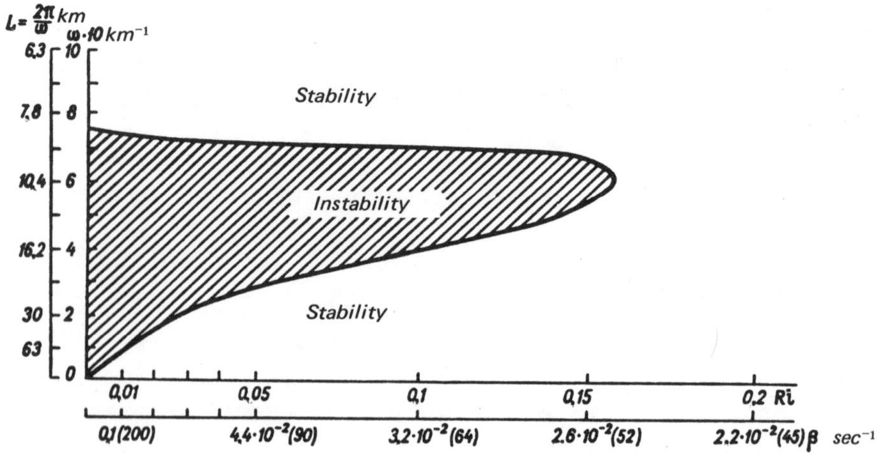

Figure 10.5. Region of unstable waves. The figures in brackets on the lower x axis give the velocities (m/sec) on the axis of a jet with half-thickness 2 km corresponding to Ri numbers for $\gamma_a - \gamma = 0.003°$ *C/m.*

interaction of internal gravity waves (Ri $< \frac{1}{4}$) depends on the vertical temperature profile. In the case of a linear profile interaction is weak, and in the case of a distorted temperature profile (tropopause, inversion layers) the interaction is pronounced. Calculations showed that waves of length 10–30 and 1–5 km are significantly nonlinear; the nonlinear interaction of such waves induces turbulence of the flow, and the energy of fluctuating motion is comparable with the energy observed in aircraft buffeting conditions. In the intermediate region of wavelengths (5–10 km) nonlinear interaction is much weaker.

Kao and Woods' investigations [176], based on the data of the American "Project Jet Stream," showed that the energy spectra for longitudinal and transverse wind-velocity components are very similar to one another in shape and that $S(\omega) \approx \omega^{-2.0}$. The variance of the wind-velocity fluctuations was found to depend on the angle between the predominant direction of the air flow and the direction of the flight along which the measurements were made. Table 10.2 shows these differences. The variance of the fluctuating velocity is greater in flights across the flow than in flights along the flow, and it is greater for the transverse component than for the longitudinal component in flights parallel to the flow.

A hypothesis used in turbulence investigations is that of the locally normal structure of turbulence, i.e., that there is a set of turbulent zones in each of which the variation of the fluctuating-velocity components with time is a stationary normal process with variance varying from zone to zone. In principle, for the region of turbulence scales in which there is a turbulence source or sink due to external factors, this hypothesis is inapplicable. These effects can be demonstrated by using data indicating the distribution of the spectral density $S(\omega)$ for some fixed turbulence scales. The distribution of $S(\omega)$ can also be used not only to reveal influxes of external energy into the spectrum, but also turbulent energy sinks. The distribution of $S(\omega)$ for several wave numbers is shown in Fig. 10.6, and the numerical characteristics of these distributions are given in Table 10.3.

The distribution of $S(\omega)$ for scale 0.5 km is trimodal, for scales 1, 20, and 50 km it is bimodal, and for 2, 5, and 10 km it is unimodal. The distribution of $S(\omega)$ for $\omega = 10^{-4}$ m^{-1} is distinguished by the greatest variance. The secondary maxima of the frequency of $S(\omega)$ for scales 0.5, 1, and 20 km indicate a comparatively high probability of appearance of these disturbances, distin-

TABLE 10.2. Characteristics of Wind-Velocity Fluctuations according to [176]

Direction of flight	Mean wind velocity (m/sec)		Variance of fluctuating velocity (m^2/sec)	
	longitudinal component	transverse component	longitudinal component	transverse component
Along flow	59.0	6.1	1.0	1.1
Across flow	62.3	6.9	2.6	2.03

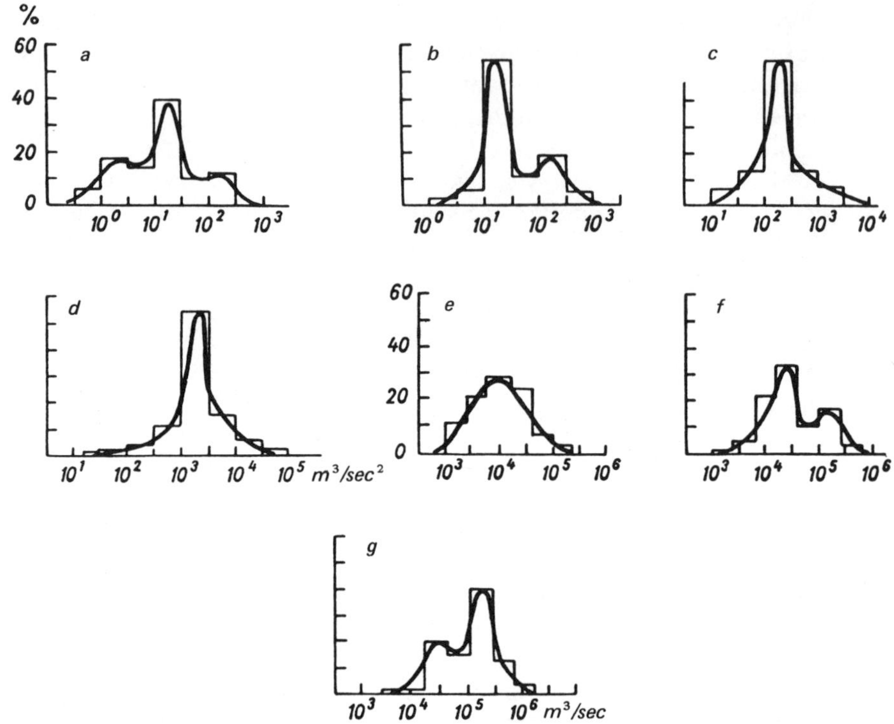

Figure 10.6. Distribution curves of spectral density $S(\omega)$ for different wave numbers ω. (a) $2 \cdot 10^{-3}$; (b) 10^{-3}; (c) $5 \cdot 10^{-4}$; (d) $2 \cdot 10^{-4}$; (e) 10^{-4}; (f) $5 \cdot 10^{-5}$; (g) $2 \cdot 10^{-5}$ m^{-1}.

TABLE 10.3. Parameters of Statistical Distribution of $S(\omega)$

ω (m^{-1})	L (km)	$S(\omega)$ (m^3/sec^2)	Mode $S(\omega)$ (m^3/sec^2)		Mode (%)	
			main	secondary	main	secondary
$2 \cdot 10^{-3}$	0.5	$5 \cdot 10^{-1} - 10^3$	$10^1 - 5 \cdot 10^1$	$10^0 - 5 \cdot 10^0$	40.0	18.0
				$10^2 - 5 \cdot 10^2$		12.4
10^3	1.0	$10^0 - 5 \cdot 10^3$	$10^1 - 5 \cdot 10^1$	$10^2 - 5 \cdot 10^2$	54.8	18.6
$5 \cdot 10^{-4}$	2.0	$10^1 - 5 \cdot 10^4$	$10^2 - 5 \cdot 10^2$	—	55.0	—
$2 \cdot 10^{-4}$	5.0	$2 \cdot 10^1 - 10^5$	$10^3 - 5 \cdot 10^3$	—	55.0	—
10^{-4}	10.0	$10^3 - 4 \cdot 10^5$	$8 \cdot 10^3 - 2 \cdot 10^4$	—	29.9	—
$5 \cdot 10^{-5}$	20.0	$10^3 - 2 \cdot 10^6$	$2 \cdot 10^4 - 6 \cdot 10^4$	$10^5 - 4 \cdot 10^5$	35.0	18.0
$2 \cdot 10^{-5}$	50.0	$4 \cdot 10^3 - 2 \cdot 10^6$	$10^5 - 4 \cdot 10^5$	$2 \cdot 10^4 - 6 \cdot 10^4$	42.0	20.5

guished by high energies, and it can be inferred that their appearance is due to an influx of energy into the turbulence spectrum. On the other hand, the secondary maxima of frequency of $S(\omega)$ for scales 0.5 (left secondary maximum) and 50 km reflect not an influx of energy, but loss of energy in the turbulence spectrum.

3. Vertical Distributions of Wind Gusts and Their Spectra

The statistical characteristics of the wind-velocity fluctuations and their changes in a horizontal plane were discussed above. Of no less interest is the vertical distribution of wind-velocity fluctuations. If sufficiently detailed data on the vertical distribution of such fluctuations are available, they can be used to obtain correlation and spectral functions, as is done for horizontal and time distributions. Although such a vertical distribution of wind-velocity fluctuations can be regarded only provisionally as a realization of the random process, the spectral decomposition has a quite definite physical meaning, which allows determination of the vertical scales of wind-velocity fluctuations and the spectral distribution of the energy of these fluctuations.

We will call such spectra energy spectra, by analogy with the spectra of real random processes, thus incorporating the above-mentioned meaning in the term.

Kozlov *et al.* [55] gave autocorrelation functions of mesoscale (from a few meters to hundreds of meters) wind-velocity fluctuations, calculated from the results of measurements of wind-velocity fluctuations in ascents of automatic stratosphere balloons equipped with special apparatus. Such functions were obtained for altitudes from the ground to 10–12 km in layers 1 km thick. An analysis showed that the best approximation for these empirical autocorrelation functions is the expression

$$R(\Delta z) = e^{-\alpha \Delta z} \cos \varphi \, \Delta z \,, \tag{10.1}$$

to which corresponds the spectral density

$$S(\omega) = \frac{\alpha}{\pi} \frac{\omega^2 + \alpha^2 + \varphi^2}{(\omega^2 - \varphi^2 - \alpha^2)^2 - 4\alpha^2\omega^2} \,. \tag{10.2}$$

The vertical disturbance scales $\Delta_0 z$, determined by the correlation radius corresponding to $R(\Delta z) = 0$, varied from 10 to 350 m. In 62% of the cases they were less than 100 m, and in 16% of the cases they were greater than 200 m. The value of α also varied in a wide range: 0.005–0.15 m^{-1}; but in 89% of the cases it was 0.005–0.05 m^{-1}. It was found also that the turbulence intensity

$\overline{u'^2}$ and the disturbance scales $\Delta_0 z$ increase with reduction of Ri. The reduction of relative turbulence intensity ψ with altitude is inversely proportional on the average to the square root of the altitude.

Measurements of wind velocity and direction by means of a system consisting of an FRS-16 precision radar set and a helium-filled rigid Mylar metallized sphere with a spiky surface have been carried out in the U.S.A. for several years. The space coordinates of the sphere, which ascends at a speed of approximately 5 m/sec, are determined every 0.1 sec. These data are automatically averaged over 50-m height intervals. The root-mean-square error of determination of the wind velocity is about 0.5 m/sec. The obtained vertical wind-velocity profile is then averaged over approximately 600-m layers, and then it becomes similar to the wind profile obtained by ordinary rawinsonde probing of the atmosphere. The difference between the initial and averaged vertical wind profile gives the vertical profile of the fluctuating wind-velocity component. The spectral density function is calculated from the data of this profile.

Daniels *et al.* [158] give averaged spectra of wind-velocity fluctuations from 210 vertical profiles obtained on Cape Canaveral for the atmospheric layer 2–16 km. These spectra are represented satisfactorily by a power law with exponent equal to 2.47 for the zonal component of the wind velocity, 2.70 for the meridional component, and 2.46 for the modulus of the wind velocity. As an example, we cite that the spectral density of the fluctuations of the wind vector modulus for the wave number $\omega = 5 \cdot 10^{-4}$ m^{-1} was less than $1.2 \cdot 10^4$ m^3/sec^2 in 99% of the cases, and less than $4.8 \cdot 10^3$ m^3/sec^2 in 50% of the cases.

TABLE 10.4. Seasonal Values of Spectral Density of Wind-Velocity Fluctuations

ω (km^{-1})	L (km)	Winter		Spring		Summer		Fall		Year	
		$S_u(\omega)$	$S_v(\omega)$	$S_u(\omega)$	$S_v(\omega)$	$S_u(\omega)$	$S_v(\omega)$	$S_u(\omega)$	$S_v(\omega)$	$S_u(\omega)$	$S_v(\omega)$
0.02	50	145.36	75.44	82.80	31.60	24.88	36.00	21.72	46.84	68.69	47.47
0.025	40	132.60	69.40	94.52	33.32	29.88	32.84	42.36	41.80	74.84	44.34
0.030	33.3	110.80	59.64	98.64	35.36	35.04	28.56	62.12	36.52	76.65	40.02
0.035	28.5	86.84	49.56	96.88	37.88	39.80	24.72	77.84	33.00	75.34	36.29
0.040	25.0	65.16	41.16	89.96	40.76	43.52	22.48	86.92	32.44	71.39	34.21
0.050	20.0	38.48	32.96	64.32	46.16	46.36	25.26	79.68	40.92	57.21	36.40
0.060	16.6	35.04	33.60	36.52	47.48	42.72	38.12	17.32	53.16	40.40	39.09
0.070	14.3	35.96	32.20	24.36	40.84	36.44	49.72	21.60	51.24	29.59	43.50
0.080	12.5	24.84	24.68	25.48	28.92	31.76	47.80	21.52	33.24	25.90	33.66
0.090	11.1	11.00	18.96	21.16	19.12	28.12	33.80	26.80	18.72	21.77	22.65
0.10	10.0	7.08	18.00	8.60	14.64	23.56	19.28	17.84	17.08	14.27	17.25
0.12	8.3	6.88	10.00	3.80	11.76	13.56	10.28	2.00	11.44	6.56	10.87
0.14	7.1	3.64	10.04	3.64	6.08	3.36	11.68	6.88	5.36	5.13	8.29
0.16	6.3	4.32	8.32	3.12	7.16	8.72	9.16	1.32	11.44	4.37	9.02
0.20	5.0	3.52	7.64	1.68	4.04	4.24	6.92	1.52	7.48	2.74	6.52
0.25	4.0	0.72	2.52	2.08	4.84	2.44	3.24	2.12	2.72	1.84	3.33
0.30	3.3	0.92	3.00	0.64	4.52	3.68	3.24	2.48	1.64	1.93	3.10
0.35	2.8	1.24	3.36	0.16	2.60	1.20	0.64	0.52	1.24	0.78	1.96
0.40	2.5	1.40	2.56	0.96	1.60	1.28	0.60	0.40	1.48	1.01	1.56

TABLE 10.5. Seasonal Values of $\overline{u'^2}$ and $\overline{v'^2}$

Season	$\overline{u'^2}$ (m^2/sec^2)	$\overline{v'^2}$ (m^2/sec^2)	$\overline{u'^2}/\overline{v'^2}$
Winter	73.62	29.57	0.39
Spring	56.61	20.96	0.37
Summer	23.06	18.67	0.83
Autumn	48.68	24.09	0.49
Year	50.49	23.32	0.46

Kao and Sands [178] obtained spectra of wind-velocity fluctuations for the atmospheric layer from the ground to an altitude of 50 km from the data of 210 rocket experiments accompanied by measurements of wind velocity. They calculated the spectra for scales from 2.5 to 50 km for each vertical wind-velocity profile. The calculations were made separately for the zonal and meridional wind-velocity components. The individual spectra were then averaged for each season.

Table 10.4 gives values of the normalized spectral density for different wave numbers ω, while Table 10.5 gives the variance of the zonal and meridional wind-velocity components for different seasons. In addition, Fig. 10.7 gives curves of the normalized mean annual values of $S(\omega)$. As Fig. 10.7 shows, the spectral density curves for fluctuations of the zonal and meridional wind-

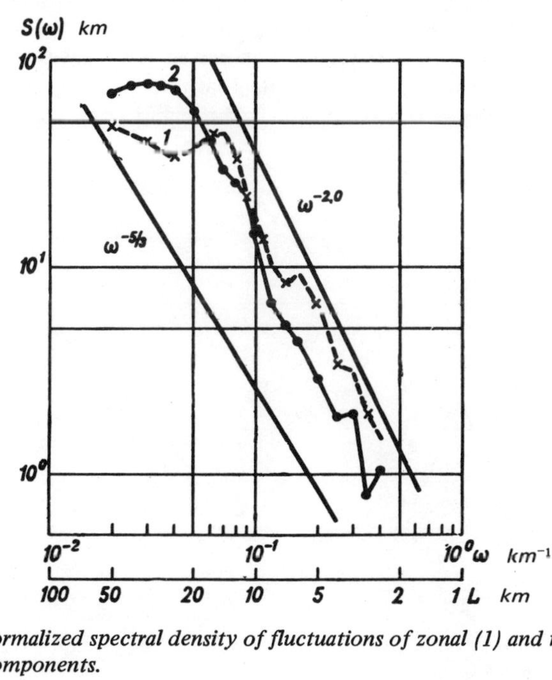

Figure 10.7. Normalized spectral density of fluctuations of zonal (1) and meridional (2) wind-velocity components.

velocity components have approximately the same shape and can be represented by the power law $S(\omega) \approx \omega^{-n}$. According to Kao and Sands' estimate the spectral density is inversely proportional to the square of the wave number ω.

4. Mesometeorological Energy Minimum in Turbulence Spectrum

It was shown in Chapter 1 that the spectral density increases with reduction of frequency. In the inertial subrange this relation is characterized by the "minus five-thirds" law. The wavelength region adjoining the inertial subrange on the low-frequency side contains the bulk of the kinetic energy of turbulent motions. This part of the spectrum contains the energy maximum corresponding to a certain range of scales of turbulent disturbances—the so-called energy-bearing subrange of the spectrum of atmospheric turbulence. At present there are no theoretical estimates of the form of the spectral density function in this spectral region. The low-frequency region adjoining the energy spectrum maximum on the low-frequency side and which depends significantly on the conditions of onset of turbulence in the flow has been investigated to an even smaller extent.

4.1. Energy Spectrum of Wind-Velocity Fluctuations at Low Altitudes. Attempts to calculate the spectrum near the ground in a wide frequency range have been made by Van der Hoven and Kolesnikova and Monin [56].

Figure 10.8 shows the energy spectrum [in coordinates $fS_u(f)$ and f] of fluctuations of the horizontal wind-velocity component in the frequency range from 10^3 to 10^{-3} h^{-1}, constructed by Van der Hoven from the results of measurements on a 125-meter meteorological mast in Brookhaven (U.S.A.). The high-frequency part of the spectrum was obtained from measurements in a period of strong winds.

Figure 10.8 shows that this spectrum has two maxima: a main maximum at frequency $f \approx 10^{-2}$ h^{-1} and a secondary maximum corresponding to $f \approx 10^{-1}$ h^{-1}. The maximum in the "synoptic region" ($f \approx 10^{-2}$ h^{-1}) is due to the passage of large-scale pressure systems, while the maximum in the micro-

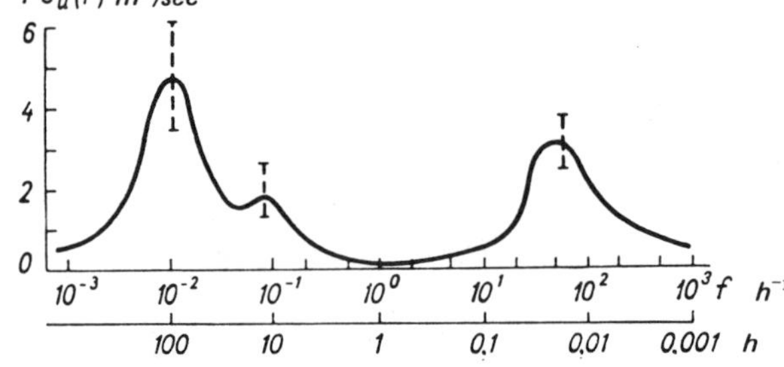

Figure 10.8. Energy spectrum of fluctuations of horizontal wind-velocity component (according to Van der Hoven).

meteorological region ($f \approx 6 \cdot 10^1\ h^{-1}$) is due to small-scale turbulence of dynamic origin. The most significant feature of the considered spectrum is the deep minimum (dip) in the mesometeorological region ($f \approx 2 \cdot 10^{-1}$ to 10^1 h^{-1}).

According to Van der Hoven's data, this feature is typical of many series of measurements.

At present there is still no theoretical explanation of the appearance of the mesometeorological minimum in the spectrum. We will discuss the qualitative explanation of this effect. Investigations of the statistical characteristics of atmospheric turbulence are based on the idea that the atmosphere is a dynamic unstable system and that periodic wave motions are always present in it. If these motions in turn become unstable, motions of a turbulent nature arise. Hence, turbulent motions are the result of "secondary" instability of the atmosphere.

The synoptic maximum is due to secondary instability of the large-scale dynamic processes in the atmosphere resulting from the difference in the heating of polar and tropic zones by the sun, and also from the difference in the nature of the underlying surface (continents, oceans). The eddy "breakup" process transfers turbulent energy from this maximum over the spectrum from large scales to small scales and ultimately to those where the energy of the turbulent motions is dissipated as heat. In the small-scale region of the turbulence spectrum turbulent energy can be produced owing to hydrodynamic stability in the wind field, particularly in the planetary boundary layer of the atmosphere, and also owing to thermal instability of the atmosphere and disintegration of short gravity waves when they become unstable. These local influxes of turbulent energy can form the second maximum in the energy spectrum.

The turbulence scales at which the second maximum of turbulent energy occurs depend not only on the vertical profiles of wind velocity and air temperature, but also on the distance from the ground. As observations have shown, the second maximum in the energy spectrum is typical of the boundary and, in particular, the surface layer of the atmosphere. In addition, since the vertical scale of turbulent disturbances increases with altitude due to thermal instability of the atmosphere, there is a shift of the turbulent energy maximum into the region of lower frequencies with increase in altitude. On the other hand, since the thermal stability of air in the troposphere increases on the average with altitude, the role of the thermal factors responsible for the appearance of the second maximum decreases. In the middle and upper troposphere, apart from the zone of the tropopause, jet streams, and atmospheric fronts, there are no dynamic conditions for direct generation of turbulence in the microscale region.

Hence, we can postulate than in the free atmosphere there need not be a maximum of turbulent energy at microscales and the associated mesometeorological energy minimum in the energy spectrum of atmospheric turbulence.

4.2. Energy Spectra of Wind-Velocity Fluctuations in Troposphere and Stratosphere. Most of the published works contain the results of experimental investigations of turbulence in the boundary layer of the atmosphere, whereas there are much fewer data on turbulence in the free atmosphere. Moreover, these data relate only to particular regions of the spectrum of atmospheric turbulence. For instance, Wan Chang Chin investigated the temporal energy spectra of fluctuations of the horizontal wind-velocity component at levels of 700 to 50 mb for two points above the U.S.A. He used the results of rawinsonde observations carried out once a week over the course of a year. These investigations clearly revealed the synoptic energy maximum in the spectrum of the wind-velocity fluctuations.

The spectra of the fluctuations of the horizontal wind-velocity component obtained by Kao and Woods [176] for the jet-stream region provide a picture of the special features of the energy spectra only for mesoscale turbulence.

Pinus [88] investigated spectra from synoptic scales to scales of a few tens of meters. He used rawinsonde observations, and also aircraft measurements with the aid of a Doppler navigation system and a hot-wire anemometer. The aircraft measurements provided one-dimensional spatial spectra of fluctuations of the horizontal wind-velocity component, and the rawinsonde measurements provided time spectra. The "frozen" turbulence hypothesis was used in the comparison of these spectra.

The investigations of Pinus showed that in the troposphere there may be energy maxima in the microscale region. Usually, however, particularly in the upper troposphere and lower stratosphere, the microscale turbulence region of the spectrum shows a loss of turbulent energy due to work against negative buoyancy forces and, hence, the modulus of the slope of the spectral density curve on this section of turbulence scales exceeds $\frac{5}{3}$.

Some idea of the possible frequency of spectra with a second energy maximum is given by the data for the frequency of aircraft buffeting (see Chapter 9). The frequency of buffeting in the upper troposphere in middle latitudes of the USSR is about 12%, including approximately 6% moderate and severe buffeting, while in the stratosphere at altitudes of 10–20 km the frequency of buffeting (irrespective of intensity) is about 1%. We should note one very important circumstance, viz., that in the free atmosphere the energy in the region of the micrometeorological maximum is only a tenth of that in the region of the "synoptic" maximum [88]. This is confirmed by the energy spectrum for altitudes 3–20 km for the U.S.A. (Vinnichenko [223]). Yet, according to Van der Hoven's data, at an altitude of about 100 m (see Fig. 10.8), the energies at the micrometeorological and synoptic maxima are approximately equal.

5. Energy of Turbulent Motions of Different Scales in Relation to Altitude

The kinetic energy of turbulent motions depends on the altitude and time of the year. The question of the role of turbulent motions of different scales in this relation is important.

In 1966–1967 the Central Aerological Observatory conducted two series of experiments (in summer and winter) in the region of Kharkov. Each series lasted a month (July and January), during which radiosondes were launched at intervals of 2 h. In the same region and at the same time there were flights of an IL-18 airplane equipped with a Doppler navigation system, hot-wire anemometer, and other scientific apparatus. During the month there were 8–10 flights at altitudes of 0.5 to 10 km over horizontal areas 200–250 km in extent.

The data of the aircraft and radiosonde measurements were used to calculate the spectral density functions of the wind-velocity fluctuations. Individual spectra for altitudes of 1–3 km are given in [88], and for altitudes of 5–18 km in [22]. The spectra of wind-velocity fluctuations from radiosonde measurements and the averaged spectra from aircraft measurements revealed several features of the distribution of turbulent energy in relation to ω and altitude [92]. The energy in relation to wave numbers was estimated in [92] as $E_\omega = \omega F(\omega)/2$.

These investigations showed that the distribution of E_ω in the bottom kilometer layer of the atmosphere is distinguished by great inhomogeneity, especially in summer. At some scales there are increased values of kinetic energy of turbulent motions. At all scales the greatest values of E_ω on the whole are obtained at low altitudes; E_ω decreases with altitude and reaches a minimum at a certain level. The altitude of this level depends significantly on the disturbance scale. For disturbances with $L \approx 0.5$–2.0 km the level of the minimum values of E_ω in summer is located at an altitude of approximately 6.0–6.5 km. With increase in the disturbance scale this level drops, and for disturbance scales $L \approx 10^3$ km the level is located at an altitude of about 1.0–1.5 km. This behavior of the level of minimum values of E can be accounted for as follows: The greater the scale of the turbulent formation, the flatter it is and the more its effect decreases with increase in altitude.

Above the level of minimum E_ω the energy of turbulent motions increases with altitude (particularly sharply in the region of large-scale turbulence), reaching a maximum at altitudes of 8–10 km. This maximum of E_ω can easily be traced for turbulence scales $L \geqslant 50$ km. It is mainly these disturbance scales that determine the total energy of turbulent motions recorded by investigators under the level of the tropospheric wind-velocity maximum [89, 176, 186].

In winter the minimum energy level for turbulence of small spatial scales is located at an altitude of 2–3 km; under this level the energy decreases more sharply with altitude than in summer. In the region of large scales there is initially an increase in energy with altitude (the maximum of E_ω is situated at an altitude of about 1 km), and it then decreases to an altitude of approximately 3 km. The maximum of turbulent energy in the upper troposphere is more pronounced in winter than in summer.

Thus, in the troposphere there are two types of vertical profiles of E_ω. On the vertical profiles of E_ω for $\omega < 10^{-4}$ m^{-1} in summer and winter there are two maxima: The first is at altitudes of 0.5–2.0 km and the second is in the 8-

to 10-km layer. The first maximum is due to turbulent friction in the planetary boundary layer, and the second is due to baroclinicity of the atmosphere and the large vertical gradients of the mean wind velocity in the upper troposphere.

The vertical profiles of E_ω for $\omega > 10^{-4}$ m^{-1} are usually distinguished by the fact that the greatest values of E_ω are found at low altitudes, and the smallest values at altitudes of 3-6 km with a slight irregular increase with altitude in the upper troposphere.

In each specific case the vertical distribution of E_ω depends on the nature of the vertical profiles of air temperature and mean wind velocity. As an example, Fig. 10.9 shows the results of a flight experiment carried out on May 26, 1972 in the region of Alma-Ata. The vertical profiles of the temperature T and wind velocity u from the data of temperature and wind sounding above Alma-Ata are shown on the right of this figure, and the layers and zones in which aircraft buffeting was observed are shown on the left (the numerator gives the maximum positive load increment in g, and the denominator gives the maximum negative load increment).

A jet stream was observed in the region of the flight investigations. The axis of this stream, oriented toward the northeast, was situated at an altitude of 12 km above Alma-Ata. The wind velocity here was about 50 m/sec. To the northwest of Alma-Ata the wind velocity decreased sharply. At an altitude of 12 km

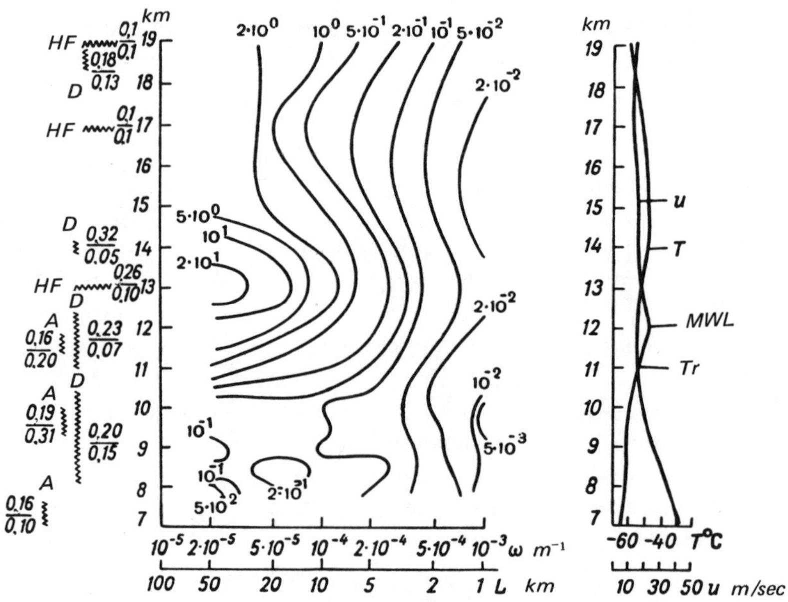

Figure 10.9. Distribution of E_ω (m^2/sec^2). Alma-Ata, May 26, 1972. HF–horizontal flight; A–ascent; D–descent; Tr–tropopause; MWL–maximum wind level.

the temperature difference above Alma-Ata and Balkhash was 3.5°C. Above Alma-Ata the temperature decreased with altitude in the 15–19 km layer, and above this layer there was strong temperature inversion. In this upper troposphere, where light winds were recorded, the value of E_ω was small; the small maximum ($E_\omega > 2 \cdot 10^{-1}$ m^2/sec^2) was associated with scales of 10–20 km. Above 10 km E_ω increased with altitude, and a maximum was observed at an altitude of about 13 km, where E_ω for $\omega = 10^{-4}$ m^{-1} was 2 m^2/sec^2, and for $\omega = 2 \cdot 10^{-5}$ m^{-1} it was $2.6 \cdot 10^1$ m^2/sec^2. The minimum value of E_ω was observed at an altitude of about 17 km and this was followed by an increase in E_ω with altitude, particularly for $\omega > 5 \cdot 10^{-5}$ m^{-1}.

6. Energy Balance in Clear-Air Turbulent Zones

As we have seen, the spatial and temporal changes in the structure of the air flow and the concomitant changes in atmospheric turbulence are of a fairly complex nature. The mechanism of formation of atmospheric turbulence can be most fully clarified by analysis of the energy-balance equation and its individual components (see Chapter 1). A knowledge of the numerical values of the components of the turbulent energy balance is of great scientific and practical value. Aircraft investigations at present can provide data on fluctuations of the temperature (T') and wind-velocity components (u', v', w'). The greatest difficulties from the experimental aspect are entailed in measurements of the air-density fluctuations and the correct determination of the components of the vertical gradient of mean wind velocity. The procedure in flight experiments in turbulent zones and possible methods of calculating the components of the vertical gradient of mean wind velocity are described in [162, 191].

Empirical data for u', v', T' and $d\bar{u}/dz$, $d\bar{v}/dz$ allow the calculation of the balance components

$$-\overline{u'w'}\frac{\partial \bar{u}}{\partial z}, \quad -\overline{v'w'}\frac{\partial \bar{v}}{\partial z}, \quad \frac{g}{T}\overline{w'T'}, \quad \frac{\partial}{\partial z}\overline{w'E},$$

where

$$E = \frac{1}{2}(\overline{u'^2} + \overline{v'^2} + \overline{w'^2}).$$

The rate of dissipation of turbulent energy (ϵ) is usually calculated from the spectral density of fluctuations of the wind-velocity components for the high-frequency part of the spectrum, where the "minus five-thirds" law is valid.

There are very few published experimental data that can be used to calculate the individual components of the turbulent energy balance in clear air. As an example we consider the results of calculations of the turbulent energy balance,

TABLE 10.6. Components of Clear-Air Turbulent Energy Balance

Altitude (km)	Type of airplane	$b^{(i)}$	E ($m^2 sec^2$)	$-\overline{u'w'}\frac{\partial \bar{u}}{\partial z}$	$-\overline{v'w'}\frac{\partial \bar{v}}{\partial z}$	$\frac{g}{T}\overline{w'T'}$	ε	Δ
				cm^3/sec^2				
9.45	T-33	$b^{(2)}$	0.78	27.7	31.6	−11.6	−45.5	−2.2
				100%		19.6%	76.8%	3.6%
19.00	U-2	$b^{(3)}$	10.30	28.5	478.0	−176.0	−367.0	+36.5
				100%		32.4%	67.6%	7.2%

given in [162, 191], for turbulent zones in which aircraft buffeting was observed.

Table 10.6 gives data on the turbulent energy balance for two turbulent zones: One of them was in the upper troposphere [191] and the other in the stratosphere [162].

The quantity Δ characterizes either the inaccuracy of determination of elements of the turbulent energy balance, or the horizontal inhomogeneity of the turbulent zone. It should also be noted that the fluctuations of u', v' and w' relate to spatial turbulence scales from a few tens of meters to 3–4 km.

Delay *et al.* [159] give data on the turbulent energy balance for five clear-air turbulent zones in which buffeting of a U-2 airplane occurred. They took into account the generation of turbulent energy due to horizontal wind-velocity gradients (Table 10.7). It follows from Table 10.7 that for four out of the five turbulent zones the value of this generation component did not exceed a few percent of the amount of turbulent energy generated by the vertical wind shears.

TABLE 10.7. Turbulent Energy Balance in Clear-Air Turbulent Zones

Altitude (km)	$b^{(i)}$	E (m^2/sec^2)	$-\overline{u'w'}\frac{\partial \bar{u}}{\partial z} - \overline{v'w'}\frac{\partial \bar{v}}{\partial z}$	$-\overline{u'^2}\frac{\partial \bar{u}}{\partial x} - \overline{u'v'}\frac{\partial \bar{v}}{\partial x}$	$\frac{g}{T}\overline{w'T'}$	ε	Δ
			cm^2/sec^3				
18.7	$b^{(2)}$	4.9	223	72	6	−246 81.7%	55 18%
18.7	$b^{(1-2)}$	3.2	148	−12	11	−158 107.5%	−11 7.5%
18.7	$b^{(2-3)}$	2.3	89	4	0	−85 90%	8 8.6%
18.6	$b^{(1-2)}$	4.2	151	12	−59 36.2%	−97 60.0%	7 4.5%
18.4	$b^{(1)}$	2.6	129	2	−72 55.0%	−67 51.0%	−8 6.1%

Table 10.7 also indicates that the remainder Δ was 4–18% of the total amount of turbulent energy generated.

An analysis of the spectra of fluctuations of the wind-velocity components showed that the spectral densities $S_u(\omega)$, $S_v(\omega)$, and $S_w(\omega)$ were approximately equal for $\omega \geqslant 10^{-3}$ m^{-1}. For smaller wave numbers $S_u(\omega)$ and $S_v(\omega)$ were much greater than $S_w(\omega)$. For the stratospheric turbulent layer (see Table 10.6) $\omega S_u(\omega)$ and $\omega S_v(\omega)$ increased, while $\omega S_w(\omega)$ decreased, with reduction of ω after $\omega \approx 10^{-3}$ m^{-1}. These features were reflected in the distribution of energy among the components of the wind-velocity fluctuations.

Table 10.8 gives data [191] on the distribution of turbulent energy among the wind vector components in the turbulent zone at an altitude of 9.45 km, where moderate buffeting of a T-33 occurred. This table also gives the turbulent energy components averaged for nine turbulent zones in the lower stratosphere, where buffeting of a U-2 occurred [180].

As can be seen, the energy of fluctuations of the longitudinal and transverse wind-velocity components at an altitude of 9.45 km, where the T-33 was subject to moderate buffeting, was almost three times greater than the energy of fluctuations of the vertical component. In the lower stratosphere the energy of fluctuations of the transverse wind-velocity component, particularly at mesoscales, was decisive (83.3%).

Investigations of other authors also show that in the stratosphere, for instance, the fluctuations of the transverse wind-velocity component are greater than the fluctuations of the vertical component and that with increase in scale the anisotropy of turbulence increases.

We can postulate that this suppression of fluctuations of the vertical wind-velocity component is due to the high thermal stability peculiar to the stratosphere.

If one assumes that the rate of turbulent energy dissipation becomes constant and the influx of turbulent energy ceases, the dissipation time $t_\epsilon = E/\epsilon$ can be calculated. Calculations show that the time for dissipation of clear-air turbulence is a few minutes. In particular, for the cases given in Table 10.6, t_ϵ = 172–280 sec. Since clear-air turbulent zones can sometimes exist for many hours (see Chapter 9), it follows that the loss of turbulent energy is made up

TABLE 10.8. Distribution of Turbulent Energy among Wind-Velocity Components

Altitude (km)	For range of frequencies ω (m^{-1})	E	$\overline{u'^2}$	$\overline{v'^2}$	$\overline{w'^2}$
		m^2/sec^2			
9.45	$10^{-2}-3\cdot10^{-3}$	0.780	0.667 42.7%	0.652 41.8%	0.241 15.5%
15.7–18.2	$10^{-3}-6\cdot10^{-5}$	99.05	27.3 14.3%	165.1 83.3%	5.7 2.4%

by the generation of energy by instability of atmospheric mesoscale processes. We can also fairly confidently conclude that when the generation of turbulent energy ceases, turbulent zones are completely dissipated in a short time (a matter of minutes).

We turn to a consideration of Table 10.6 and 10.7. These tables show that up to 20–30% of the turbulent energy generated by Reynolds stresses and vertical wind shears (for moderate and severe turbulence) is spent on overcoming the effect of buoyancy forces, and up to 70–80% is dissipated as heat. A high level of turbulent energy dissipation can greatly affect the nature of the mean temperature field in the free atmosphere and can affect not only local mesoscale processes, but also (from the set of turbulent zones) processes of planetary scale.

In fact, in a turbulent zone in which severe buffeting of a U-2 occurred and the rate of turbulent energy dissipation as heat was 367 cm^2/sec^3, the heating of the air could reach approximately 0.13°C/h, i.e., values quite comparable with heating due to radiative heat transfer [113]. As was shown above, the spectral density $S(\omega)$ for $\omega = 2 \cdot 10^{-3}$ m^{-1} usually relates to the inertial subrange of the turbulence spectrum. Hence, data on the statistical distribution of $S(\omega)$ for this wave number can be used to calculate the value and frequency of ϵ, and also the rate of air heating $\Delta T/\Delta t$ in the troposphere and lower stratosphere in clear air.

An idea of this is given by Table 10.9, which gives the values of $F(\omega) = 2S(\omega)$ and ϵ corresponding to the values found for $\omega = 2 \times 10^{-3}$ m^{-1}. The main frequency maximum (36.6%) occurs at $F(\omega) = 10\text{–}50$ m^3/sec^2, $\epsilon = 0.96\text{–}11.0$ cm^2/sec^3, and $\Delta T/\Delta t = 3.5 \cdot 10^{-4}$ to $4 \cdot 10^{-3}$°C/h. In approximately 3% of the cases $F(\omega) = 5 \cdot 10^2$ to $2 \cdot 10^3$ m^3/sec^2, $\epsilon = 340\text{–}2700$ cm^2/sec^3, and $\Delta T/\Delta t = 0.12\text{–}0.97$°C/h.

Data on the dissipation of turbulent energy in the layer 7.6–12.2 km above

TABLE 10.9. Frequency $F(\omega)$ for $\omega = 2 \cdot 10^{-3}$ m^{-1}, ϵ, and $\Delta T/\Delta t$ in Clear-Air Turbulent Zone (USSR, Temperate Latitudes)

$F(\omega)$	$10^0 - 5 \cdot 10^0$	$5 \cdot 10^0 - 10^1$	$10^1 - 5 \cdot 10^1$	$5 \cdot 10^1 - 10^2$
ε cm^2/sec^3	$3 \cdot 10^{-2} - 3.4 \cdot 10^{-1}$	$3.4 \cdot 10^{-1} - 9.6 \cdot 10^{-1}$	$9.6 \cdot 10^{-1} - 1.1 \cdot 10^1$	$1.1 \cdot 10^1 - 3 \cdot 10^1$
$\frac{\Delta T}{\Delta t}$ °C/h	$1.1 \cdot 10^{-5} - 1.2 \cdot 10^{-4}$	$1.2 \cdot 10^{-4} - 3.5 \cdot 10^{-4}$	$3.5 \cdot 10^{-4} - 4 \cdot 10^{-3}$	$4 \cdot 10^{-3} - 1.1 \cdot 10^{-2}$
P%	11.0	10.4	36.6	18.5
ΣP%	11.0	21.4	58.0	76.5
$F(\omega)$	$10^2 - 5 \cdot 10^2$	$5 \cdot 10^2 - 10^3$	$10^3 - 2 \cdot 10^3$	
ε cm^2/sec^3	$3 \cdot 10^1 - 3.4 \cdot 10^2$	$3.4 \cdot 10^2 - 9.6 \cdot 10^2$	$9.6 \cdot 10^2 - 2.7 \cdot 10^3$	
$\frac{\Delta T}{\Delta t}$ °C/h	$1.1 \cdot 10^{-2} - 1.2 \cdot 10^{-1}$	$1.2 \cdot 10^{-1} - 3.5 \cdot 10^{-1}$	$3.5 \cdot 10^{-1} - 9.7 \cdot 10^{-1}$	
P%	20.5	2.5	0.5	
ΣP%	97.0	99.5	100.0	

TABLE 10.10. Dissipation of Turbulent Energy in Upper Troposphere

H km	7.6–9.1	9.1–10.4	10.4–12.2	7.6–12.2
$\bar{\varepsilon}$ W/m²	0.59	0.36	0.37	1.32

the U.S.A. are given in [217]. Assuming the following mean values $\bar{\epsilon}$ for assessment of the intensity of aircraft buffeting:

$b^{(i)}$	$b^{(0)}$	$b^{(1)}$	$b^{(2)}$	$b^{(3)}$
$\bar{\epsilon}$, cm²/sec³	1.5	30	85	675

Trout and Panofsky [217] used information about the frequency of aircraft buffeting at different altitudes above the U.S.A., published in [152], and calculated the integral dissipation of turbulent energy for the upper troposphere. The results of the calculations of $\bar{\epsilon}$ are given in Table 10.10.

The dissipation in the 7.6-12.2 km layer is a little greater than 1 W/m², i.e., it is a quantity of almost the same order as dissipation in the boundary layer [186].

The solution of the question of the role of clear-air turbulence in large-scale atmospheric processes requires further experimental investigations that will provide data for calculation of all the components of the turbulent energy balance. It is necessary, in addition, to know the approximate volume of the atmosphere on the planetary scale occupied at the same time by clear-air turbulence.

References

1. V. N. Arkhangel'skii and N. Z. Pinus, Basic principles in the design of a model of atmospheric turbulence, *Tr. TsAGI*, No. 1342, pp. 3–9 (1971).
2. V. N. Arkhangel'skii, A. V. Bashinskii, and A. M. Ershov, Investigation of effective velocities of vertical gusts in routine flights, *Tr. IsAGI*, No. 1342, pp. 10–21 (1971).
3. V. N. Barakhtin, Some features of aircraft-buffeting zones at altitudes of 8–10 km, *Meteorol. Gidrol.*, No. 5, pp. 3–8 (1963).
4. V. N. Barakhtin, Aerosynoptic conditions for aircraft buffeting in upper troposphere and lower stratosphere over the Asiatic USSR, *Tr. TsIP*, No. 157, pp. 17–23 (1966).
5. V. P. Belyaev, Airborne meter for average values and fluctuations of air temperature, *Tr. TsAO*, No. 84, pp. 19–27 (1969).
6. V. P. Belyaev, V. I. Silaeva, and S. M. Shmeter, Microstructure of wind field involving convection, *Tr. TsAO*, No. 97, pp. 39–52 (1970).
7. A. M. Borovikov and I. P. Mazin, Orientation of cloud bands, *Tr. TsAO*, No. 79, pp. 90–95 (1967).
8. G. S. Buldovskii, Aircraft buffeting in stratosphere in relation to horizontal distribution of temperature and wind, *Tr. Gidromettsentra SSSR*, No. 70, pp. 93–103 (1970).
9. G. S. Buldovskii, Relation between turbulence in stratosphere causing aircraft buffeting and vertical distribution of meteorological parameters, *Tr. Gidromettsentra SSSR*, No. 70, pp. 79–93 (1970).
10. G. K. Batchelor, *The Theory of Homogeneous Turbulence*, Cambridge University Press, London (1959).
11. A. A. Vasil'ev, Wind distribution over Crimean mountains and features of helicopter buffeting in various synoptic situations, *Tr. GGO*, No. 171, pp. 51–62 (1965).
12. A. A. Vasil'ev, Analysis of clear-air turbulence from radiosondo data, *Tr. Gidromettsentra SSSR*, No. 35, pp. 13–28 (1968).
13. A. A. Vasil'ev, Assessment of effectiveness of synoptic criteria of existence of clear-air turbulence, *Tr. Gidromettsentra SSSR*, No. 35, pp. 29–35 (1968).
14. A. A. Vasil'ev *et al.*, "Clear-air turbulence," *Tekh. Zap. VMO*, No. 227, pp. 256–268 (1969).
15. I. V. Vasil'chenko and A. A. Ledokhovich, Some results of aircraft sounding in the Golodnaya Steppe, *Tr. GGO*, No. 135, pp. 55–59 (1962).
16. N. F. Vel'tishchev, Interpretation of mesostructure of cloud field, *Tr. MMTs*, No. 8, pp. 45–54 (1965).
17. E. K. Verle and V. P. Zhabunina, Some data on atmospheric turbulence in upper troposphere and lower stratosphere, *Tr. DVNIGMI*, No. 15, pp. 83–112 (1963).
18. N. K. Vinnichenko, Experience in using a hot-wire anemometer on an airplane, *Tr. TsAO*, No. 54, pp. 85–99 (1964).

19. N. K. Vinnichenko, Clear-air turbulence at altitudes of 6–12 km, *Izv. Akad. Nauk SSSR, Fiz. Atmos. Okeana*, 2(11):1135–1141 (1968).
20. N. K. Vinnichenko and G. N. Shur, Magnetic recording system for slowly varying meteorological parameters, *Tr. TsAO*, No. 42, pp. 147–154 (1962).
21. N. K. Vinnichenko and G. N. Shur, An on-board multichannel code converter, *Tr. TsAO*, No. 63, pp. 96–104 (1965).
22. N. K. Vinnichenko, N. Z. Pinus, and G. N. Shur, Some results of experimental investigations of turbulence in the troposphere, in: *International Colloquium on the Fine-Scale Structure of the Atmosphere and Its Relation to Radio-Wave Propagation*, Nauka, Moscow (1965), pp. 65–75.
23. N. K. Vinnichenko, N. Z. Pinus, and G. N. Shur, Investigations of clear-air turbulence in the stratosphere, *Tr. TsAO*, No. 100, pp. 86–98 (1970).
24. F. Ya. Voit, E. E. Kornienko, and S. B. Khusid, Statistical characteristics of structure of vertical motions in cumulus clouds, *Izv. Akad. Nauk SSSR, Fiz. Atmos. Okeana*, 7(11):1206–1208 (1971).
25. V. I. Vorob'ev, *Jet Streams in Temperate and High Latitudes* [in Russian], Gidrometeoizdat, Leningrad (1960).
26. E. F. Vorob'eva and B. N. Trubnikov, Cellular convection associated with anisotropic turbulence and presence of a relative eddy field, *Tr. TsAO*, No. 75, pp. 39–50 (1967).
27. P. A. Vorontsov, *Turbulence and Vertical Streams in the Boundary Layer of the Atmosphere* [in Russian], Gidrometeoizdat, Leningrad (1966).
28. P. A. Vorontsov and A. A. Vasil'ev, Questions of meteorological support for flights on helicopter routes in mountain regions, *Tr. GGO*, No. 171, pp. 122–129 (1965).
29. N. I. Vul'fson, *Investigation of Convective Motions in the Free Atmosphere* [in Russian], Izd. AN SSSR, Moscow (1961).
30. L. S. Gandin *et al.*, *Fundamentals of Dynamic Meteorology* [in Russian], Gidrometeoizdat, Leningrad (1955).
31. N. F. Helmholtz [Gel'mgol'ts], Atmospheric turbulence and calmness of flights in region of tropopause and jet streams, *Tr. KazNIGMI*, No. 19, pp. 3–30 (1963).
32. M. A. German, Turbulent exchange in clouds, *Metéorol. Gidrol.*, No. 10, pp. 15–22 (1963).
33. M. A. German, Some results of an experimental investigation of the structural and energy characteristics of turbulence in clouds, *Tr. GGO*, No. 154, pp. 46–57 (1964).
34. F. A. Gisina, Effect of gradients of mean velocity and temperature on spectral characteristics of turbulence, *Izv. Akad. Nauk SSSR, Fiz. Atmos. Okeana*, 2(8):804–814 (1966).
35. A. G. Gorelik and A. A. Chernikov, Some results of a radar investigation of the structure of the wind field at altitudes of 50–700 m, *Tr. TsAO*, No. 57, pp. 3–19 (1964).
36. A. G. Gorelik, V. V. Kostarev, and A. A. Chernikov, Coordinate Doppler method of wind observation and some results of investigation of wind-field inhomogeneities in the atmosphere, *Meteorol. Gidrol.*, No. 10, pp. 12–20 (1965).
37. N. I. Davydov, E. G. Lomonosov, and I. A. Cherenkova, Synoptic-statistical method of diagnosis and prognosis of clear-air turbulence, *Meteorol. Gidrol.*, No. 12, pp. 45–54 (1968).
38. G. M. Jenkins and D. G. Watts, *Spectral Analysis and Its Applications*, Holden-Day, San Francisco (1968).
39. V. A. Georgii [Dzhordzhio] and N. A. Petrenko, Jet streams in the atmosphere, in: *Manual of Short-Term Weather Forecasting* [in Russian], Part 2, Gidrometeoizdat, Leningrad (1965), pp. 252–309.
40. Yu. P. Dobrolenskii, *Flight Dynamics in a Calm Atmosphere* [in Russian], Mashinostroenie, Moscow (1969).

41. A. S. Dubov, Determination of vertical wind gusts from oscillations of airplane in flight with allowance for pilot action, *Tr. GGO*, No. 121, pp. 109–124 (1961).
42. A. S. Dubov and M. A. German, Spectral density of vertical wind gusts in clouds, *Izv. Akad. Nauk SSSR, Fiz. Atmos. Okeana*, 1(7):670–676 (1965).
43. G. K. Dudko and G. B. Reznikov, *Doppler Measuring Equipment for Aircraft Speed and Drift Angle* [in Russian], Sovetskoe Radio, Moscow (1964).
44. A. S. Zaitsev, Fluctuations of wind direction in boundary layer of atmosphere, *Tr. GGO*, No. 254, pp. 70–76 (1971).
45. S. S. Zilitinkevich, D. L. Laikhtman, and A. S. Monin, Dynamics of boundary layer of atmosphere, *Izv. Akad. Nauk SSSR, Fiz. Atmos. Okeana*, 3(1):297–333 (1967).
46. S. L. Zubkovskii, Frequency spectra of fluctuations of horizontal wind-velocity component in the ground layer of air, *Izv. Akad. Nauk SSSR, Ser. Geofiz.*, No. 10, pp. 1425–1433 (1962).
47. S. L. Zubkovskii, Experimental investigations of spectra of fluctuations of vertical wind-velocity component in the free atmosphere, *Izv. Akad. Nauk SSSR, Ser. Geofiz.*, No. 8, pp. 1285–1295.(1963).
48. V. N. Ivanov, Turbulent energy and its dissipation, *Izv. Akad. Nauk SSSR, Ser. Geofiz.*, No. 9, pp. 1405–1413 (1964).
49. V. N. Ivanov and A. E. Ordanovich, *Structure of Boundary Layer of Atmosphere with Unstable Stratification* [in Russian], Part 3, Izd. Otd. Nauchno-tekhn. Informatsii Gidromettsentra SSSR, Obninsk (1969).
50. A. I. Kartsivadze, G. M. Makharadze, and A. A. Ordzhonikidze, Experimental investigation of velocities of vertical motions in convective clouds, *Tr. Inst. Geofiz. Akad. Nauk GruzSSR*, 28:196–309 (1972).
51. I. F. Kvartskheliya, Regular features of relative position of levels of tropopause and maximum wind velocity, *Collection of Works of Tbilisi Hydrometeorological Observatory* [in Russian] (1969), pp. 125–132.
52. I. A. Kibel', Some kinds of wave motions in the free atmosphere, *Tr. MMTs*, No. 6, pp. 3–7 (1965).
53. Z. N. Kogan, Instability of internal waves in stratified jet streams, *Meteorol. Gidrol.*, No. 5, pp. 12–21 (1972).
54. Z. N. Kogan and N. P. Shakina, Instability of disturbances of finite amplitude in the jet stream, *Tr. TsAO*, No. 112, pp. 49–59 (1973).
55. V. N. Kozlov, N. Z. Pinus, and L. V. Shcherbakova, Some statistical characteristics of wind-velocity fluctuations in the troposphere, *Tr. TsAO*, No. 63, pp. 37–46 (1965).
56. V. K. Kolesnikova and A. S. Monin, Spectra of oscillations of meteorological fields, *Izv. Akad. Nauk SSSR, Fiz. Atmos. Okeana*, 1(7):653–670 (1965).
57. K. Ya. Kondrat'ev, E. P. Borisenkov, and A. A. Morozkin, *Practical Utilization of Data of Meteorological Satellites* [in Russian], Gidrometeoizdat, Leningrad (1966).
58. B. M. Koprov, Spectra of turbulent fluctuations of vertical wind-velocity component in boundary layer with developed convection, *Izv. Akad. Nauk SSSR, Fiz. Atmos. Okeana*, 1(11):1151–1160 (1965).
59. B. M. Koprov and L. R. Tsvang, Direct measurements of turbulent heat flux on board an airplane, *Izv. Akad. Nauk SSSR, Fiz. Atmos. Okeana*, 1(6):643–648 (1965).
60. D. L. Laikhtman and É. K. Byutner, Main criteria of the intensity of turbulence in a mountain region, *Tr. GGO*, No. 171, pp. 32–37 (1965).
61. D. L. Laikhtman and Yu. Zh. Al'ter-Zalik, Use of aerological data for determination of aircraft buffeting in the free atmosphere, *Izv. Akad. Nauk SSSR, Fiz. Atmos. Okeana*, 2(5):534–536 (1966).
62. J. L. Lumley and H. A. Panofsky, *The Structure of Atmospheric Turbulence*, Interscience, New York (1964).

63. L. V. Landau and E. M. Lifshits, *Mechanics of Continuous Media* [in Russian], Gostekhizdat, Moscow (1953).
64. V. D. Litvinova, Diurnal course of turbulence coefficient above level terrain, *Tr. TsAO*, No. 78, pp. 57–62 (1967).
65. E. G. Lomonosov and Yu. N. Lunin, Determination of clear-air turbulent zones by computer, *Meteorol. Gidrol.*, No. 4, pp. 41–47 (1969).
66. V. F. Makarov and G. N. Shur, Some features of turbulence structure in mountain regions, *Tr. TsAO*, No. 112, pp. 91–97 (1973).
67. L. T. Matveev, *Fundamentals of General Meteorolgy. Physics of the Atmosphere* [in Russian], Gidrometeoizdat, Leningrad (1965).
68. Yu. V. Mel'nichuk, Measurement of turbulence in precipitation by means of a Doppler radar station, *Izv. Akad. Nauk SSSR, Fiz. Atmos. Okeana*, 2(7):695–704 (1966).
69. Yu. V. Mel'nichuk and A. A. Chernikov, Radar measurements of divergence in vertical flows in the atmosphere, *Tr. TsAO*, No. 110, pp. 36–46 (1973).
70. Yu. V. Mel'nichuk and A. A. Chernikov, Measurements of wind and turbulence from radar echoes from clear air, *Tr. TsAO*, No. 110, pp. 22–35 (1973).
71. Yu. V. Mel'nichuk and A. A. Chernikov, An operative method of detecting turbulence in clouds and precipitation, *Tr. TsAO*, No. 110, pp. 3–11 (1973).
72. V. E. Minervin, Turbulence in lower layer of atmosphere and in low-level clouds, *Tr. TsAO*, No. 71, pp. 76–92 (1966).
73. A. S. Monin, Spectrum of turbulence in a thermally inhomogeneous atmosphere, *Izv. Akad. Nauk SSSR, Ser. Geofiz.*, No. 3, pp. 397–407 (1962).
74. A. S. Monin, Effect of thermal stratification of medium on turbulence, in: *International Colloquium on the Fine-Scale Structure of the Atmosphere and Its Relation to Radio-Wave Propagation*, Nauka, Moscow (1965), pp. 113–127.
75. A. S. Monin and A. M. Yaglom, *Statistical Hydromechanics* [in Russian], Part 1, Nauka, Moscow (1965); Part 2, Nauka, Moscow (1967).
76. Sh. A. Musaelyan, *Obstacle Waves in the Atmosphere* [in Russian], Gidrometeoizdat, Leningrad (1962).
77. R. V. Ozmidov, Some data on large-scale characteristics of field of horizontal velocity components in ocean, *Izv. Akad. Nauk SSSR, Ser. Geofiz.*, No. 11, pp. 1708–1719 (1964).
78. S. Panchev, Statistical structure of wind-velocity field in the free atmosphere, *Meteorol. Gidrol.*, No. 7, pp. 25–28 (1963).
79. L. A. Pakhomov, Aircraft equipment for wind-vector measurements, *Tr. TsAO*, No. 41, pp. 91–105 (1962).
80. V. A. Patsaeva, Investigation of orographic disturbances of atmosphere by means of balanced sounding balloons, *Tr. TsAO*, No. 59, pp. 84–92 (1964).
81. E. M. Pekelis, Numerical methods of solving the nonlinear problem of a flow of air over ground irregularities, *Tr. MMTs*, No. 6, pp. 57–64 (1965).
82. E. M. Pekelis, Flow of air over an isolated obstacle (nonlinear problem), *Meteorol. Gidrol.*, No. 10, pp. 24–30 (1966).
83. N. Z. Pinus, Structure of wind-velocity field in upper atmosphere and lower stratosphere, *Meteorol. Gidrol.*, No. 4, pp. 7–13 (1962).
84. N. Z. Pinus, Statistical characteristics of horizontal wind-velocity component at altitudes of 6–12 km, *Izv. Akad. Nauk SSSR, Ser. Geofiz.*, No. 1, pp. 177–188 (1963).
85. N. Z. Pinus, Vertical motions in thunder clouds, *Dokl. Akad. Nauk SSSR*, 150(4): 788–790 (1963).
86. N. Z. Pinus, Some results of investigations of meso- and microstructure of wind field at altitudes of 6–12 km, *Tr. TsAO*, No. 53, pp. 4–21 (1964).
87. N. Z. Pinus, Some features of development of turbulence above level terrain, *Izv. Akad. Nauk SSSR, Fiz. Atmos. Okeana*, 1(3):266–274 (1965).

88. N. Z. Pinus, Energy spectra of wind-velocity fluctuations in the free atmosphere, *Meteorol. Gidrol.*, No. 4, pp. 3–11 (1966).
89. N. Z. Pinus, Energy of macroturbulent motions in the atmosphere, *Izv. Akad. Nauk SSSR, Fiz. Atmos. Okeana*, 4(8):803–811 (1968).
90. N. Z. Pinus, Physical model of clear-air turbulence, *Meteorol. Gidrol.*, No. 6, pp. 56–66 (1971).
91. N. Z. Pinus, Physical model of clear-air turbulence, in: *Proceedings of the All-Union Conference on Questions of Meteorological Support for Supersonic Aviation* [in Russian], Izd. LGMI (1971), pp. 131–140.
92. N. Z. Pinus, Energetics of turbulent motions in the troposphere, *Izv. Akad. Nauk SSSR, Fiz. Atmos. Okeana*, 8(8):810–817 (1972).
93. N. Z. Pinus, Statistical characteristics of spectra of wind-velocity fluctuations in upper troposphere and lower stratosphere, *Meteorol. Gidrol.*, No. 1, pp. 82–84 (1974).
94. N. Z. Pinus and S. M. Shmeter, *Atmospheric Turbulence Affecting Flight of Aircraft* [in Russian], Gidrometeoizdat, Moscow (1962).
95. N. Z. Pinus *et al.*, Basic principles of automatic processing of aircraft meteorological information, *Meteorol. Gidrol.*, No. 9, pp. 3–9 (1964).
96. N. Z. Pinus and L. V. Shcherbakova, Structure of wind field in a thermally stratified atmosphere, *Izv. Akad. Nauk SSSR, Fiz. Atmos. Okeana*, 2(11):1126–1134 (1966).
97. N. Z. Pinus and E. F. Shumanskaya, Statistical properties of thermal stratification of troposphere and lower stratosphere, *Meteorol. Gidrol.*, No. 5, pp. 3–13 (1970).
98. C. H. B. Priestly, *Turbulent Transfer in the Lower Atmosphere*, University of Chicago Press, Chicago (1959).
99. I. G. Pchelko, *Aerosynoptic Conditions of Aircraft Buffeting in the Upper Troposphere and Lower Stratosphere* [in Russian], Gidrometeoizdat, Leningrad (1962).
100. I. G. Pchelko, Clear-air turbulence, *Meteorol. Gidrol.*, No. 12, pp. 10–18 (1966).
101. A. A. Reshchikova, Clear-air turbulence in jet streams, *Tr. TsAO*, No. 54, pp. 53–65 (1964).
102. A. A. Reshchikova, Spatial distribution of orographic turbulent zones on Alder–Novorossiisk section during the bora, *Tr. TsAO*, No. 3, pp. 56–70 (1965).
103. S. V. Solonin and O. G. Bogatkin, Atmospheric turbulence causing buffeting of supersonic aircraft in stratosphere and problems of predicting it, *Tr. LGMI*, No. 34, pp. 71–83 (1968).
104. S. V. Solonin, O. G. Bogatkin, and L. I. Tolstobrova, Aerosynoptic conditions and prediction of buffeting of supersonic aircraft in stratosphere, in: *Proceedings of the All-Union Conference on Questions of Meteorological Support for Supersonic Aviation* [in Russian], Izd. LGMI (1971), pp. 150–159.
105. G. K. Sulakvelidze, *Rain and Hail Showers* [in Russian], Gidrometeoizdat, Leningrad (1967).
106. G. K. Sulakvelidze, A. Sh. Bibilashvili, and V. F. Lapcheva, *Formation of Precipitation and Effect of Hail Processes* [in Russian], Gidrometeoizdat, Leningrad (1965).
107. V. I. Tatarskii, *Wave Propagation in a Turbulent Atmosphere* [in Russian], Nauka, Moscow (1967).
108. B. N. Trubnikov, Investigations of air flows over mountain regions with due regard to thermal inhomogeneity of underlying surface, *Izv. Akad. Nauk SSSR, Ser. Geofiz.*, No. 2, pp. 150–160 (1964).
109. B. N. Trubnikov, Transport of mesoscale tropospheric disturbances in upper atmosphere, *Tr. TsAO*, No. 69, pp. 36–43 (1966).
110. B. N. Trubnikov, Some questions of the theory of free (cellular) and forced convection, *Tr. TsAO*, No. 75, pp. 3–39 (1967).
111. R. A. Fisher, *Statistical Methods for Research Workers*, Oliver and Boyd, Edinburgh (1954).

112. J. O. Hinze, *Turbulence*, McGraw-Hill, New York (1959).
113. A. Kh. Khrgian, *Physics of the Atmosphere* [in Russian], Gidrometeoizdat, Leningrad (1969).
114. S. B. Khusid, Results of experimental investigations of characteristics of vertical gusts in cumulus clouds, *Tr. UkrNIGMI*, No. 86, pp. 121–126 (1970).
115. Yu. V. Chernov, Investigations of ascending flows by means of gliders, *Tr. TsAO*, No. 63, pp. 70–76 (1965).
116. V. I. Chernysh, An aircraft digital memory for meteorological information, *Tr. TsAO*, No. 63, pp. 85–96 (1965).
117. N. P. Shakina, Internal waves and onset of turbulence in the free atmosphere, *Tr. TsAO*, No. 112, pp. 28–48 (1973).
118. N. S. Shishkin, *Clouds, Precipitation, and Thunderstorm Electricity* [in Russian], Gostekhizdat, Moscow (1964).
119. S. M. Shmeter, Features of the wind field in the vicinity of cumulonimbus clouds, *Meteorol. Gidrol.*, No. 11, pp. 20–27 (1962).
120. S. M. Shmeter, Stages of development of cumulonimbus clouds and features of the distribution of meteorological parameters in their zone, *Tr. TsAO*, No. 53, pp. 54–79 (1964).
121. S. M. Shmeter, Interaction of cumulonimbus clouds and the wind field, *Izv. Akad. Nauk SSSR, Fiz. Atmos. Okeana*, 2(10):1026–1032 (1966).
122. S. M. Shmeter, Structure of fields of meteorological elements in the region of cumulonimbus clouds, *Tr. TsAO*, No. 8, p. 120 (1969).
123. S. M. Shmeter, Spectral structure of turbulence in the region of thick convective clouds, in: *Turbulent Flows* [in Russian], Nauka, Moscow (1971), pp. 223–228.
124. S. M. Shmeter, Structure of thermal turbulence causing aircraft buffeting, in: *Proceedings of the All-Union Conference on Questions of Meteorological Support for Supersonic Aviation* [in Russian], Izd. LGMI (1971), pp. 168–177.
125. S. M. Shmeter, *Physics of Convective Clouds* [in Russian], Gidrometeoizdat, Leningrad (1972).
126. S. M. Shmeter and V. I. Silaeva, Vertical flows within cumulonimbus clouds, *Meteorol. Gidrol.*, No. 10, pp. 7–13 (1966).
127. V. A. Shnaidman, Vertical distribution of characteristics of wind fluctuating velocity in the upper troposphere, *Izv. Akad. Nauk SSSR, Fiz. Atmos. Okeana*, 5(4):200–212 (1969).
128. G. N. Shur, Experimental investigations of energy spectrum of atmospheric turbulence, *Tr. TsAO*, No. 43, pp. 79–91 (1962).
129. G. N. Shur, Universal self-balancing bridge for an airplane resistance thermometer, *Tr. TsAO*, No. 63, pp. 104–108 (1965).
130. G. N. Shur, Turbulence spectrum of a stable stratified atmosphere, *Tr. TsAO*, No. 78, pp. 100–108 (1967).
131. G. N. Shur, Energy transfer across turbulence spectrum in the free atmosphere, in: *Turbulent Flows* [in Russian], Nauka, Moscow (1970), pp. 228–233.
132. G. N. Shur, Aircraft investigations of clear-air turbulence in the troposphere and stratosphere, in: *Proceedings of the All-Union Conferences on Questions of Meteorological Support for Supersonic Aviation* [in Russian], Izd. LGMI (1971), pp. 141–149.
133. M. I. Yudin, Questions of turbulence theory and wind structure in application to the problem of aircraft oscillations, *Tr. NIU GUGMS, Ser. 1*, No. 35 (1946).
134. L. Yu. Yarovskaya, Author's Abstract of Dissertation: Atmospheric turbulence in the stratosphere and questions of diagnosis and prediction of STS buffeting, LGMI (1974).
135. R. W. Astheimer, The remote detection of clear-air turbulence, *Clear Air Turbulence and Its Detection*, New York (1968), pp. 436–448.

136. D. Atlas *et al.*, The birth of CAT and microscale turbulence, *J. Atmos. Sci.*, 27(6):903–913 (1970).
137. A. Auer and J. Marwitz, Estimates of air moisture flux into hailstorms on the High Plains, *J. Appl. Meteorol.*, 7(2):196–198 (1968).
138. A. H. Auer and W. Sand, Updraft measurements beneath the base of cumulus and cumulonimbus clouds, *J. Appl. Meteorol.*, 5(4):461–466 (1966).
139. R. B. Blackman and J. M. Tukey, *The Measurement of Power Spectra*, Dover, New York (1958).
140. R. Bolgiano, Structure of turbulence in stratified media, *J. Geophys. Res.*, 67(8):3015–3023 (1962).
141. R. J. Boucher and K. R. Glover, Radar, aircraft, and meteorological observations on shear-induced turbulence, in: *Proceedings of the International Conference on Atmospheric Turbulence*, May 18–21, London (1971), pp. 655–657.
142. A. M. Bradbury, Glider flight in the lower stratosphere above cumulonimbus clouds, *Meteorol. Mag.*, 102(1209):110–120 (1973).
143. J. Briggs and W. T. Roach, Aircraft observations near jet-streams, *Q. J. Roy. Meteorol. Soc.*, 89(5):225–247 (1963).
144. P. Brochet and H. Gland, *Contribution à l'étude statistique de la turbulence en clair air, zone Europe-Méditerranée*, Ministère des Travaux publics et des Transports (1968).
145. K. G. Brundidge and L. J. Clodman, Model cross-sections across the jet-streams, *J. Appl. Meteorol.*, 3:303–317 (1962).
146. A. J. Brunstein, Lessons to be learned from accidents attributed to turbulence, in: *Proceedings of the International Conference on Atmospheric Turbulence*, May 18–21, London (1971), p. 12.
147. A. Burns, *Power Spectra of Low Level Atmospheric Turbulence Measured from an Aircraft*, Ministry of Aviation, C.P. 733 (1964).
148. H. R. Byers and R. R. Braham, *Thunderstorm*, Washington (1949).
149. W. T. Cochran *et al.*, What is the Fast Fourier Transform?, *IEEE Trans.*, AU-15(2):45–56 (1967).
150. R. T. Collis, Lidar detection of CAT, *Astronaut. Aeronaut.*, 2(12):52–54 (1964).
151. D. Colson, Analysis of clear-air turbulence for March 1962, *Mon. Weather Rev.*, 91(2):73–82 (1963).
152. D. Colson, Clear-air turbulence and upper level meteorological patterns, *CAT and Its Detection*, Plenum Press, New York (1969), pp. 337–360.
153. D. Colson and H. A. Panofsky, An index of clear-air turbulence, *Q. J. Meteorol. Soc.*, 91:390 (1965).
154. C. D. Cone, Thermal soaring of birds, *Am. Scientist*, No. 50, pp. 180–209 (1962).
155. J. W. Cooley *et al.*, Historical notes on the Fast Fourier Transform, *IEEE Trans.*, AU-15(2):76–79 (1967).
156. T. H. Cooke, A smoke-trail technique for measuring wind (Plate III), *Q. J. Roy. Meteorol. Soc.*, 88(375):83–88 (1962).
157. W. Crooks *et al.*, Project HICAT. High altitude clear-air turbulence measurements and meteorological correlations, Tech. Rep. AEFDL-TR-68-127 (1968).
158. G. S. Daniels, J. R. Scoggins, and O. E. Smith, Terrestrial environment (climatic) criteria for use in space vehicle development, 1966 revision, NASA TN 53328 (1966).
159. R. D. Delay and J. A. Dutton, An analysis of conditions associated with an occurrence of stratospheric CAT, *J. Atmos. Sci.*, 28:1272–1279 (1971).
160. H. Dietz, Turbulenz im volkenfreien Raum in der Nähe von Gewittern, *Berl. Wetterkarte*, No. 8, p. 5 (1968).
161. K. Doi, Formation and movement of precipitation clouds, *Geophys. Mag.*, 31(2):313–330 (1962).

161a. C. Dousset, R. Joatton, and P. Stuckelberger, Vol Stratosphérique et metéorologie. Premières données du Concorde, *L'Aéronaut. l'Astronaut.*, No. 38, pp. 81–87 (1972).

162. J. A. Dutton, An energy budget for a layer of stratospheric CAT, *Radio Sci.*, 4(12): 1137–1142 (1969).

163. L. J. Ahernberger and R. J. Wilson, Subjective evaluation of the XB-70 airplane response to turbulence and controlled inputs, in: *Proceedings of the International Conference on Atmospheric Turbulence*, May 18–21, London (1971), p. 18.

164. R. M. Endlich, The detailed structure of the atmosphere in regions of clear-air turbulence, Final Report, Contr. CWB-10324, Stanford Res. Inst., Cal. (1963).

165. R. M. Endlich and C. S. McLean, Jet-stream structure over the central United States determined from aircraft observations, *J. Appl. Metereorol.*, 4(1):83–90 (1965).

166. F. Fiedler and H. A. Panofsky, Atmospheric scales and spectral gaps, *Bull. Am. Meteorol. Soc.*, 51(12):1114–1120 (1970).

167. Final report of analysis in the ICAO short-term high level turbulence report, Aviation Weather Service Japan Meteorol. Agency, Nov. 1967.

168. H. P. Foltz, Prediction of clear-air turbulence, Dept. Atm. Sci., Colorado State University, Fort Collins (1967), p. 158.

169. T. D. Foster, Onset of manifest convection in a layer of fluid with a time-dependent surface temperature, *Phys. Fluids*, 12(12): 2482–2487 (1969).

170. J. N Garrison, An assessment of atmospheric turbulence data for aeronautical applications, *Proceedings of the International Conference on Atmospheric Turbulence*, May 18–21, London (1971), p. 18.

171. W. Gray, Calculation of cumulus draft velocities in hurricanes from aircraft observations, *J. Appl. Meteorol.*, 4(4): 463–474.

172. K. Haman, On some causes of clear-air turbulence, *Acta Geophys. Pol.*, 10(4):335–357 (1962).

173. H. E. Hart and L. W. Cooper, Thunderstorm airflow studies using radar transponders and superpressure balloons, in: *Proceedings of the XIII Radar Meterological Conference*, August 20–23 (1968).

174. W. Hesse, *Handbuch der Aerologie*, Leipzig (1961).

175. J. Hicks and E. Angell, Radar observations of breaking gravitational waves in visually clear atmosphere, *J. Appl. Meteorol.*, 7(1):114–121 (1968).

176. S. K. Kao and H. D. Woods, Energy spectra of mesoscale turbulence along and across the jet-stream, *J. Atmos. Sci.*, 21:513–519 (1964).

177. S. K. Kao and A. H. Sizoo, Analysis of clear-air turbulence near the jet-stream, *J. Geophys. Res.*, 71(16):3799–3805 (1966).

178. S. K. Kao and E. E. Sands, Energy spectrums, mean and eddy kinetic energies of the atmosphere between surface and 50 kilometers, *J. Geophys. Res.*, 71(22):5213–5220 (1966).

179. S. K. Kao and V. B. Taylor, Mean kinetic energies of eddy mean currents in the atmosphere, *J. Geophys. Res.*, 69(6):1037–1051 (1964).

180. S. K. Kao and J. B. Globhard, An analysis of heat-, momentum-transport and spectra for clear-air turbulence in mid stratosphere, *Pure Appl. Geophys.*, 88(5):180–185 (1971).

181. Kimura Ryuiji, Tsu Hiroji, and Yagihashi Akiko, Convective patterns in a plane Couette flow, *J. Meteorol. Soc. Jpn.*, 49(4):249–260 (1971).

182. G. E. King, Some characteristics of severe turbulence, CAADRP, Tech. Rep. No. 13 (1969).

183. Z. N. Kogan and N. R. Shakina, Numerical investigations of internal waves in jet-streams including nonlinear effects, *Boundary-Layer Meteorol.*, No. 5, pp. 79–93 (1973).

184. E. E. Kordes and B. J. Love, Preliminary evaluation of XB-70 airplane encounters with high altitude turbulence, NASA, TND-4209 (1967).

185. J. P. Kuettner and D. K. Lilly, Lee waves in the Colorado Rockies, *Weatherwise*, 21(5):180–186 (1968).

186. E. C. Kung, Kinetic energy generation and dissipation in the large-scale atmospheric circulation, *Mon. Weather Rev.*, 94(2):67–82 (1966).

186a. D. K. Lilly, D. E. Waco, and S. J. Adelfang, Stratospheric mixing estimated from high-altitude turbulence measurements, *J. Appl. Meteorol.*, 13:488–493 (1974).

187. G. H. Ludlam, Severe local storms: a review, *Meteorol. Monogr.*, 5(27):1–32 (1962).

188. J. L. Lumley, The spectrum of nearly inertial turbulence in a stably stratified fluid, *J. Atmos. Sci.*, 21(1):99–102 (1964).

189. J. S. Malkus and H. Riehi, *Cloud Structure and Distributions over the Tropical Pacific Ocean*, Berkeley (1964).

190. H. Mantis, The structure of winds of the upper troposphere at mesoscale, *J. Atmos. Sci.*, 20(1):94–106 (1963).

191. G. K. Mather, Flight studies of severe clear-air turbulence—the current position and further research needs, in: *Proceedings of the International Conference on Atmospheric Turbulence*, May 18–21, London (1971), p. 12.

192. J. J. MacPherson and E. G. Morrissey, Stratospheric turbulence and temperature change measurements from the COLDSCAN project, in: *Proceedings of the International Conference on Atmospheric Turbulence*, May 18–21, London (1971), p. 12.

193. L. O. Myrup, Atmospheric measurements of the buoyant subrange of turbulence, *J. Atmos. Sci.*, 25:1160–1164 (1968).

194. L. O. Myrup, Turbulence spectra in stable and convective layers in the free atmosphere. *Tellus*, 21(3):341–354 (1969).

195. C. W. Newton, Dynamics of severe convective storms, *Meteorol. Monogr.*, 5(27):33–58 (1963).

196. E. Palm, Two-dimensional and three-dimensional waves, *Geophys. Publ. Norv.*, 20(3): 25 (1958).

197. S. Panchev and D. Syrakov, Spectra of thermally stratified turbulent flow with no shear, *Tellus*, 23(6):500–505 (1971).

198. H. A. Panofsky, Spectra of atmospheric variables in the boundary layer, *Radio Sci.*, 4(12):1101–1105 (1969).

199. S. Penn, Investigation of the mesoscale atmospheric structure relative to the life cycle of clear-air turbulence, in: *Proceedings of the International Conference on Atmospheric Turbulence*, May 18–21, London (1971), p. 14.

200. S. Penn and T. A. Pisinsku, Mesoscale structure of the atmosphere in regions of clear-air turbulence, AFCRL-67-0115 (1965).

201. O. M. Phillips, On the Bolgiano and Lumley–Shur theories of the buoyancy subrange, *International Colloquium on the Fine-Scale Structure of the Atmosphere and Its Relation to Radio-Wave Propagation*, Nauka, Moscow (1965).

202. C. J. Readings and D. R. Rayment, The high-frequency fluctuation of the wind in the first kilometer of the atmosphere, *Radio Sci.*, 4(12):1127–1131 (1969).

203. E. R. Reiter, *Meteorologie der Strahlströme (Jet-Streams)*, Springer-Verlag, Vienna (1961).

204. E. R. Reiter and A. Burns, Atmospheric structure and clear-air turbulence, *International Colloquium on the Fine-Scale Structure of the Atmosphere and Its Relation to Radio-Wave Propagation*, Nauka, Moscow (1965).

205. E. R. Reiter and A. Burns, The structure of clear-air turbulence derived from "Topcat" aircraft measurements, *J. Atmos. Sci.*, 23:206–212 (1966).

206. J. H. Richter, High resolution tropospheric radar soundings, *Radio Sci.*, 4(12):1261–1268 (1969).

207. C. Rider, M. R. Thomson, and F. E. Verinder, Measurement of extreme mechanical turbulence during low level flights by Mirage A3-76, in: *Proceedings of the International Conference on Atmospheric Turbulence*, May 18–21, London (1971), p. 18.
208. W. T. Roach, Some aircraft reports of high-level turbulence, *Meteorol. Mag.*, 98(1160): 65–68 (1969).
209. W. T. Roach, On the influence of synoptic development on the production of high level turbulence, in: *Proceedings of the International Conference on Atmospheric Turbulence*, May 18–21, London (1971), p. 18.
210. R. S. Scorer, Causes and consequences of standing waves, in: *Proceedings of the Symposium on Mountain Meteorology*, Colorado (1967), pp. 75–101.
211. R. A. Steiner, Review of NASA high-altitude clear-air turbulence sampling programs, *J. Aircraft*, 3(1):48–52 (1966).
212. R. Steiner and R. H. Rhyne, Some measured characteristics of severe storm turbulence, Nat. Severe Storm Project, Rep. No. 10, p. 17 (1962).
213. R. Steiner and R. H. Rhyne, Atmospheric turbulence and airplane response: convective-type clouds, *J. Aircraft*, 1(1):13–15 (1964).
214. R. J. Taylor, *Manual on Aircraft Loads*, Pergamon Press (1965).
215. R. J. Taylor, J. Warner, and N. E. Bacon, Scale length in atmospheric turbulence as measured from an aircraft, *Q. J. Roy. Meteorol. Soc.*, 96:750–755 (1970).
216. D. S. Treddenick, Turbulence measurements in and near thunderstorms, Aeronautical Rep. LR-534, Ottawa (1970).
217. D. Trout and H. R. Panofsky, Energy disspiation near tropopause, *Tellus*, 21:355–358 (1969).
218. K. Tsuchiya and T. A. Fujita, A satellite meteorological study of evaporation and cloud formation over western Pacific under the influence of the winter monsoon, *J. Meteorol. Soc. Jpn.*, 45:232–250 (1967).
219. C. F. Van Thullenar *et al.*, Severe storm detection and circumnavigation, Final Report, Federal Aviation Agency, Contr. ARDS-A-176, U.S. Weather Bureau, Washington (1963).
220. J. Warner, The microstructure of cumulus cloud. Part 1, *J. Atmos. Sci.*, 26(5):1049–1059 (1969).
221. J. Warner, The microstructure of cumulus cloud. Part 3. The nature of the updraft, *J. Atmos. Sci.*, 27(4):682–688 (1970).
222. R. J. Wilson, B. Y. Love, and R. R. Larson, Evaluation of effects of high-altitude turbulence encounters on the XB-70 airplane, NASA, TN D-6457 (1971).
223. N. K. Vinnichenko (Winnichenko), The kinetic energy spectrum in the free atmosphere–1 second to 5 years, *Tellus*, 22:158–166 (1970).